The Development of Chemical Principles

COOPER H. LANGFORD
The University of Calgary

RALPH A. BEEBE
Late Professor, Amherst College

DOVER PUBLICATIONS, INC.
New York

Bibliographical Note

This Dover edition, first published in 1995, is an unabridged and corrected
republication or the work first published in the *Addison-Wesley Series in
Chemistry* in 1969 by the Addison-Wesley Publishing Company, Inc., Reading,
Massachusetts.

Library of Congress Cataloging-in-Publication Data

Langford, Cooper Harold, 1934–
    The development of chemical principles / Cooper H. Langford, Ralph A.
Beebe.
        p.   cm.
    Originally published: Reading, Mass. : Addison-Wesley, © 1969, in series:
Addison–Wesley series in chemistry.
    Includes index.
    ISBN 0-486-68359-1 (pbk.)
    1. Chemistry.   I. Beebe, Ralph A. (Ralph Alonzo), b. 1898.   II. Title.
0031.2.L34   1995
541.2—dc20   94-40923
                                                                    CIP

Manufactured in the United States of America
Dover Publication, Inc., 31 East 2nd Street, Mineola, N.Y. 11501

Preface

If longer titles were convenient, this book might be accurately called *The Development of Structural Principles in Chemistry: An Introductory Study*. We like the term "development" because it suggests the two main things we are trying to do in this book.

We try to record experimental observations, identify creative insights, and pursue deductive consequences in the manner characteristic of the development of a scientific concept. For example, we quote the early data on conductance and colligative properties of electrolyte solutions, discuss Arrhenius' synthesis, and pursue the deductive consequences of his idea of partial ionization to identify its flaw and reveal the origins of the Debye-Hückel theory. We hope that the reader can participate in this development to "rediscover" the theory of ionization for himself and, at least vicariously, practice science. We also hope it will be apparent that the last word in the story has probably not been said.

We also try to follow the longer-range development of fruitful ideas to show that a good theoretical idea evolves. We try to show how the meaning of a term changes as the idea it represents is found to be applicable in more and more new areas. The evolution of the concept of molecule from the micro-billiard ball of Avogadro through the architectural notions of Kekulé, Le Bel, and Lewis to the complex of electron-nuclear interactions of Mulliken is a striking example of this type of development.

Obviously, both senses of the term "development" often commit us to examination of the history of our subject, including some of its false steps and discarded ideas. We do this mainly with science students in mind. Our conviction is that involving the student as a *critical* participant in the development of ideas should be given first priority. Science students, especially, should be encouraged early to practice science.

The topical focus of the book is structure. We follow the suggestion of the Westheimer report that chemistry is naturally divided by differences of intellectual style into study of structure, synthesis, and dynamics. A study of concept development implies a study of intellectual style. Historically, structural thinking matured first in chemistry and seems an appropriate emphasis for a first course. Clearly, a balanced treatment of modern structural theory cannot ignore some of dynamics and synthesis.

The plan of the book is to explore three lines of the development of contemporary chemical structural theory: the classical theory of bonding in molecules; the ionic interpretation of electrolyte solutions; and the physical theory of atomic structure. Important characteristics of each of these somewhat independent genetic strains are evident in our current hybrid. The first chapter reviews the foundations of atomic-molecular theory and stoichiometry. The next two chapters show how isomer relationships and reaction patterns lead to representation of molecules by geometrical diagrams with bond lines. Chapters 4 and 5 explore the properties of solutions and the origin of the concept of ion. Energetic relationships in chemistry are introduced in Chapter 6 on the basis of an "empirical" discovery of the Nernst equation. (This is intended to give concrete chemical foundation to a first contact with thermodynamic quantities and to make reasonable demands on the student's mathematical preparation.) The final three chapters treat the development of a theory of atomic structure and its significance for interpretation of the chemist's earlier theoretical entities, bonds and ions.

Accomplishments that are not strictly required for the development of the basic theme are treated in appendixes to several chapters. Such topics are kinetic theory of gases, a survey of organic functional group reactivity and nomenclature, and applied equilibrium calculations. The book as a whole has an appendix introducing the dynamics of chemical reaction from the new fundamental process (molecular-beam) viewpoint.* Only this appendix presumes a background in calculus.

---

*See R. Wolfgang, "The Revolution in Chemical Dynamics," *J. Chem. Ed.*, **45** (1968), 359.

Since the development of modern chemistry depends so much on concepts from mechanics, electricity, and wave physics, appendixes on these subjects are included. We are especially indebted to Professor A. B. Arons for permitting us to include his introductory chapter on waves as Appendix 3. The basic wave ideas are usually less familiar to the introductory chemistry student than are the elementary notions of mechanics.

If active participation in the development of ideas is the objective of the book, its problem and assignment materials bear a major responsibility for accomplishing this aim. We have placed problems in the body of the text at appropriate points, either to illustrate or to advance the argument. Probably the ideal procedure would be to solve these problems as they arise. We have also prepared some larger, more independent assignments, called "seminars" for lack of a better term. They present some specific experimental information as a challenge to theory building, usually after some outside reading. They provide an opportunity for articulation of extended critical arguments tied to specific results and for the development of a sense of the rules of scientific "debate." We have used them as assignments of the "term paper" type, which students have found rewarding. They might also serve as the basis for interesting discussion seminars.

We have not tried to write a comprehensive text. We are not sure what the content of introductory (general?) chemistry should be at present. However, we do enthusiastically endorse the conclusion of the report of the 1957 conference on "Improving the Quality of Introductory Physics Courses"* that the magnitude of the subject requires radical paring to allow time for deep exploration of principles and methods of reasoning. The detailed content of the introductory course should probably be adapted to the overall curricular situation in each institution. This procedure is certainly feasible since there are now

---

*Am. J. Phys. **25**, (1957), 417.

available in inexpensive editions many good specialized monographs written for beginners.

We believe that emphasis on development is relevant to a wide variety of curricula, and we have used the materials of this book in several situations. The book originated in a one-semester course for well-prepared science and liberal arts students at Amherst College. That course was terminal for the liberal arts students but led to a second-semester course in thermodynamics and equilibrium for the science students. With some supplements, the material was used for the first-year course for science students at Carleton University. Parts of it have also been useful in a first-year course at Voorhees College, Denmark, South Carolina, and in a summer enrichment institute for secondary school students.

We owe unacknowledgeable debts to our predecessors; we have learned from many earlier texts and benefited from conversations with many teachers. However, a few of our most important debts can be acknowledged here. Professor Jay A. Young read the entire manuscript and gave us lively criticism and valuable encouragement. Parts of the manuscript were commented upon by Professors Francis Bonner, Robert H. Weightman, Richard D. Fink, Peter Kruus, James M. Holmes, and Paul M. Laughton. We owe a special debt to our teachers. Our scientific careers have a common "grandpaternity" (C. H. L.) and "paternity" (R. A. B.) in the inspiration engendered by Professor H. S. Taylor. Professor R. L. Burwell, Jr., a Taylor student, taught C. H. L. and in the process advocated a healthy skepticism toward excesses of pedagogic theory. (That lesson may not have been sufficiently well learned.) Both of us have been deeply interested in the universally required introductory "Science 1" course at Amherst College, and C. H. L. taught on its staff for several years. In the period that the course was directed by Professor A. B. Arons, it offered a lesson in the best science teaching. Our introductory chemistry course at Amherst was undoubtedly a response to the challenge of Science 1. We can only hope we have met the challenge.

We warmly acknowledge the expert contributions of Mrs. Madelaine deFriesse, Mrs. H. N. Stassen, Mrs. Ann Holt, and Mrs. Martha Langford in the preparation of the manuscript. We have to admit that the problems were checked by first-year students at Amherst and Carleton. We apologize to them and thank them. The patience of our wives in support of this project certainly must also be recognized.

Ottawa, Ontario
Amherst, Massachusetts
January, 1969

C. H. L.
R. A. B.

Contents

CHAPTER 9 THE ELECTRONIC THEORY OF VALENCE

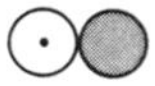

"Olefiant gas"
(ethylene) molecule
John Dalton, 1808

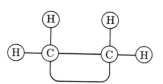

Ethylene molecule after
Crum Brown, 1864

The structure diagram is more than a convenience to the chemist. It is an important tool of reasoning. Here one molecule is shown as it has been represented at five stages in the development of chemical theory. Interestingly, all but the first are still "correct" in the sense that they completely explain some aspects of experiments and are in use today in only slightly revised form.

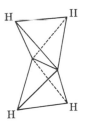

Ethylene molecule in stereochemical representation of J. H. van't Hoff, 1874

H . . H
. C : : C .
H . . H

Ethylene in electron dot representation of G. N. Lewis, 1916

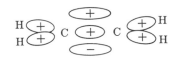

Ethylene molecule in current molecular orbital electron cloud representation after theory of R. S. Mulliken

The Periodic System of the Elements

Group 0
H 1
He 2

0	I	II	III	IV	V	VI	VII
He 2	Li 3	Be 4	B 5	C 6	N 7	O 8	F 9
Ne 10	Na 11	Mg 12	Al 13	Si 14	P 15	S 16	Cl 17

0	I	II	III	IVa	Va	VIa	VIIa	VIII			Ib	IIb	IIIb	IV	V	VI	VII	0
Ar 18	K 19	Ca 20	Sc 21	Ti 22	V 23	Cr 24	Mn 25	Fe 26	Co 27	Ni 28	Cu 29	Zn 30	Ga 31	Ge 32	As 33	Se 34	Br 35	Kr 36
Kr 36	Rb 37	Sr 38	Y 39	Zr 40	Nb 41	Mo 42	Tc 43	Ru 44	Rh 45	Pd 46	Ag 47	Cd 48	In 49	Sn 50	Sb 51	Te 52	I 53	Xe 54
Xe 54	Cs 55	Ba 56	La 57 *	Hf 72	Ta 73	W 74	Re 75	Os 76	Ir 77	Pt 78	Au 79	Hg 80	Tl 81	Pb 82	Bi 83	Po 84	At 85	Rn 86
Rn 86	Fr 87	Ra 88	Ac 89 ◆	Th 90	Pa 91	U 92	Np 93	Pu 94										

* Lanthanons

Ce 58	Pr 59	Nd 60	Pm 61	Sm 62	Eu 63	Gd 64	Tb 65	Dy 66	Ho 67	Er 68	Tm 69	Yb 70	Lu 71

◆ Actinons

Th 90	Pa 91	U 92	Np 93	Pu 94	Am 95	Cm 96	Bk 97	Cf 98	Es 99	Fm 100	Md 101	No 102	Lr 103	Kh 104

CHAPTER 1

The Atomic-Molecular Theory

1.1 THE DIFFERENTIATION BETWEEN SIMPLE SUBSTANCES AND MIXTURES; THE STARTING POINT FOR CHEMICAL SCIENCE

Consider a random assortment of natural phenomena: an apple falls, iron rusts, water boils, soda fizzes, a dye fades, a wheel rotates, a green leaf develops, a rocket soars, an electric bulb glows, silver is electroplated onto another metal. Which of these phenomena are the special province of chemical science? Which are within the domain of physics? Which are of concern to both?

Roughly speaking, physics is not primarily concerned with the kind of "stuff" under observation. The law of acceleration under gravitation is meant to apply equally well to a pound of lead or a pound of feathers, and does, provided air resistance is properly considered (e.g., eliminated by conducting experiments in a vacuum). In an event such as "falling," no new "stuff" is formed. Except for possible dents, the piece of lead seems pretty much the same after falling as it did before. On the other hand, the rusting of iron, the electrical deposition of silver, and the fading of a dye produce what would obviously be called changes in the sort of "stuff" present. These are chemical changes. It will be our task here to develop a systematic logical explanation* of the nature of such events.

To begin to develop an explanation of chemical change, it is first important to decide what are simple chemical materials and what are complicated ones. This is not an easy task. In Herbert Butterfield's important book, *The Origins of Modern Science*, the chapter on the origin of scientific chemistry is titled "The Postponed Scientific Revolution in Chemistry." Chemistry was late to bloom, largely because the most "common" materials are not necessarily simple.

---

* "Explanation" is a tricky concept. In this book we shall try to illustrate what counts as a scientific explanation and show something of the process by which some specific explanations arise. We shall not have occasion to attempt an analysis of the idea of explanation. The reader is urged to explore this fascinating question. A good starting point might be S. Toulmin's small book, *Foresight and Understanding*, Harper & Row, New York, 1963 (Harper Torchbook TB 564).

There is an obvious temptation to try to begin the study of chemistry with the assumption that the most common entities are the *simple* constituents of matter. Aristotle suggested that all matter was composed of varying proportions of earth, air, fire, and water. This idea dominated Greek science, and was influential until almost 1800. It was finally disproved after the development of increasingly sophisticated chemical laboratory procedures which permitted the study of processes that *resolved the supposed elements* into components. For example, the development of pumps and "pneumatic troughs" made possible the collection, measurement, and study of gases. Only then did it become clear that all gases were not air and that, in fact, the air we breathe is a mixture of gases, each having its own distinctive properties. Precise gas handling also led to a new understanding of the role of weight in chemistry. When *all* the products of a reaction (including gases) could be collected and weighed, it was soon realized that the weight of all the products was exactly equal to the weight of all the reactants, that is, weight is conserved in chemical reactions. In modern terminology, the idea of weight as being due to the force of gravity and as being proportional to the *mass* of the body leads to expression of the result as the law of *conservation of mass*. Conservation of mass, or constancy of weight in reactions, suggests a decision to regard as *matter* only that which has weight or mass. From this point of view "fire" clearly leaves the list of elements (see Problem 1.3). We shall see that the decision to focus on weighing led to more progress in chemistry in the following century than in the previous two millenia. This result has abundantly justified the decision.

Out of these historical circumstances grew a more sophisticated notion of the composition of substances. It has rarely been carefully formulated, and it is not easy to give precise and at the same time concise descriptions of the procedures by which simple chemical substances are recognized.* Most chemists learn the rules more by example than by precept.

The first distinction which must be made is that between a *chemical* and a *physical* change, because the definition of pure substance depends on this distinction. The arguments are inevitably somewhat circular, since the distinction may be made only after some idea of what is a *pure* chemical substance has been given.

To put the matter as briefly as possible, *a pure substance is one that has invariant properties.* The properties intended must be listed. The list includes the following simple properties:

melting point (temperature of melting),

boiling point (temperature of boiling at normal pressure),

density (weight per unit volume),

---

* The process of giving the definition of a concept in terms of procedures and operations, which must be carried out in order to observe or measure the entity in question, is known as giving an "operational definition." Giving operational definitions is certainly crucial in experimental science, which purports to deal with the observable and measurable aspects of nature.

refractive index (the degree to which a beam of light is bent on passing from air into the substance),

electrical conductivity,

viscosity (a measure of the ease with which a liquid or a gas will flow),

solubility in a particular liquid (solvent).

An arbitrary sample of matter will have a definite numerical value for each of these properties. For a sample to be *pure*, the value must remain *constant* when the sample is subjected to any *physical* process designed to purify it which does *not* cause a *chemical* change. A few key examples of these *physical* processes are:

   1. distillation,   2. crystallization,   3. chromatography.

Each of these processes is important to both laboratory chemistry and the chemical industry. They are best described by considering examples.

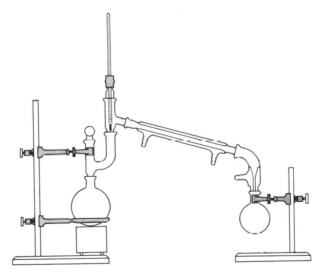

Fig. 1.1. Simple distillation apparatus. (Adapted from A. Corwin and M. Bursey, *Elements of Organic Chemistry: As Revealed by the Scientific Method*, Addison-Wesley, Reading, Mass., 1966.)

1. *Distillation.* When a mixture of grain alcohol and ether is boiled, and a thermometer is placed in the vapor (in such a way that droplets of liquid will just condense on its bulb), the temperature of boiling may be recorded. As the sample boils away, the boiling temperature rises. If the apparatus is designed so that the vapor may be condensed to liquid again and collected (see Fig. 1.1), fractions boiling over restricted temperature ranges may be collected and subjected to redistillation. Finally, one may collect a sample which boils over a very narrow temperature range or (to the degree of precision of the measurement being made) a sample, all of which distills at a "constant" temperature. Such a sample is considered pure. (Obviously, the significance of purity is somewhat variable according to the sensitivity and precision of the temperature measurement.) The higher-boiling component of the alcohol-ether mixture boils at 78°C.* The lower-boiling component boils at 34°C. By appropriate combination of fractions and redistillations, *all* the original mixture may be separated into the 78° fraction and the 34° fraction.

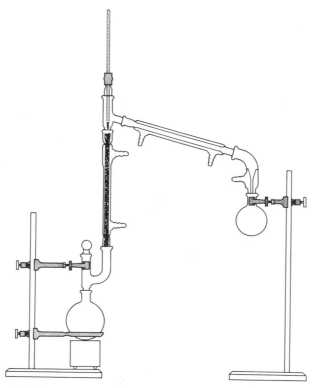

Fig. 1.2. Fractional-distillation apparatus. The lower-boiling fraction tends to rise to the top as vapor. The higher-boiling fraction tends to flow back (reflux to the bottom) as liquid. (Adapted from A. Corwin and M. Bursey, *Elements of Organic Chemistry: As Revealed by the Scientific Method*, Addison-Wesley, Reading, Mass., 1966.)

In practice, it is convenient to arrange for a series of successive condensations and revaporizations in the distillation apparatus so that the lower-boiling component works its way to the collection flask first in an essentially pure state, and fraction collection and redistillation is eliminated. Figure 1.2 shows such an apparatus, called a *fractional-distillation apparatus*.

---

* The symbol C denotes the centigrade (or Celsius) temperature scale on which water boils at 100° and ice melts at 0°.

2. *Crystallization.* One of the properties that may be considered a constant characteristic of pure substances is their *solubility* in suitable liquids at specific temperatures. The fact that different solid substances display different solubilities in a liquid is extremely useful for purification purposes. Perhaps crystallization from solution is the most widely used method of purification. Characteristically, when a chemical reaction is run with the aim of preparing a substance, the material first isolated is impure, as evidenced, for example, by the fact that the whole of a sample does not melt at a single well-defined temperature (melting point) but over a range of 10° or 20°. Frequently, the entire sample may be dissolved in a solvent such as hot water, but upon cooling, crystals separate again. (This is an instance of the common phenomenon of greater solubility at higher temperature.) If only relatively small quantities of the "impurities" are present, their solubility may not be exceeded at the lower temperature, and they remain in solution. The crystals isolated from this "recrystallization" then melt over a smaller temperature range. Finally, after successive recrystallizations, a sample melting sharply at a well-defined temperature may be obtained and regarded as pure.

Problem 1.1   When a small quantity of "pure" water boiling at 100°C is added to "pure" ethyl alcohol (grain alcohol) boiling at 78°C, a "mixture" is prepared which boils at a fixed temperature of 95°C. Determine whether a new "substance" has been formed. Explain how you would tell. Suppose that the test designed to determine if a new "substance" has been formed results in failure. Comment on the exclusive use of a *single* criterion of purity.

Problem 1.2   Two samples *A* and *B* each melt sharply at 133°C. However, all mixtures of these two samples begin to melt at lower temperatures. Are *A* and *B* samples of the same compound? Explain your reasoning.

3. *Chromatography.* A third method for the isolation and purification of substances is that of chromatography. This depended initially on the characteristic tendency of substances to be adsorbed (i.e., more or less, to "stick") on a solid surface to varying degrees. The name *chromatography* derives from the earliest application of the method: the separation of *colored* pigments in natural products. The procedure is perhaps best described by an example.

The autumn colors of maple or aspen leaves are already present at the height of the growing season in early summer, but at this time their red or yellow pigments are swamped out by an excessive amount of the green pigment, chlorophyll. The colors in the green leaves can be separated by chromatography. The pigments are extraced from the green leaves by grinding them in a mortar with a small volume of a 1:9 mixture of benzene and petroleum ether (the low-boiling fraction of petroleum). The extract is gently shaken with a drying agent such as anhydrous (water-free) calcium sulfate and then filtered. A cylindrical column is constructed by filling the vertical glass tube, shown in Fig. 1.3, with a suitable solid adsorbent such as powdered confectioner's sugar. The sugar will not dissolve in petroleum ether. Moreover, since it is finely powdered, the sugar presents a relatively extensive surface for the selective adsorption of the pigments. The solution extracted from the leaves is now poured onto the top of the column and it moves down through the column either by force of gravity or by gentle suction applied at the bottom. The petroleum ether and benzene are little attracted by the sugar at the solid-liquid interface; the chlorophyll and the other pigments are much more strongly attracted. As a result, these colored materials do not move along as fast as the main stream of petroleum ether, but rather they are retarded by adsorption on the surface of the sugar in the column. Because of their differences in adsorbability (or stick-to-itiveness) on the sugar surface, the colored components are retarded to varying degrees, and a chromatogram is developed.

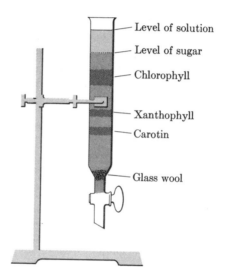

Fig. 1.3. Chromatography column.

Suppose that we chose green leaves from a tree that would eventually turn yellow in autumn owing to the disappearance of the chlorophyll at that time of year. In that case, the chromatogram might develop as shown in Fig. 1.3 with the faster moving pigments, carotin and its closely related chemical cousin xanthophyll, coming through first and producing their characteristic yellow bands. These are followed later by the characteristic green bands of chlorophyll. If we had chosen a maple leaf containing a red pigment as well as green and yellow, then we might have expected a red band with its own characteristic retention time as the solution flows through the column. By appropriate techniques we could now separate the bands and extract the colored substances. In this way we would have effected a sorting that otherwise would be very difficult indeed.

We have chosen here to cite the application of the technique involving familiar materials such as sugar and growing leaves, and the low-boiling fraction of petroleum. Although the technique of chromatography gets its name from the separation of a mixture of colored substances into bands, the term is much more generally applied. The mixture of substances to be separated may be quite devoid of color; yet we refer to the separation process as chromatography. A number of solids in powdered or granular form have been found to serve well as column materials.

Although the column chromatography, which has been described here, came first (the phenomena were reported by Tswett as early as 1906, and the process was widely used by 1930), several modifications based on similar principles have been worked out more recently.* For instance, paper chromatography has been applied with great success to the separation of the amino acid mixtures produced by the degradation of proteins from living tissue. In this case the sugar is replaced by a paper strip. Mixtures of low-boiling (and therefore highly volatile) simple organic substances are successfully separated by specially treated chromatographic columns where the "carrier" fluid is not a liquid, but rather the most inert of all gases, helium. This technique is referred to as vapor-phase chromatography or VPC, and it came into general use in the early 1950's. This technique has proved invaluable and has to a large degree displaced the more laborious technique of fractional distillation, especially when small quantities of materials are involved.

1.2 COMPOUNDS AND ELEMENTS

Having discussed the separation of pure substances from mixtures by so-called "physical" means, we may turn to a consideration of the cases in which substances undergo more profound alterations that are recognized as *"chemical"* changes *producing new substances.* Again it proves useful to identify the *simplest of pure substances.* And again the history of the subject reveals the difficulty associated with ready acceptance of too simple a notion. The notion that all substances are formed by the combination of a limited set of simplest substances—*elements*—is ancient. Ancient and premodern lists of elements were commonly limited to as few as Aristotle's four—earth, air, fire, and water. They were compiled on the basis of intuition, and procedures for experimental recognition of elements were not outlined (no "operational" definition of the concept was given). Perhaps the first useful approach to the problem was phrased by Robert Boyle in his *Sceptical Chymist* of 1661. Boyle suggested that elements should be recognized as the ultimate limit of chemical analysis. When a substance could not be caused to undergo chemical changes resolving it into two substances, then it must be considered an *element* and *not* a substance produced from the combination of elements—a *compound.*

---

* One modification known as thin-layer chromatography has been especially useful in separating mixtures of some of the less-volatile substances. A fascinating account of the application of thin-layer chromatography to the separation of pigments is given in an article entitled "The Changing Colours in the Plant World" by E. C. Grob, *Palette 21*, Sandoz Ltd., CH-4002, Basel, Switzerland.

Table 1.1

Table of simple substances belonging to all the kingdoms of nature, which may be considered as the elements of bodies*

New names	Old names
Light	Light
Caloric	Heat
	Principle or element of heat
	Fire, igneous fluid
	Matter of fire and of heat
Oxygen	Dephlogisticated air
	Empyreal air
	Vital air, or base of vital air
Azote	Phlogisticated air or gas
	Mephitis, or its base
Hydrogen	Inflammable air or gas, or .
	the base of inflammable air

Oxidable and acidifiable simple substances not metallic

New names	Old names	New names	Old names
Sulfur Phosphorus Charcoal }	The same names	Muriatic radical Fluoric radical Boracic radical }	Still unknown

* List of elements compiled by Lavoisier, which appeared in his *Elements of Chemistry*. (Adapted from A. Lavoisier, *Elements of Chemistry*, the Great Books Foundation, Chicago, 1949.)

Boyle's program is not simple and was not carried out systematically until agreement on the importance of weight as a means of measuring the amount of a substance had been reached and good techniques for manipulation of gases were available. It should be clear that no single experiment will establish the status of a substance as an element. If a substance can be resolved, it must be a compound, but the constituents into which it is resolved may only be tentatively identified as elements. Later experiments might show that they, too, can be resolved.* The line of argument is best clarified by a simple example. Suppose that iron and sulfur are heated together (reacted). A product of strikingly different properties is produced. For example, sulfur may be dissolved in the liquid carbon disulfide, but the product of the reaction cannot.

---

* To anticipate later chapters, we note that there is considerable confidence in the modern list of elements because it has proved possible to develop *out of* the chemical discovery of the elements and related developments of more modern physics a *theory* of the constitution of elements that explains the occurrence of the known elements and important relationships among them. In this context it has even proved possible to *make* "new" elements.

Table 1.1 continued

Oxidable and acidifiable simple metallic bodies

New names	Regulus of	New names	Regulus of
Antimony	Antimony	Mercury	Mercury
Arsenic	Arsenic	Molybdenum	Molybdenum
Bismuth	Bismuth	Nickel	Nickel
Cobalt	Cobalt	Platinum	Platinum
Copper	Copper	Silver	Silver
Gold	Gold	Tin	Tin
Iron	Iron	Tungsten	Tungsten
Lead	Lead	Zinc	Zinc
Manganese	Manganese		

Salifiable simple earthy substances

New names	Old names
Lime	Chalk, calcareous earth Quicklime
Magnesia	Magnesia, base of Epsom salt Calcined or caustic magnesia
Barytes	Barytes, or heavy earth
Argill	Clay, earth of alum
Silex	Siliceous or vitrifiable earth

Iron can be dissolved in acidic aqueous (water) solutions with the evolution of an odorless gas; so can the product, but now with the evolution of a foul-smelling gas. If varying quantities of iron and sulfur are used, the resultant product mixture will contain *only* the new substance and *either* excess iron or excess sulfur, the total weight of material remaining equal to the sum of the weights of iron and sulfur mixed. It is reasonable to infer that iron and sulfur have combined to form a new substance that must be a compound [iron(II) sulfide in modern nomenclature]. It may be true that iron and sulfur (the constituents) are elements, but this is not proven.

Problem 1.3 One of the key historical accomplishments in the systemization of chemistry was the explanation of the process of combustion (burning) given by Lavoisier about 1780. It represents one of the early thorough commitments to weighing. Metals are commonly found in nature as the metal oxide, which was known in the eighteenth century as a "calx." When the calx is heated with carbon (charcoal), it is converted into the metal. The metal burned in air gives the calx. According to Lavoisier's predecessors, the calx was an element that combined with a substance called phlogiston ("the fire principle") obtained from carbon to give a *compound*,

Table 1.2

List of the atomic weights of the elements

Element	Symbol	Atomic Number	Atomic Weight	Element	Symbol	Atomic Number	Atomic Weight
Actinium	Ac	89	(227)	Mercury	Hg	80	200.59
Aluminum	Al	13	26.98	Molybdenum	Mo	42	95.94
Americium	Am	95	(243)	Neodymium	Nd	60	144.24
Antimony	Sb	51	121.75	Neon	Ne	10	20.183
Argon	Ar	18	39.948	Neptunium	Np	93	(237)
Arsenic	As	33	74.92	Nickel	Ni	28	58.71
Astatine	At	85	(210)	Niobium	Nb	41	92.91
Barium	Ba	56	137.34	Nitrogen	N	7	14.007
Berkelium	Bk	97	(249)	Nobelium	No	102	(253)
Beryllium	Be	4	9.012	Osmium	Os	76	190.2
Bismuth	Bi	83	208.98	Oxygen	O	8	15.9994
Boron	B	5	10.81	Palladium	Pd	46	106.4
Bromine	Br	35	79.909	Phosphorus	P	15	30.974
Cadmium	Cd	48	112.40	Platinum	Pt	78	195.09
Calcium	Ca	20	40.08	Plutonium	Pu	94	(242)
Californium	Cf	98	(251)	Polonium	Po	84	(210)
Carbon	C	6	12.011	Potassium	K	19	39.102
Cerium	Ce	58	140.12	Praseodymium	Pr	59	140.91
Cesium	Cs	55	132.91	Promethium	Pm	61	(147)
Chlorine	Cl	17	35.453	Protactinium	Pa	91	(231)
Chromium	Cr	24	52.00	Radium	Ra	88	(226)
Cobalt	Co	27	58.93	Radon	Rn	86	(222)
Copper	Cu	29	63.54	Rhenium	Re	75	186.23
Curium	Cm	96	(247)	Rhodium	Rh	45	102.91
Dysprosium	Dy	66	162.50	Rubidium	Rb	37	85.47
Einsteinium	Es	99	(254)	Ruthenium	Ru	44	101.1

Element	Symbol	Number	Weight	Element	Symbol	Number	Weight
Erbium	Er	68	167.26	Samarium	Sm	62	150.35
Europium	Eu	63	151.96	Scandium	Sc	21	44.96
Fermium	Fm	100	(253)	Selenium	Se	34	78.96
Fluorine	F	9	19.00	Silicon	Si	14	28.09
Francium	Fr	87	(223)	Silver	Ag	47	107.870
Gadolinium	Gd	64	157.25	Sodium	Na	11	22.9898
Gallium	Ga	31	69.72	Strontium	Sr	38	87.62
Germanium	Ge	32	72.59	Sulfur	S	16	32.064
Gold	Au	79	196.97	Tantalum	Ta	73	180.95
Hafnium	Hf	72	178.49	Technetium	Tc	43	(99)
Helium	He	2	4.003	Tellurium	Te	52	127.60
Holmium	Ho	67	164.93	Terbium	Tb	65	158.92
Hydrogen	H	1	1.0080	Thallium	Tl	81	204.37
Indium	In	49	114.82	Thorium	Th	90	232.04
Iodine	I	53	126.90	Thulium	Tm	69	168.93
Iridium	Ir	77	192.2	Tin	Sn	50	118.69
Iron	Fe	26	55.85	Titanium	Ti	22	47.90
Krypton	Kr	36	83.80	Tungsten	W	74	183.85
Lanthanum	La	57	138.91	Uranium	U	92	238.03
Lawrencium	Lw	103	(257)	Vanadium	V	23	50.94
Lead	Pb	82	207.19	Xenon	Xe	54	131.30
Lithium	Li	3	6.939	Ytterbium	Yb	70	173.04
Lutetium	Lu	71	174.97	Yttrium	Y	39	88.91
Magnesium	Mg	12	24.312	Zinc	Zn	30	65.37
Manganese	Mn	25	54.94	Zirconium	Zr	40	91.22
Mendelevium	Md	101	(256)				

* Based on mass of C^{12} at 12.000 ... . The ratio of these weights to those on the older chemical scale (in which oxygen of natural isotopic composition was assigned a mass of 16.0000 ...) is 1.000050. (Values in parentheses represent the most stable known isotopes.)

the metal. They wrote as "equations"

$$\text{calx} + \text{phlogiston (from carbon)} \rightarrow \text{metal},$$

$$\text{metal} \xrightarrow{\text{burning}} \text{calx} + \text{phlogiston}.$$

Lavoisier called attention to two key aspects of this situation: first, that the metal burned in air *gained* weight on forming the calx; second, that carrying out combustion in a sealed vessel led to *no overall change in weight.*

a) On the basis of these facts, defend Lavoisier's contention that the metal is an element and the calx a compound formed by combination of the metal with a gas from the air (dubbed by Lavoisier, oxygen) according to the equation:

$$\text{metal} + \text{oxygen} \rightarrow \text{metal oxide (calx)}.$$

b) Discuss the importance of Lavoisier's second observation for the central role given to weighing in chemistry.

c) Is it possible for carbon to be considered an element by (i) the phlogistonists, (ii) Lavoisier?

d) Formulate the argument if one were willing to accept the postulate that phlogiston has *negative* weight.

In 1789, Lavoisier's textbook *Traité Elémentaire de Chimie* was published. It is the first systematic "modern" account of chemistry. Table 1.1 reproduces its list of elements [note that heat (*caloric*) and light (*luminére*) are still regarded as substances and elements]. This table is mainly based on experimental documentation of the type we have described. Table 1.2 is a modern list of the elements along with the extremely convenient shorthand symbols used for the names. The atomic weights and atomic numbers are also included. The significance of the atomic numbers will become apparent in Chapter 7.

1.3 THE LAW OF DEFINITE PROPORTIONS

The next major question in the development of chemistry is the question of the proportions in which elements may combine to form compounds. Can any weight of element A combine with any arbitrary weight of element B to form a compound of A and B? Or, are there certain privileged ratios of weights such that there are only a limited number of possible compounds of two elements? An affirmative answer to the second question set the stage for Dalton's development, on a scientific basis, of the atomic theory: the collection of concepts which asserts that matter is composed ultimately of infinitesimal particles which are the final point of possible chemical subdivision.*

---

* In a 20th century context, the adjective "chemical" modifying "subdivision" is crucial since the spectacular results of a further *subdivision* of atomic particles themselves have impinged on all our lives. We refer, of course, to nuclear disintegration.

L. K. Nash has written in his history of the atomic molecular theory:

It has long been an implicit article of faith among chemists that a given compound (substance) always contained its components in a fixed proportion by weight—for example that the weights of hydrogen and oxygen present in 1 gram of water would not vary with the source of the water.*

And yet as recently as 170 years ago at the turn of the eighteenth century, a famous dispute occurred between two French chemists, Berthollet and Proust, both highly respected scientists of their day. Let us illustrate the point of controversy. Suppose that we heat a freshly polished copper strip in contact with air. We note the production of iridescent colors on the copper surface going from orange-red to purple to black. We presume these colors are due to oxides of copper. Berthollet contended that there was an infinite series of copper oxides—with progressively variable colors and progressively variable proportions by weight of copper and oxygen. In contrast, Proust maintained that the "compounds" of continuously variable oxygen-to-copper ratio reported by Berthollet were actually nothing but *mixtures* of just two compounds in which the oxygen-to-copper ratio had two discrete and invariant values.

Proust defines his position with the following trenchant declaration.

According to our principles a compound . . . is a privileged product to which nature assigns fixed ratios. Let us recognize, therefore, that the properties of true compounds are invariable as is the ratio of their constituents . . . The cinnabar of Japan is constituted according to the same ratio (by weight) as the cinnabar of Spain. Silver is not differently oxidized or muriated in the muriate (chloride) of Peru than in that of Siberia . . .

. . . There is a balance which, subject to the decrees of nature, regulates *even in our laboratories* the ratios of compounds. . . .*

Other materials were cited by Berthollet as having continuously varying properties and composition by weight. These are the metal alloys (such as brass) and the glasses. Proust quite correctly countered that these were not true substances. Nor did they seem to be mixtures of substances. We now recognize these materials as solutions of one or more solid substances in another. Proust's position has withstood all the weapons of assault initially brought to bear by Berthollet.

The scales were gradually tipped in favor of Proust's beliefs, but the controversy had an important effect on the development of chemistry. In an attempt to support their respective positions, both contestants and doubtless many of their coworkers gathered a great body of quantitative data which set the stage for the flowering of Dalton's atomic theory.

---

* Leonard K. Nash, *The Atomic Molecular Theory* (Harvard Case Histories in Experimental Science, No. 4, pp. 32, 33, Harvard University Press, Cambridge, Mass., 1950). Quoted by permission of the publisher.

Proust's thesis has come to be known as the law of definite proportions. It may be stated: *The proportions by weight in which elements enter into any given compound are invariable.*

1.4 THE DEVELOPMENT OF THE DALTONIAN ATOMIC THEORY

The atomic theory of matter is usually attributed to the English chemist John Dalton, who published his statement of the theory in 1807. However, the concept of "atomism" (from the Greek word *atomos*, indivisible) can be traced back to 400 B.C., when certain Greek philosophers conceived and propounded the concept. At that time and for many centuries thereafter, the advantages of putting such a concept to experimental test were not fully appreciated. Now, of course, we accept the tenet that any concept which cannot stand up under experimental tests should either be modified or discarded.

The Greek atomists arrived at their notion of the atom as a development from their philosophy. They made the general *observation* that many or indeed most objects formed in nature are subject to a state of change. On the other hand, they felt that there must be something permanent associated with real and tangible materials. This conflict of ideas was resolved by postulating the reality of invisible and unalterable minute atoms, which gave permanence to matter, and postulating further that the observed changes in matter could be accounted for in terms of the motions of these atoms. Untrammeled by the demands of experimental test, they found it natural to develop and extend the idea to "account for" observed properties of matter. For instance, might not the sharp taste of vinegar be due to pointed atoms, or the smooth feel of olive oil to atoms of rounded shape? At the time, it seemed quite reasonable to put forward these ideas, and they were well received by some. However, atomism was not universally endorsed because it was as easy to raise difficulties as to explain some phenomena. How could these very tiny and inanimate particles account for living things? How could these particles, believed to be limited in variety, account for the apparently limitless variety in nature? There was no adequate way to test these questions; indeed we still wrestle with some of them today and have no complete explanation, even with the advantage of modern theory and the facilities of modern laboratories.

John Dalton is now regarded by many as the father of modern atomic-molecular theory. Certainly such outstanding scientists as Boyle and Newton, who preceded him by a century and a half, believed in, and in their writing, referred to the particulate nature of matter. Dalton, however, was able to state the theory in a way that brought out its direct connection to the experimental data of Proust, Berthollet, and others. The Daltonian theory, based on the available experimental data at the turn of the eighteenth century, may be summarized as follows.

1. All matter is composed of tiny indivisible and indestructible particles (atoms).

2. There are several different kinds of atoms characteristic of the several different elements.

3. a) Atoms of a given element all have the same *weight*.
   b) Atoms of different elements have different *weights*.

4. The atoms combine in various ratios of small whole numbers to form compounds.

The concept of characteristic weights of the atoms of the different elements turns out to be a particularly fruitful one: it leads directly to Proust's law of definite proportions. This concept appears to have been first clearly emphasized by Dalton. Today we modify the postulates somewhat because of recent subtle discoveries,* but with the data of Dalton's day, the postulates proved to offer a compelling explanation of the quantitative laws of chemical combination. We shall now present and discuss the laws in succession, recapitulating the crucial law of definite proportions.

1.5 QUANTITATIVE RELATIONSHIPS IN CHEMISTRY

In any specific compound the constituent elements are always combined in the same proportions by weight, regardless of the source or mode of preparation. This law has already been stated in the earlier discussion of Proust's contention of the definite and invariant composition of substances. According to atomic theory we may (1) regard compounds as composed of smallest characteristic units called molecules, (2) regard these molecules as aggregates of definite numbers of the different atoms corresponding to specific elements, for example, the atoms of the elements carbon, hydrogen, and oxygen present in alcohol or sugar, (3) regard the constituent atoms within the molecule as definite in relative numbers and of characteristic weights for each element. These three ideas lead directly to the conclusion that compounds have definite proportions. We must say that the law of definite proportions is entirely consistent with the Daltonian atomic theory. It is difficult even to see the direction of any argument by which we might conceive of some theory quite independent of that of Dalton which would serve as well here. This idea is illustrated in Fig. 1.4.

The Law of Multiple Proportions

If two elements form more than one compound, then the different weights of one which combine with the same weight of the other are in a ratio of small whole numbers. Unlike the law of definite proportions, this law was not recognized as generally applicable until *after* the presentation of the atomic theory in 1804, and it was an early triumph of the theory that this law was predicted.

---

* Atomic weight becomes subordinate to atomic number with the discovery of isotopes. See Chapter 8.

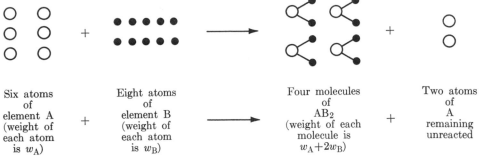

| Six atoms of element A (weight of each atom is w_A) | + | Eight atoms of element B (weight of each atom is w_B) | ⟶ | Four molecules of AB_2 (weight of each molecule is $w_A + 2w_B$) | + | Two atoms of A remaining unreacted |

Fig. 1.4. Illustration of a chemical reaction according to the atomic-molecular model, assuming that the compound formed consists of one atom of element A associated with two atoms of element B to form the triatomic molecule AB_2. (Since element A is in excess, some of its atoms are left unreacted.) The composition of the compound would have the *definite proportions:* fraction of A = $w_A/(w_A + 2w_B)$, and fraction of B = $2w_B/(w_A + 2w_B)$, where w represents the weight of an atom. (From A. B. Arons, *Development of Concepts of Physics*, Addison-Wesley, Reading, Mass., 1965.)

In fact Dalton himself in the period from 1801 to 1804 observed from his experiments on the reactions of oxygen with nitrous gas (nitrogen) that a certain volume of air containing a fixed quantity of oxygen would combine with 36 or 72 measures of nitrogen "according to certain variations in the mode of conducting the experiment." Soon afterward, Dalton observed another example of simple multiple proportions, finding, for the two gases methane and ethylene, that a fixed weight of carbon combined with twice the weight of hydrogen in the case of methane as compared to ethylene. These observations were, of course, entirely consistent with the predictions of the atomic theory. Suppose for instance that in compound I there are two atoms of element B to one atom of element A but that in compound II only one atom of B is combined with an atom of A. Then it follows that the weights of B combining with any fixed weight of A in the two compounds will be related *to each other* by a ratio of small whole numbers, in this case 2:1.

The strength of the Daltonian theory lay in its power to predict and its stimulation of new research. Chemists of his day were led to seek out additional examples of the law of multiple proportions, and a considerable number of such examples were found either in existing data or in data from freshly conceived and executed experiments.

It is not surprising that the multiple-proportion relationship was not at first noticed in data existing prior to 1804. Until one looked at the data with the relationship in mind, it was not at once obvious. For example, in earlier experiments on compounds of carbon and oxygen, an analysis by Lavoisier of carbon dioxide ("carbonic acid," "fixed air") had yielded 28% by weight of carbon and 72% of oxygen. Analysis by Clement and Desormes of carbon monoxide ("carbonic oxide," "inflammable air") had resulted in finding 44% of carbon and 56% of oxygen. The multiple-proportion relationship is not immediately evident here, and in fact for some years it was overlooked. However it becomes apparent if we fix the weight of one element, say, at 1 g for carbon and find (by recalculation from the data) the weights of the other element, oxygen, combined with this 1 g of carbon in the two compounds. From the percentage-composition data given above, these weights of oxygen turn out to be 2.57 g and 1.27 g respectively for the carbon dioxide and the carbon monoxide. These numbers then stand in the ratio 2.02 to 1.00. Taking into account the probable lack of precision in the data, we may reasonably call this ratio 2:1.

Problem 1.4 Examine quantitatively the weight percentage-composition data in the following table for agreement with the law of multiple proportions. Consider both the data available in 1808 and the modern values.

Substance		Composition by weight	
Name used by Dalton	Modern name	Quoted about 1808	Modern values
Carbonic oxide	Carbon monoxide	44% C; 56% O	42.9% C; 57.1% O
Carbonic acid	Carbon dioxide	28% C; 72% O	27.3% C; 72.7% O
Olefiant gas	Ethylene	83% C; 17% H	85.6% C; 14.4% H
Carburetted hydrogen	Methane	71% C; 29% H	74.9% C; 25.1% H

Comment on the significance of the difference between the data of 1808 and modern values with respect to the way in which the law would be discussed. Consider the problem of deciding what is experimental error.

The Law of Equivalent Proportions

Suppose that two substances A and B react with each other and with a third substance C. All three may be either elementary or compound substances. Now the weights of A and of B reacting separately with some fixed weight of C bear some ratio R to each other. Usually R is not an integer. If now we find the ratio R' by which A and B react with each other, then it is observed that R' is some simple multiple or fraction of R. Thus $R' = nR$, *where n is a small integer* or a ratio of small integers.

This statement of the law is of necessity rather complex, and the relationship is made clearer by an illustration. Nitrogen (A) and oxygen (B) react with hydrogen (C) to form ammonia and water respectively. One gram of hydrogen will react as above with 4.66 g of nitrogen and with 8.00 g of oxygen. Now the nitrogen and oxygen can form several different compounds. One of these, we shall show later, contains single atoms of nitrogen and oxygen in a molecule and hence has the formula NO. In this compound the ratio R' of the reacting weights of nitrogen (A) and oxygen (B) is found to be 0.875. On the other hand, the value of R is $4.66/8.00 = 0.583$. Since $R' = nR$, or $0.875 = 0.583n$, solving for n we get 1.501 or $\frac{3}{2}$.

Problem 1.5   Consider the significance of the difference between 1.501 and $\frac{3}{2}$ in the preceding calculation.

For other compounds of nitrogen and oxygen, n has different values, but all represent ratios of small integers within the limits of experimental error.

An analysis of these apparently rather complex relationships still finds rather straightforward explanation in the atomic theory. Thus a fixed weight of hydrogen C containing some fixed number of atoms combines with either nitrogen A or oxygen B. According to the atomic theory we may expect that the numbers of atoms of A and B reacting with the fixed number of atoms of C will bear some fairly simple integral (or integral fraction) relationship to the number of atoms of C reacting. This means on reflection that the number of atoms of A and of B reacting *with each other* may also be expected to stand in a ratio of small integers.

The three laws of definite, multiple, and equivalent proportions all lend support to the credibility of the Daltonian theory, although no one of these laws gives definitive "proof" of the theory, nor is such "proof" unequivocal if we bring to bear the combined weight of all three. It always remains possible that an alternative explanation of the laws exists. However, the Daltonian atomic theory was so successful in interpreting the apparent significance of these various weight relations that most chemists of the early decades of the nineteenth century came to use it and explore its implications as a central concern of further research. It had been, at the very least, fruitful in calling their attention to subtle but quite striking regularities. To any research scientist this is an exciting triumph.

1.6 THE PROBLEM OF ATOMIC WEIGHTS

We have witnessed the success of the atomic theory in "explaining" the quantitative weight relationships associated with the composition of compounds and with the interaction of substances in chemical changes. The next problem that fairly clamored for solution in the early nineteenth century was that of determining the weights of atoms or at least their weights relative to each other. (In fact, Dalton's first announcement of his atomic theory contained a table of relative atomic weights.)

Dalton knew that 1 g of hydrogen combined with 8 g of oxygen to form water. Adopting a *rule of greatest simplicity* (for lack of any alternative), he assumed HO as the formula for the molecule of water (i.e., one atom of hydrogen to one atom of oxygen). Then it would follow directly that the atomic weight of the oxygen atom, relative to that of the hydrogen atom (arbitrarily set at 1) must be 8. From the data already cited on carbon monoxide and carbon dioxide, and with the postulate of greatest simplicity, he assumed (in this case "correctly") that the formulas were respectively CO and CO_2. Since the weight of carbon associated with 1 g of oxygen in carbon monoxide was known to be

0.75 g, Dalton postulated that the atomic weight of carbon must be 6 based on the scale of 1 for hydrogen and 8 for oxygen. Assuming the formula CO_2 for carbon dioxide, similar reasoning would again lead to 6 for the atomic weight of carbon. We now "learn" in childhood that the assumption of HO for water is "incorrect." If Dalton had abandoned his rule of maximum simplicity and postulated H_2O for water, then all would have been well, and he would have emerged with the "correct" relative atomic weights of 16 for oxygen and 12 for carbon. But how was he to know this? What possible justification could there be for selecting one formula over another?

If we pursue the principle of greatest simplicity further we run into inconsistencies. For example, let us attempt to calculate the atomic weight of nitrogen on the basis of atomic weights of 1 for hydrogen and 8 for oxygen and the weight relationships in ammonia and nitrous oxide, the oxide of nitrogen of lowest oxygen content. The greatest simplicity rule suggests the formulas NH and NO. The known compositions of the two compounds together with the atomic weights of 1 for hydrogen and 8 for oxygen would lead to atomic weights of nitrogen of $4\frac{2}{3}$ and 14 respectively! It is obvious that we must look further for help, but in what direction?

Problem 1.6 Carry out the calculations leading to the inconsistent atomic weights for nitrogen if you adopt the rule of greatest simplicity. Suppose the compositions by weight are: Ammonia, 17.6% H, 82.4% N; nitrous oxide 36.4% O, 63.7% N.

It is easy to see that a system of relative atomic weights could be worked out if we knew the formulas and the combining weights. Further reflection will show that we can also work out the formulas provided we have at our disposal both the combining weights and the atomic weights. It seems, however, that we must find the two interdependent entities, molecular formula and atomic weights, *at the same time*. This was the problem that faced chemists in the 1830's. Dalton's theory may in fact be summarized with a simple equation. If the only experimental data are the combining weights (weights of elements A and B that enter into a compound) we can write

$$\frac{\text{weight A}}{\text{weight B}} = \frac{\text{number of atoms of A} \times \text{atomic weight A}}{\text{number of atoms of B} \times \text{atomic weight B}}. \tag{1.1}$$

Only the factor on the left is experimentally accessible. The term on the right (number of atoms of A/number of atoms of B) represents an aspect of the theory, the "formula" of a molecule. The term on the right (atomic weight of A/atomic weight of B) represents an aspect of the theory, the relative atomic weights of A and B. We have a "triangular difficulty." If we can know two vertices of the triangle in Fig. 1.5, we can calculate the third. So far experiment gives only one.

It turns out that help was to be found from the study of *gases* and in particular from their *volume* ratios in chemical reactions studied in a series of

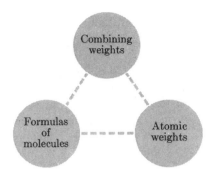

Fig. 1.5. The atomic-molecular theory "triangle."

meticulous experiments by the French chemist Joseph Louis Gay-Lussac. But the way was tortuous and long and culminated only after several decades in the interpretation of the work of Dalton, Gay-Lussac, and Amedo Avogadro by Cannizzaro.

1.7 THE GAS LAWS

In 1643, the Italian physicist Evangelista Torricelli suggested that we are living at the bottom of a "sea of air" whose weight exerts a pressure on us and everything around us. He also invented the barometer to measure this pressure (by measuring the height of a column of mercury that this pressure would support). Less than twenty years later, Robert Boyle had completed a careful quantitative study of the compressibility of gases—the relationship between the volume occupied by a confined gas and the pressure exerted on the walls of its confining container. This relationship, known as *Boyle's law*, states that if a *given weight* of gas is trapped in a vessel of variable volume, and the *temperature is kept constant*, then any decrease in volume increases pressure proportionally. That is,

$$P \cdot V = \text{constant} \quad \text{(for fixed weight of gas at fixed temperature),}$$

where P is the pressure exerted on the walls of the vessel. (Pressure is expressed in units of the normal pressure of the atmosphere, i.e., in atmospheres, or in terms of the height of a column of mercury which the pressure can support, in mm of Hg,* or in terms of the force per unit area, i.e., in newtons per square meter.) V is the corresponding volume of the gas (expressed, for example, in liters or cubic centimeters).

Problem 1.7 A balloon is filled with helium at ordinary atmospheric pressure (1 atm). Its volume is $0.14 \ m^3$. It is released (e.g., lost). As it rises, the atmospheric pressure decreases. Find the volume when the balloon has reached a height at which the confining pressure is now only 0.7 atm.

---

* 1-atm pressure supports an Hg column 760 mm or 76 cm high.

Problem 1.8 A sample of gas was found to occupy 540 milliliters (ml) when the barometer in the laboratory read 756 mm of mercury. A few hours later, the barometer had fallen to 729 mm, but the temperature was unchanged. Find the new volume of the gas.

Boyle's law applies at constant temperature. Now suppose that we warm or cool a sample of gas while maintaining the pressure constant (as for example by confining at atmospheric pressure). How will the volume vary? The answer to this question appeared more than one hundred years after the enunciation of Boyle's law. A relation between volume and temperature was discovered independently by Jacques Charles (1746–1832) and Joseph Louis Gay-Lussac (1778–1850). At constant pressure, they observed that the increase of volume was proportional to the increase of temperature. This is usually called Charles' law.

There is a surprising feature in the presentation of Boyle's and Charles' laws. *It has not been necessary* to specify the chemical composition of the gas. For example, when heated from 0° to 1°C, almost all gases expand about $\frac{1}{273}$ (0.37%) in volume. No such unique distinctive property exists for solids or liquids. So called "thermal-expansion coefficients" must be evaluated for each chemically different solid. A temperature increase of 1°C leads to volume changes of 0.01% for ice, 0.0025% for quartz, and about 0.02% for mercury.

It was also observed that the thermal expansion of gases remained nearly linear with respect to temperature over most of the readily accessible temperature range. Noting that a gas contracts $\frac{1}{273}$ of its volume when cooled from 1° to 0°C, one might imagine that upon cooling the gas to approach -273°C (that is, 273° below the melting point of ice), its volume would approach zero. Actually, all gases liquefy before this point is reached, but the idea may be employed to construct a new temperature scale. Instead of designating the melting point of ice as the zero of temperature, one may designate the point at which gas volume is predicted to vanish. On such a scale (called the "absolute" or "Kelvin" scale with 0°K being -273°C, or more precisely -273.15°C), it follows that the volume of a gas is directly proportional to the temperature. Denoting "absolute" temperatures by the symbol T, we have

$$V \propto T \quad \text{or} \quad V/T = \text{a constant} \quad \text{(for a fixed weight of gas at constant pressure).}$$

The law of Gay-Lussac and Charles is simpler and more useful in the form employing the absolute temperature scale.

Problem 1.9 In the paragraph above we have expressed the relationship between the volume of a gas and its centigrade temperature. The algebraic expression for this relationship is

$$V_t = V_0 + \frac{t}{273} \times V_0,$$

where V_0 and V_t are the volume of a gas sample at 0° and t°C respectively. Clearly from the above discussion T (in degrees Kelvin) is simply $t + 273$. Now carry out the algebra to show that $V_t/V_0 = T/T_0$, or more generally that $V_t/V'_t = T/T'$.

Equations (1.2a) and (1.2b) summarize the two experimental laws which provide at least a good first approximation to the behavior of all gases. Before proceeding we must admit that, for a fixed weight of gas,

$$PV = \text{a constant} \quad \text{(at constant } T\text{)}, \tag{1.2a}$$

$$V/T = \text{a constant} \quad \text{(at constant } P\text{)}. \tag{1.2b}$$

More precise study shows these laws describe the behavior of gases accurately only at low pressures. However, they provide a useful first approximation to the actual behavior of gas at higher pressures (near 1 atm). The approximation is very valuable in both theory and practice. Let us now try to combine the two laws into a single equation. A simple mathematical step serves to combine the two laws into a general law. Since we have $V \propto 1/P$ (for constant T) and $V \propto T$ (for constant P), it follows that V is *jointly* proportional to $1/P$ and to T or $V \propto T/P$. Any statement of proportionality can be converted to an equation by introducing a proportionality constant. In the present case, let this constant be represented by the symbol r. We have $V = rT/P$ or

$$PV = rT. \tag{1.3}$$

Previously, we noted the important condition of the constancy of quantity of gas. (The "constant" r depends on the quantity and type of the gas under consideration.)

Problem 1.10   It is known that 2.02 g of hydrogen gas occupy 22.4 liters when under so-called standard conditions (1 atm pressure, 0°C or 273°K).

a) Find the value of r in liter atm/deg.
b) Find the volume that the 2.02 g sample will occupy at 9.0 atm and "room temperature" (21°C).
c) The density of oxygen is 1.43 g/liter under standard conditions. Find the weight of oxygen that yields the *same* value of r as 2.02 g of hydrogen.

In the next section we shall be concerned with the relationships among the volumes of reacting gases. The first point (of several!) that will concern us in this section is that attention must be given to comparing gas volumes at the same temperature and pressure.

1.8 GAY-LUSSAC'S LAW OF COMBINING VOLUMES*

During the first years of the nineteenth century concurrently with the development of Dalton's atomic theory, substantial progress was being made in labora-

---

* In preparing this section we have relied heavily on section 29.1, A. B. Arons, *The Development of Concepts of Physics*, Addison-Wesley, Reading, Mass., 1965.

tory techniques for purifying and measuring different gases. This made possible for the first time quantitative study of reactions between gases. In 1809, Gay-Lussac (who was associated with the group surrounding Berthollet*) published a report in which he described his intention to make known some new properties of gases, the effects of which are regular, by showing that these substances combine amongst themselves in very simple proportions, and that the contraction of volume which they experience on combination also follows a regular law. Note in this remark the concern for volume. The relation between volume of a gas and quantity of a substance in weight terms was explored in Section 1.7.

The famous naturalist Alexander von Humboldt and Gay-Lussac had performed experiments on the formation of water vapor by passing sparks through mixtures of hydrogen and oxygen. In these experiments they had observed that, for any given volume of oxygen *completely consumed* in the reaction, exactly *twice* this volume of hydrogen was required. (The measurements were precise to ±0.19%!) The volume of water vapor produced, recalculated by means of Eq. (1.3) to the temperature and pressure prevailing before the reaction, was found to be equal to that of the hydrogen consumed. We can write an equation describing these results as:

2 volumes hydrogen + 1 volume oxygen → 2 volumes water vapor

if all volumes are measured at, or recalculated to, the same temperature and pressure.†

The experiments reveal a *law of definite proportions by volume.* But what is more, they indicate that the volumes themselves bear *small-whole-number-ratios to each other.* The integers appear immediately and no special theoretical viewpoint is required to reveal the regularity. This law was "easily" discovered. Note the contrast to the law of multiple proportions by weight, which was formed only after Dalton had thought of atoms.

Gay-Lussac investigated volume relations in other gas reactions and also recomputed on a volume basis the gravimetric data on gas reactions previously reported by other investigators. Some of his results, expressed in modern language, are presented in Table 1.3.

Gay-Lussac's results were rounded off to the integer values; in very few instances was the evidence as precise as that in the hydrogen-oxygen reaction. In cases such as the oxides of nitrogen, which are difficult to handle experimentally, the experimental ratios departed from integer values by as much as 6 or 8%. It is evidence of Gay-Lussac's faith (was it justified at the time

---

* Keep in mind the influence Berthollet might be expected to exert on any work bearing on Dalton's atomic theory since he rejected the law of definite proportions.
† The gas volumes here must have been measured at some temperature above 100°C if the pressure was one atmosphere. At any temperature less than 100°, the water vapor would condense.

Table 1.3

**Gay-Lussac's summary of volume relations in a number
of reactions between gaseous substances**

2 volumes hydrogen + 1 volume oxygen → 2 volumes water vapor	(1)
3 volumes hydrogen + 1 volume nitrogen → 2 volumes ammonia	(2)
2 volumes sulfur dioxide + 1 volume oxygen → sulfur trioxide (solid)	(3)
1 volume ammonia + 1 volume hydrogen chloride → ammonium chloride (solid)	(4)
2 volumes carbon monoxide + 1 volume oxygen → 2 volumes carbon dioxide	(5)
1 volume nitrogen + 1 volume oxygen → 2 volumes nitric oxide	(6)
1 volume oxygen + carbon (solid) → 2 volumes carbon monoxide	(7)

by logic?) in the correctness of a simple law, that he chose to assume that such large discrepancies arose from experimental uncertainties.

Gay-Lussac summarized the significance of the data in the law of combining volumes which we now name in his honor. This law states that the volumes of the gases entering into a chemical reaction and of the gaseous products are in the ratio of small integers.

We have the summary in Gay-Lussac's own words:

Thus it appears evident to me that gases combine in the simplest proportions when they act on one another. . . . In all the preceding examples the ratio of combination is 1 to 1, 1 to 2 or 1 to 3. It is very important to observe in considering weights that there is no simple relation between the elements of any one compound; it is only when there is a second compound between the same elements that the new proportion of the element that has been added is a multiple of the first quantity. Gases, on the contrary, in whatever proportions they may combine, always give rise to compounds whose [constituents] by volume are multiples of each other. Not only, however, do gases combine in very simple proportions . . . but the apparent contraction of volume which they experience on combination has *also* a simple relation to the volume of the gases. . . .

These ratios by volume are not observed with solid or liquid substances, nor when we consider weights, and they form a new proof that it is only in the gaseous state that substances are in the same circumstances and obey regular laws.*

Problem 1.11   In what sense does the law of combining volumes seem to lend support to Dalton's theory? How would you interpret the significance of the small whole numbers in Gay-Lussac's data? Given that the volume of a gas at a given temperature and pressure is directly related to the number of particles present, what seems to be implied concerning the number of particles present in equal volumes of different gases? Interpreted in this way, what does the two-to-one ratio in the hydrogen-oxygen reaction suggest concerning the binary or ternary character of the water molecule?

---

* Alembic Club Reprint No. 4, p. 15, 1899 (Simpkin, Marshall, Hamilton, Kent & Co., Ltd. Reissue Ed., E. & S. Livingstone Ltd, 16 and 17 Teviot Place, Edinburgh, 1960).

1.9 AVOGADRO'S MODEL FOR GASES

The previous section developed a broad hint that the interpretation of Gay-Lussac's law lies in the idea that equal volumes of gases contain equal numbers of particles (molecules, in most cases). If the fundamental unit of chemical reaction involves a small integral number of atoms or molecules reacting to give products, the appearance of small integers relating volumes in reactions strongly suggests that the volumes contain the same numbers of particles. For reasons connected with his meteorological researches and a misreading of the implications of an earlier interpretation of Boyle's law by Newton, Dalton *rejected* Gay-Lussac's results *precisely because they seemed to imply the equal numbers idea*. Dalton attacked the experiments of Gay-Lussac and presented some somewhat sloppy data of his own to show the incorrectness of the combining-volumes regularities. However, all the evidence points to the fact that Gay-Lussac's work was a marvel of achievement when one considers the apparatus available to him at the time, and subsequent experimentation settled the issue of the validity of Gay-Lussac's work. Its interpretation in terms of atomic theory was offered by the Italian physicist Amedeo Avogadro in 1811 in a rather obscurely phrased essay. It was to be half a century before the full impact of Avogadro's theory was generally accepted. Until its significance was more completely apprehended, the problem of atomic weights remained apparently unsolved. In fact, some reputable scientists of the era were skeptical about the possibility of its ultimate solution and, as a result, were also skeptical about the foundations of the atomic theory.

Avogadro's model for gases may conveniently be presented under four headings.

a) The particles of a gas are not in contact. Dalton's concept that the gas particles were in contact was now abandoned. Rather it was postulated by Avogadro, and indeed by others as well, that a gas consisted largely of empty space and that any individual particle occupied only a tiny fraction of its share of the total volume. This meant that, unlike the weights, the absolute *volumes* of the gas *particles* themselves had little significance.

b) *Avogadro's hypothesis.* The focal feature of Avogadro's proposed model is referred to as Avogadro's hypothesis: Under equal conditions of pressure and temperature, equal volumes of different gases contain equal numbers of particles. This is true whether the gas is an elementary or a compound substance; it is true whether the particles are monatomic or polyatomic, and it even applies to mixtures of gases. This law ultimately turned out to have a high utility, but at first it presented some rather puzzling implications, especially in its application to the law of combining volumes. Let us consider the volume relationships between the reactant gases hydrogen and oxygen and the resultant product water vapor, if the whole system is kept at a temperature well above 100°C, so that water vapor will not condense under normal atmospheric pressure. (See Fig. 1.6.)

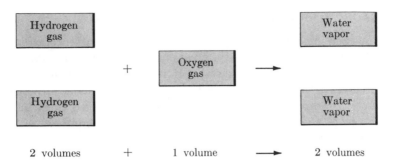

Fig. 1.6. Volume relation in Gay-Lussac's experiment on the formation of water vapor.

By Avogadro's hypothesis, these equal volumes contain equal numbers of particles. On the basis of Dalton's principle of maximum simplicity we may conclude that the ultimate particle (molecule) of water would contain one atom of hydrogen and one of oxygen and we would write the equation

$$H + O \rightarrow HO.$$

But there is an inconsistency here, because this would suggest that, in contrast to the findings of Gay-Lussac, one volume of hydrogen plus one volume of oxygen produces one volume of water vapor. To dodge this difficulty let us consider the third part of Avogadro's gas model, which is an imaginative departure from the current thinking of that period.

c) *Molecules of elementary gases.* Prior to Avogadro, Dalton and everyone else who had written about the problem, had assumed that the ultimate particles characterizing an elementary gas were always composed of single atoms. Avogadro now proposed that for some gaseous elements at least, the smallest particles were not single atoms, but rather assemblies of two or more like atoms. Up to this point in our discussion we have applied the term molecule only to compound substances in which the molecule would contain two or more *different* atoms of the different elements entering into combination. Now we consider Avogadro's suggestion that the ultimate particles characteristic of certain elements may consist of groups of two or more *like* atoms. Thus we conceive of molecules of elementary substances. When we refer to the concept "equal volumes, equal numbers" we refer to numbers of molecules characterizing a given compound *or* elementary substance.

Using this concept of polyatomic molecules for certain elementary gases, we need no longer encounter an inconsistency in the interpretation of the law of combining volumes. Returning to Fig. 1.6, for instance, we may now postulate a diatomic molecule for oxygen and thus we emerge with Fig. 1.7, which gives the pictorial representation that two monatomic molecules of hydrogen

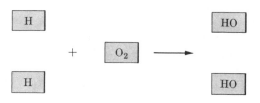

Fig. 1.7. Scheme for a consistent hypothesis.

may react with one diatomic molecule of oxygen to produce two molecules of HO. There is nothing inconsistent between Fig. 1.7 and the experimental facts of hydrogen-oxygen reaction.

However, in postulating polyatomic molecules perhaps we have opened a Pandora's box. We might equally well postulate diatomic hydrogen as well as diatomic oxygen, leading to the formula H_2O for water, or triatomic hydrogen, giving us H_3O, or monatomic hydrogen with O_4 leading to

$$2H + O_4 \rightarrow 2HO_2.$$

Have we merely compounded the confusion?

Avogadro's gambit was to make one further postulate which seems to smack of Dalton's unsuccessful maneuver in dealing with the atomic weight problem.

d) *Avogadro's rule of simplicity.* Out of the multiplicity of possible formulas for water which become possible if we accept the notion of polyatomic molecules of elements, how shall we choose the uniquely correct formula? Faced with the problem, Avogadro added a final postulate: if several processes leading to several different formulas (for water in this case) are in harmony with both the experimental observations and the other postulates of the theory, then *choose as correct the one that is simplest.* On this basis he rejected the HO formula as leading to an inconsistency, but accepted H_2O as the simplest formula which gave a consistent interpretation of *all* of the experiments (see also Section 1.6).

It must be noted that we have not reached a unique solution of the problem here, because we have not ruled out the possibility that either hydrogen or oxygen molecules might be more complex than diatomic. The final attempt to focus on a unique solution for the formula of a gaseous substance was that of Cannizzaro, and the results were presented to the chemical community in 1860. Thus Avogadro's work, unlike that of Dalton, lay fallow for almost half a century.

In addition to the failure of Avogadro's model for gases to lead to unique molecular formulas, there were several other objections or at least sources of reservation which militated against its immediate acceptance. Avogadro could give no physical reason for the concept of equal volumes, equal numbers, although it may have seemed plausible enough. He could offer no mechanism to explain why like atoms in elementary gas molecules could cling together in clusters of two or more. His model worked well for the four common elementary gas molecules H_2, N_2, O_2, and Cl_2, *but* with some other gases the molecules were not diatomic, *but* instead the data and reasoning led to Hg, P_4, As_4, and at successively lower temperatures to S, S_2, S_6, and S_8. This departure from simplicity lacked appeal. And finally, if the molecules are tiny, then how tiny? It would have been satisfying to know, but the answer to this problem of size did not come until late in the nineteenth century. For these reasons scientists were reluctant to accept Avogadro's postulates without reservation, and even if they did, no satisfactorily unequivocal path to the relative atomic weights was indicated.

In default of an adequate system of relative atomic weights it was customary in the middle decades of the nineteenth century to rely on a system of equivalent weights, a system which gained in acceptability because of Faraday's electrochemical experiments in the 1830's. Electrochemistry is treated in Chapter 6. Although the equivalent weight concept really could contribute nothing directly to the solution of the atomic weight problem, it was used with telling effect to find exact atomic weights once the way was found to discovering molecular formulas and even rough atomic weights.

For a time, molecular formulas seemed somewhat unimportant, but the rapid growth of a new field, organic chemistry (or the chemistry of carbon compounds), revived the demand for a resolution of the problem. In this period hundreds of carbon compounds consisting of apparently simple molecules were isolated from natural or synthetic sources. The need for a clear and unequivocal knowledge of molecular formulas became one of paramount importance, since the same percentage composition by weight was found to occur in clearly different compounds (see Chapter 2).

1.10 THE METHOD OF CANNIZZARO

Stanislao Cannizzaro, Avogadro's countryman, succeeded in making Avogadro's hypothesis attractive by employing it in an extremely clever way. This resolution of the problem of atomic weights and molecular formulas may be understood by examining the examples listed in Table 1.4.

Suppose, as Cannizzaro did, that we set ourselves the problem of finding the atomic weights of the set of elements listed by symbol in column 3 of Table 1.4. We proceed as follows:

1. Prepare in pure form both the elementary gases and as many as possible of their compounds, which either are gaseous or can be vaporized without decomposition at elevated temperatures.

2. Analyze each compound thus prepared for percent by weight of its constituent elements.

3. Now find the weight of each element in an arbitrarily fixed volume of the elementary substances and of each of the prepared compounds, say in 1 liter at standard conditions (0°C and 1-atm pressure).*

4. Consider what should be expected of the table if Avogadro's hypothesis is correct. *All of the weights of an element should be related by a set of integers— the number of atoms in the molecules, since the calculation based on equal volumes means equal numbers of molecules.* In our list of the compounds for each element, it is probable that we shall have selected at least one having only one atom of that element per molecule, or at least that the list will contain two compounds differing by only one atom per molecule. In that case, the greatest common divisor of the set of weights in (3) will reduce them to integers corresponding to the number of atoms in the molecule.

We now examine our data in column 3 and select the greatest common divisor for each element. The ratios of these values will be the ratios of the atomic weights of the different elements considered. Preserving these ratios we may convert the row of greatest common divisors to the accepted atomic weights of the final row, which are based on O = 16.000 (or ^{12}C = 12.000†). As we have noted, division of the numbers listed for each element in column 3 by the greatest common divisor now tells us the number of atoms per molecule and hence we may obtain the molecular formulas and the relative molecular weights given in column 4.

Closing the circle, note that the existence of the required integral ratios which we predict on the basis of Avogadro's hypothesis impressively *confirms* the hypothesis!

Problem 1.12 The derivation of formulas and molecular weights from the data of column 3 will be better understood after completion of column 4 for the lower half of Table 1.4. Do this as an exercise.

1.11 THE MOLE CONCEPT

On the basis of the atomic-molecular theory it becomes evident that atoms and molecules are the reactive units in chemical change. The simplest chemical events that we could conceive of would involve the union of small numbers of atoms or small molecules into more complex molecules or other processes such as dissociation or exchange of partners also involving small numbers.

---

* Because some of the compounds may well be liquids rather than gases at the above conditions of temperature and pressure, the observed volumes are "reduced" using the gas laws to the volume which the compound would occupy if it were a gas at conditions chosen. Clearly the choice of a particular temperature and pressure is arbitrary.

† Twentieth century results indicate that there are in fact chemically equivalent atoms of varying weight (isotopes) which occur mixed in nature to give an element its characteristic atomic weight. The most modern atomic weight scale is constructed on the basis of assigning to one isotope of carbon (^{12}C) the value 12.000. See Chapter 7.

Table 1.4

Illustration of Cannizzaro's analysis of the density and composition data for various gases*

| (1) | (2) | (3) Grams/liter of each constituent element, based on composition data | | | | | | | | (4) |
Name of gas or vapor	Density ρ, recalculated to 0°C and 1-atm pressure, g/liter	H	O	S	P	Cl	N	Hg	C	Molecular formula and relative molecular mass, Scale O ≡ 16
Hydrogen	0.090	0.090								H_2, 2.02
Oxygen	1.43		1.43							
Sulfur (<1000°C)	8.59			8.59						
Sulfur (>1000°C)	2.86			2.86						
Phosphorus	5.53				5.53					P_4, 124
Chlorine	3.16					3.16				
Nitrogen	1.25						1.25			
Mercury	8.96							8.96		
Water	0.803	0.090	0.713							H_2O, 18.02
Hydrogen chloride	1.63	0.045				1.58				
Ammonia	0.760	0.135					0.625			NH_3, 17
Phosphine	1.52	0.135			1.38					

Hydrogen sulfide	1.52	0.090		1.43					
Hydrogen cyanide	1.20	0.045					0.625		0.53
Chloride of phosphorus	6.13				1.38	4.74			
Calomel	10.54					1.58		8.96	
Corrosive sublimate	12.12					3.16		8.96	
Nitrous oxide	1.96		0.713				1.25		
Nitric oxide	1.34		0.713				0.625		
"Carbonic oxide"	1.25		0.713						0.54
"Carbonic acid"	1.96		1.43						0.53
Oxide of sulfur	2.86		1.43	1.43					
Ethylene	1.25	0.180							1.07
Alcohol	2.05	0.270	0.71						1.07
All numbers in the vertical column are integral multiples of:		0.045	0.713	1.43	1.38	1.58	0.625	8.96	0.535
Converted to scale O ≡ 16, numbers in the preceding row become:		1.01	16.000	32.1	31.0	35.5	14.0	201	12.0

* From A. B. Arons, *Development of Concepts of Physics* (Addison-Wesley, Reading, Mass., 1965).

Unfortunately, such simple situations, although they can be studied, are inconvenient to investigate. Hence we resort to "counting out" large numbers of atoms or molecules in sufficient quantity so that their relative weights or volumes (if gaseous) can be conveniently measured. In this procedure of "counting out" it is convenient to define a *unit* of weight or mass for each substance that has a *fixed* number of atoms or molecules; comparing quantities of substances in such units directly compares the number of atoms or molecules available for reaction. It is the natural unit system for discussion of chemical reactions. How should it be chosen?

Suppose that we set up a scale of molecular weights of 16.00 units for the oxygen atom; then, relative to the arbitrary base O = 16.00, the weight of the O_2 molecule is 32.00 units, the H_2 molecule 2.02, and the H_2O molecule 18.02. Then it is clear that weights of these substances chosen in these ratios will contain the same number of molecules. In particular note that 32.00 g of O_2, 2.02 g of H_2, and 18.02 g of H_2O must contain the same number of molecules. Similarly, a number of grams of any substance equal to its molecular weight must have the same number of molecules as our 32.00 g of O_2, 2.02 g of H_2, or 18.02 g of H_2O. For any substance, the weight in grams that is numerically equal to its molecular weight is called *a mole*. Note that the number of moles in any sample is simply the number of grams divided by the molecular weight of the substance:

$$\text{moles} = \text{grams}/MW,$$

where MW is the molecular weight. Elements like mercury exist as monatomic species (Hg not Hg_2 in the gaseous state). It is conventional to call a mole of atoms a *gram-atom* rather than a mole, since the term mole is clearly related to the word molecule. The authors feel this to be an unnecessary complication of terminology. It will not be consistently followed in this book. We shall speak of a "mole of atoms" as well as a "mole of molecules."

Consider, from another point of view, what the mole unit system accomplishes: 2.02 g of hydrogen react with exactly 16.00 g of oxygen to form water. Reexpressed in moles, 1.00 mole of hydrogen reacts with 0.50 mole of oxygen. The mole units bring to the surface the 2:1 relationship, the simple integer relationship at the root of atomic theory.

If a mole of any substance is the same number of molecules, what is that number? First it is important to say that it is arbitrary. It is fixed by using an atomic weight scale based on O = 16.00 and measuring samples in grams. However, it *is* fixed, and the number of molecules in a mole has been christened Avogadro's number (designated N_0). It is not necessary to know it for chemical application of the Dalton-Avogadro atomic-molecular theory. However, a significant final vindication of the theory has come from the fact that several independent methods developed in later years for indirectly counting atoms yield a consistent value for N_0; namely, 6.023×10^{23} atoms or molecules per mole.

It should be noted that N_0 represents a number regardless of the entities being counted; this number is called a mole. Just as we refer to a dozen oranges, or a gross of bolts, we may refer to a mole of atoms or a mole of molecules or a mole of elephants for that matter. While the mole of atoms is a highly useful quantity, the mole of elephants is scarcely a useful concept. N_0 elephants of average weight, let us say 5 tons each, would equal the weight of more than 700 earths!

1.12 THE SIGNIFICANCE OF CHEMICAL EQUATIONS

Having established molecular formulas and a system of atomic and molecular weights, we are now in a position to apply the law of conservation of mass to chemical reactions in the convenient form of "chemical equations."

If we pass a stream of hydrogen gas over a bed of black, powdered copper oxide at high temperature, say 450 to 500°C, we note that the black powder turns to a reddish-orange color, and a clear liquid condenses in the outlet tube. Upon examination, this liquid turns out to be water. We may conclude that the following reaction has occurred:

$$\text{copper oxide} + \text{hydrogen} \rightarrow \text{copper} + \text{water}.$$

By using the symbols for the elements and the appropriate formulas, the story of any chemical event can be expressed more compactly than in words. Moreover, the mole concept and the knowledge of the atomic and molecular weights make it possible for us to give the chemical equation quantitative meaning. Thus we write:

1. copper oxide + hydrogen $\rightarrow$ copper + water,
$$CuO + H_2 \quad \rightarrow \quad Cu + H_2O,$$
1 molecule* + 1 molecule $\rightarrow$ 1 atom + 1 molecule,
1 mole + 1 mole $\rightarrow$ 1 mole + 1 mole,
79.5 g + 2.0 g $\rightarrow$ 63.5 g + 18.0 g;

2. alcohol + oxygen $\rightarrow$ carbon dioxide + water,
$$C_2H_6O + 3O_2 \quad \rightarrow \quad 2CO_2 + 3H_2O,$$
1 molecule + 3 molecules $\rightarrow$ 2 molecules + 3 molecules,
1 mole + 3 moles $\rightarrow$ 2 moles + 3 moles,
46.0 g + 3 × 32.0 g $\rightarrow$ 2 × 44.0 g + 3 × 18.0 g.

Reaction (2) is conveniently demonstrated by applying a flame to a pool of the liquid alcohol in an open dish. It burns quietly, combining with the oxygen of the air to form gaseous carbon dioxide and the condensable vapor

---

* It would be difficult to prove in this case that any particular oxygen atom is exclusively bound to a specific copper atom to form a molecule of copper oxide. However, it is convenient to refer to a pair of copper and oxygen atoms, CuO, as a molecule. This we shall do.

of water. Note that the "law of conservation of mass" (see Problem 1.3) is obeyed here since the sum of the masses of the reactants (46.0 g + 3 × 32.0 g) is equal to that of the products (2 × 44.0 g + 3 × 18.0 g). We have, in all cases, the same numbers of *atoms* before and after reaction but the numbers of molecules on the left and right of the equation are not necessarily the same; for instance, they are (1 + 3) and (2 + 3) respectively for the reaction of alcohol with oxygen. If the numbers of *molecules* used and produced as in reaction (1) are the same, this is a fortuitous circumstance. Keep firmly in mind, henceforth, the *three* meanings of a chemical equation, i.e., ratios of (a) molecules, (b) moles, (c) grams.

1.13 DETERMINATION OF MOLECULAR WEIGHTS.   THE GAS LAW AGAIN

Reconsider the gas law of Section (1.7) in the light of Avogadro's hypothesis. Avogadro's hypothesis asserts that a given volume at a fixed temperature and pressure contains the same number of molecules of any gas. The gas law relates volume, temperature, and pressure by the equation

$$PV = rT, \tag{1.3}$$

where r is taken to be a function of the quantity and nature of the given gas. However, if Avogadro's hypothesis is correct, *the same number of moles of any gas must give the same P, V, T relation.* Thus the proportionality constant r may be factored into the number of moles of gas, n, times a constant which is characteristic of all gases. This universal constant is designated by R. For a given sample of any gas $r = nR$. Let us reexamine the result of Problem 1.10(c). We may now rewrite Eq. (1.3) as

$$PV = nRT, \tag{1.4}$$

which applies to all gases. The value of R is 0.082 liter-atm/mole-deg.

Problem 1.13   What experimental data would be required to find the value of R? How could you demonstrate that R as used in Eq. (1.4) is a constant for *all conditions* ordinarily encountered in the laboratory and that it is truly a *universal* constant for all gases?

If n is the number of moles, it is, of course, just the weight of gas divided by the molecular weight, $n = w/MW$. Thus we may calculate molecular weights from P, V, T data for a given weight of gas by the equations

$$n = \frac{w}{MW} = \frac{PV}{RT} \tag{1.5}$$

or

$$MW = \frac{wRT}{PV}. \tag{1.6}$$

Returning to Problem 1.10, we note that 2.02 g of hydrogen (which is one mole) occupy 22.4 liters at standard temperature and pressure (0°C, 1 atm). Since one mole of any gas (N_0 molecules) occupies the same volume at the same temperature and pressure, this means that 22.4 liters of any gas at standard temperature and pressure will contain one mole. Therefore the molecular weight of any gaseous substance may be obtained by finding the weight of gas which occupies 22.4 liters at standard temperature and pressure. This of course represents a special application of Eq. (1.6), where $R = 0.082$ liter-atm/mole-deg, $V = 1.0$ liter, $P = 1.0$ atm, and $T = 273°$K.

Problem 1.14 A hydrocarbon, a compound of C + H, has a density of 1.87 g/liter at 273°K and 1 atm. What is the molecular weight?

With the technique for calculation of molecular weight in mind and armed with the concept of mole units, we shall illustrate the way in which the formula of a molecule is determined now that the atomic-molecular theory is complete. That is, the problem now is to calculate the formula of a molecule given (1) the elemental composition data, (2) data on gas density, and (3) a table of atomic weights.

Consider a compound that is 52.2% C, 13.0% H, and 34.8% O. Its vapor has a density of 2.05 g/liter at standard temperature and pressure. We may begin by thinking of a 100-g sample of this substance. It contains 52.2 g C, 13.0 g H, and 34.8 g O. How many "moles" of each does this represent? (Moles = g/atomic weight.) We have

$$\text{C;} \quad \frac{52.2}{12.0} = 4.35 \text{ moles,}$$

$$\text{H;} \quad \frac{13.0}{1.00} = 13.0 \text{ moles,}$$

$$\text{O;} \quad \frac{34.8}{16.0} = 2.18 \text{ moles.}$$

These numbers of moles, of course, represent the relative numbers of each of the three types of atoms in the substance. If we can reduce the ratios 4.35:13.0:2.18 to a ratio of integers, we will have the simplest formula of the compound. Divide each of the numbers of moles by the smallest: 4.35/2.18 = 1.99, 13.0/2.18 = 5.97, and 2.18/2.18 = 1. Let us assume that the deviation from integral ratios here is due to the experimental errors involved in measuring the data as given. Then the ratios of numbers of moles of the elements involved may be expressed as 2:6:1. The simplest formula of the compound is C_2H_6O. If this were also the *molecular* formula, the molecular weight would be $(2 \times 12) + (6 \times 1) + (16) = 46$. According to Eq. (1.6):

$$MW = \frac{wRT}{PV}, \tag{1.6}$$

but density d is just weight per unit volume or w/V. So the equation may be written

$$MW = d\,\frac{RT}{P} = (2.05)\,\frac{(0.082)(273)}{(1.0)} = (2.05)(22.4) = 46.$$

The molecular weight confirms the molecular formula C_2H_6O (as opposed to, say, $C_4H_{12}O_2$ with a molecular weight of 92).

Problem 1.15   A 0.596-g sample of a gaseous compound containing only boron and hydrogen occupies 484 ml at 273°K and 1 atm. When the compound is burned in oxygen, all the hydrogen is converted to 1.17 g of water. Find the molecular weight, simplest formula, and molecular formula of the boron hydride.

1.14 STOICHIOMETRIC CALCULATIONS

In Section 1.12, we saw that the balanced chemical equation for a reaction has quantitative significance with respect to the ratio of moles reacting or produced. This can be very useful in working out the weights or volumes of all the substances involved in a reaction, provided we have the means of finding the number of moles of any one of them. This type of calculation, known as a stoichiometric calculation, is illustrated by the following problem.

Consider the following sample calculation which may now be carried out. The salt, ammonium dichromate, dissociates when heated according to the equation,

$$(NH_4)_2Cr_2O_7 \rightarrow N_2 + 4H_2O + Cr_2O_3.$$

a) Find the weight of chromium(III) oxide, Cr_2O_3, that can be obtained from 10.00 g of the salt.
b) Find the volume of N_2 gas that can be obtained at 25° and 78 cm of mercury pressure.

Solution

MW of $(NH_4)_2Cr_2O_7{}^* = (2 \times 14.0) + (8 \times 1.0) + (2 \times 52.0) + (7 \times 16.0) = 252,$

$$\text{number of moles of } (NH_4)_2Cr_2O_7 = \frac{10.00\text{ g}}{252.0\text{ g/mole}} = 0.0397.$$

a) Number of moles of $Cr_2O_3{}^* = 0.0399,$
   weight of $Cr_2O_3 = 0.0399$ mole $\times$ 152 g/mole $= 6.03$ g.

---

* Neither $(NH_4)_2Cr_2O_7$ or Cr_2O_3 can be regarded as consisting of molecules. It might be better therefore to substitute the term gram formula weight for the term mole in referring respectively to 252 g or 152 g of these substances. However, it is customary to use the term mole in such cases remembering that we are really referring, say for Cr_2O_3, to $2N_0$ atoms of chromium and $3N_0$ atoms of oxygen. This point will be elaborated in Chapter 5.

b) Number of moles of $N_2 = 0.0399 \times 2 = 0.0794$

$$PV = nRT$$

$$V = \frac{nRT}{P} = \frac{(0.0794)(0.082)(298)}{(\frac{78}{76})} = 1.9 \text{ l.}$$

SUMMARY PROBLEMS

Problem 1.16 Applying the conservation of mass principle in the atomic-theory form of conservation of atoms, copy the following chemical equations and balance them (i.e., supply the correct numbers of atoms or molecules—or, equally well, moles—which must be involved).

a) __ BCl_3 +   __ P_4 +   __ H_2 →   __ BP +   __ HCl

b) __ IBr +   __ NH_3 →   __ NI_3 +   __ NH_4Br

c) __ NH_3 +   __ O_2 →   __ N_2 +   __ H_2O

d) __ FeS_2 +   __ O_2 →   __ Fe_2O_3 +   __ SO_2

e) __ $C_7H_6O_2$ +   __ O_2 →   __ CO_2 +   __ H_2O

Problem 1.17 A simple method for obtaining hydrogen for laboratory purposes employs the reaction of CaH_2 with water to produce $Ca(OH)_2$ and H_2.

a) Write the equation for the reaction.
b) Find the number of moles of H_2 that may be produced from 100 g of CaH_2.
c) Find the number of grams of H_2.
d) Determine what volume this H_2 will occupy at 1 atm pressure and 23°C (room temperature).

Problem 1.18 Exactly 500 ml (milliliters) of a gas at standard temperature and pressure weighs 0.581 g. Elemental analysis reveals that the composition of the gas is C = 92.24%, H = 7.76%. Find the simplest formula and the molecular formula of the gas.

Problem 1.19 The density of NO is found to be 0.2579 g/liter at a temperature and pressure at which the density of O_2 is 0.2750g/liter. Find the atomic weight of N.

Problem 1.20 An oxide of nitrogen is 63.7% N. Its vapor density is 1.98 g/liter. Find its molecular formula. A chloride of sulfur is 52.51% Cl. Its vapor density is 6.03 g/liter. Find its molecular formula.

Problem 1.21 The following table contains approximate molecular weights (from vapor density) and weight-percentage composition data for a series of compounds.

Name	MW	%C	%H	%O
Methane	16	75	25	
Heptane	100	84	16	
Formaldehyde	30	40	6.7	53.3
Propylene	42	85.7	14.3	
Acetic acid	60	40	6.7	53.3
Ether	74	64.9	13.5	

Taking the atomic weight of oxygen as 16, develop an argument to show that the atomic weight of hydrogen is 1. Find the molecular formulas. Suppose that the compounds methane and formaldehyde had been omitted from the table. Could you then prove that *none* of the compounds listed contained only *one* atom of carbon?

Problem 1.22   A certain compound of manganese and oxygen is a solid substance which cannot be vaporized by heating without undergoing decomposition. A 2.39-g sample of the compound contains 0.64 g of oxygen and 1.65 g of manganese. Consult Table 1.2 for the atomic weights.

  a) Can you find (i) the molecular formula, (ii) the simplest formula of the compound? State in brief your reasons for your answer.
  b) If finding either (i) or (ii) of Part (a) is possible, perform the operations required to get a numerical answer.

READINGS AND EXTENSIONS

The subject of this chapter is called in modern chemical practice *stoichiometry*. A good treatment of the theory and applications of stoichiometry is to be found in L. K. Nash, *Stoichiometry*, Addison-Wesley, Reading, Mass., 1966. The treatment is thorough and sophisticated. Clearly, a key concept introduced in this chapter is that of *mole*. The way this idea weaves through all of modern chemical thinking is explored in a simple and straightforward book by W. F. Kieffer, *The Mole Concept in Chemistry*, Reinhold Publishing Co., New York, 1962. Both of these books are inexpensive paperbacks.

The historical themes of this chapter are given a full exposition in an intriguing case study by L. K. Nash, *The Atomic-Molecular Theory*, Harvard University Press, Cambridge, Mass., 1950. The events which led chemistry up to the point at which Chapter 1 of this book begins are briefly but effectively summarized in J. R. Partington's *A Short History of Chemistry*, Harper Torchbooks, New York, 1960. Both of these are paperbacks too. Perhaps the authoritative history of chemistry now available, which provides interesting background to this chapter and all that follow, is A. J. Ihde's *The Development of Modern Chemistry*, Harper and Row, New York, 1966.

APPENDIX TO CHAPTER 1

Molecules in Motion

1A.1 DERIVATION OF THE KINETIC THEORY EQUATION

Chapter 1 has traced the *chemical* origins of the concepts of atom and molecule. The only crucial property with which our atom or molecule was endowed was the property of weight (or the more fundamental physical property *mass* from which weight arises by the action of gravity). But, a *model* conception like that of molecule or atom is likely to carry with it additional features that are not immediately used. The molecule proposed by Avogadro may be thought of as a sort of small-scale billiard ball. It should be more than just weighable; it should "do the things billiard balls do." It is very interesting that just this extension of the concept of atom can lead to an explanation of the gas laws introduced in Section 1.7. The laws are explained as a consequence of molecules in motion by the kinetic theory of gases, which developed shortly after the Dalton-Avogadro theory.

We have noted the *experimental fact* that the pressure, volume, and temperature of a gas are interrelated, to a close approximation, according to the equation

$$PV = nRT. \qquad (1A.1)$$

Here n is the number of moles of gas and R is a constant, the numerical value of which depends on the P, V, and T units. If the units are liters, atmospheres, and degrees Kelvin, then R is found experimentally to be 0.082 liter-atmospheres per mole per degree.

It becomes a matter of interest to see if we can *derive* this equation on the basis of postulates about the physical behavior of gas molecules. We begin by adding to our earlier models of molecules the ideas that in a gas (1) they are in rapid, random motion, (2) to a close approximation, they may be regarded as point centers of mass (their volumes are small compared to the volume in which they are confined), and (3) any collisions between a gas molecule and the walls of a containing vessel are perfectly elastic. As a result of postulate 2, we may initially disregard the possibility of intermolecular collisions. Postu-

39

late 3 is required to eliminate the concept that the gas molecules finally "come to rest" as a bouncing rubber ball finally does.

The presumed relatively low volume of the actual gas molecules is consistent with the fact that a gas, which has been condensed to the liquid or solid state, then occupies only a few thousandths of the volume which it filled when in the gaseous state.

It is not difficult to see in a *qualitative* sense how these ideas may be used to explain the experimental behavior of gases. Confining a gas in a smaller volume means that molecules moving at a constant speed will strike the wall more often. It seems very reasonable to connect the pressure exerted by the gas on the wall (force per unit wall area) with the forces exerted during these collisions. We suggest a simple relation between P and V. We know that at a given T that relation is $P \times V = $ constant. But, since the constant RT rises as temperature does, our model may force us to connect temperature to the molecular speeds.

To achieve a *quantitative* understanding, we shall attempt to relate the measurable PVT quantities to the important physical characteristics of a collection of microscopic billiard balls (gas molecules), i.e., their number and also their masses and velocities.

Now suppose we consider the pressure exerted by a moving "ball" on one wall of a cubical container, say the wall in the yz-plane as represented in Fig. 1A.1.

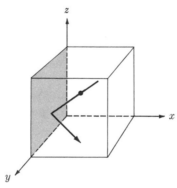

Fig. 1A.1. A gas molecule colliding elastically with the yz side of a cubical box.

We shall try to calculate the pressure (force per unit area) by computing the *average* force exerted on the wall by the molecules in collision with the wall. We face an averaging problem because, first, not all the molecules are in contact with the wall and are exerting a force at any given time, *and* it would be presumptuous to assume that all molecules would exert the same force during a collision. (This last point turns out to correspond to the observation that the molecules may be moving at different velocities.)

Now let's see how we might calculate a force on the wall if we knew molecular velocities. When a molecule hits a wall, the "reaction" principle (Newton's third law) asserts that the force exerted on the molecule is equal and opposite to the force exerted by the molecule on the wall. When the molecule bounces off the wall, its direction of motion will be "reversed." This means that a force has acted because velocity is a vector (has magnitude and direction), and a change of direction means a velocity change, which, according to Newton's second law, is the evidence of a force acting. Newton's second law defines force as equal to mass times acceleration, where acceleration is *just the rate of change of velocity* when velocity is written as $\mathbf{v}$, the symbol Δ means the change in a quantity, and F_{av} the average force:

$$F_{av} = m \frac{\Delta \mathbf{v}}{\Delta t} = \frac{\Delta(m\mathbf{v})}{\Delta t}.$$

The average force is the change in $m\mathbf{v}$ (called momentum) for a given time period Δt. If we find changes in momentum for the molecules, we can calculate average forces on the wall.

At any given instant we may regard one molecule as moving with a velocity $\mathbf{v}$. Each molecule has a component of velocity ($\cos \theta \mathbf{v}$) along the axes x, y, and z. Let us designate these components of velocity of molecule 1 as $\mathbf{v}_x$, $\mathbf{v}_y$, and $\mathbf{v}_z$ respectively. From the Pythagorean theorem these components of velocity are related to $\mathbf{v}$ by the equation

$$\mathbf{v}_x^2 + \mathbf{v}_y^2 + \mathbf{v}_z^2 = \mathbf{v}^2. \tag{1A.2}$$

Let us consider the collisions of one typical molecule with one wall of the vessel in the yz-plane, shaded in Fig. 1A.1. The force due to a collision will depend only on the component of velocity $\mathbf{v}_x$ along a line parallel to the x-axis. The momentum of the molecule as it strikes the wall will be $m\mathbf{v}_x$. Since the collision is assumed to be perfectly elastic, the molecule will rebound with a component of velocity $-\mathbf{v}_x$ and a corresponding momentum $-m\mathbf{v}_x$. Hence the change in momentum is $2m\mathbf{v}_x$. Before striking the specified face a second time, the molecule must traverse the cubical space to the opposite wall, rebound, and return to the same face. If L is the length of the side of the cube, the time required for this return trip between impacts with the face will be $2L/\mathbf{v}_x$ sec. Then the *rate* of change of momentum, $\Delta m\mathbf{v}_x/\Delta t$, is:

$$\frac{2m\mathbf{v}_x}{2L/\mathbf{v}_x} = \frac{m\mathbf{v}_x^2}{L}. \tag{1A.3}$$

This is the average force exerted from within by molecule 1 on the chosen face of the cubic box.

Problem 1A.1 Would it make any difference to the calculation just completed if the molecule followed a path between two collisions with the yz-wall as shown on p. 42?

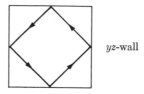

yz-wall

If you can't see the answer to this problem, consider the meaning of the fact that a bullet fired from a gun held exactly horizontal strikes the ground at the same time as a bullet that is simultaneously dropped from the gun.

So far, we have considered only the effect of one molecule which was assumed to be moving at constant velocity. It is necessary now to take into account all N gas molecules present in the box and to adopt some suitable method of averaging their various speeds. We may define a quantity $\overline{v^2}$ called the *mean-square velocity* as the sum of the squares of all the velocities divided by N, the number of molecules. Then

$$\overline{v_x^2} = \frac{v_{x_1}^2 + v_{x_2}^2 + \cdots + v_{x_N}^2}{N}. \tag{1A.4}$$

From Eq. (1A.2) we can see that the mean square velocity along any one of the three coordinates such as the x-axis will be $\frac{1}{3}\overline{v^2}$, since molecular motion is equally probable along each of the three axes, and hence

$$\overline{v_x^2} = \frac{v_{x_1}^2 + v_{x_2}^2 + \cdots + v_{x_N}^2}{N} = \frac{1}{3}\overline{v^2}. \tag{1A.5}$$

From Eq. (1A.3) we see that the force due to one molecule striking a surface of the cube is $m v_{x_1}^2/L$, and for molecule 2 it is $m v_{x_2}^2/L$, etc. Hence the force due to all N molecules striking the specified wall of the cube will be

$$\text{force} = \frac{m v_{x_1}^2}{L} + \frac{m v_{x_2}^2}{L} + \cdots + \frac{m v_{x_N}^2}{L} \tag{1A.6}$$

$$= \frac{m}{L}(v_{x_1}^2 + v_{x_2}^2 + \cdots + v_{x_N}^2). \tag{1A.7}$$

From Eq. (1A.5) it is seen that the sum of the squares of the velocities in Eq. (1A.7) must be $\overline{v_x^2}$ or $\frac{1}{3}\overline{v^2}$. Then it follows that

$$\text{force} = \frac{1}{3}\frac{mN}{L}\overline{v^2}. \tag{1A.8}$$

Pressure is force per unit area. The *area* of the wall is L^2, and hence the pressure P is

$$P = \frac{1}{3} \frac{mN}{LL^2} \overline{\mathbf{v}^2} = \frac{1}{3} \frac{mN}{L^3} \overline{\mathbf{v}^2} = \frac{1}{3} \frac{mN}{V} \overline{\mathbf{v}^2}, \tag{1A.9}$$

where V is the volume of the cubic box. This leads to the gas equation derived from the kinetic theory since the pressure on all walls is presumably the same:

$$PV = \tfrac{1}{3}mN\overline{\mathbf{v}^2}. \tag{1A.10}$$

We call this the kinetic-molecular equation.

1A.2 RELATIONSHIPS WHICH DEVELOP FROM THE KINETIC-MOLECULAR EQUATION

Boyle's Law. In Eq. (1A.10) the terms m and N are constant. At constant temperature, the average squared speed of the molecules is a constant independent of the number of collisions with the wall if the collisions are elastic. Thus Eq. (1A.10) is *equivalent to* Boyle's law: the product of the pressure times the volume of a gas is a constant which depends only on the number of molecules in the sample under consideration.

The Relationship to Temperature. There is no way to derive a value for the term $\overline{\mathbf{v}^2}$ except by resorting to experimental results. We do know that, *by experiment*, $PV = nRT$ (see Eq. 1.4), where n is the number of moles of gas. Now we may express N of Eq. (1A.10) as nN_0, where N_0 is Avogadro's number. Then we may write

$$PV = n\tfrac{2}{3}N_0(m\overline{\mathbf{v}^2}/2),$$

and hence we may *suggest the interpretation*

$$RT = \tfrac{2}{3}N_0(m\overline{\mathbf{v}^2}/2).$$

The quantity $m\overline{\mathbf{v}^2}/2$ expresses the *kinetic energy* per molecule and hence $N_0(m\overline{\mathbf{v}^2}/2)$ is the *kinetic energy of a mole of gas*. Then

$$N_0(m\overline{\mathbf{v}^2}/2) = \tfrac{3}{2}RT. \tag{1A.11}$$

From this we may conclude that the *temperature* is a parameter which is directly *proportional to the kinetic energy of the gas molecules*. Dividing through by N_0, we get

$$m\overline{\mathbf{v}^2}/2 = \tfrac{3}{2}(R/N_0)T = \tfrac{3}{2}kT. \tag{1A.12}$$

Here $k = R/N_0$ is the gas constant for a single molecule, called *Boltzmann's constant*. From Eq. (1A.12) we see that the absolute temperature is directly proportional to the average value of the kinetic energy of a single molecule.

We see from Eq. (1A.11) that RT must be expressed in units of energy and hence, since $PV = nRT$, it follows that PV must be an energy term. We have expressed PV in terms of liters times atmospheres, which may not at first appear to represent energy. However, it is true that pressure $\times$ volume = (force/area)(area $\times$ length). Cancelling area, we have the units force $\times$ length. Work or energy is defined in terms of product of force times distance, and thus we see that PV does in fact have the units of energy.

Graham's Law of Diffusion. So far, of course, all this is interpretation. It is plausible but it is by no means certain that other approaches wouldn't also allow us to derive an equation equivalent to Boyle's law. The present model would be much more convincing if it could derive consequences directly relating to its fundamental idea: molecules in random motion.

Over a century ago, the Scottish scientist Thomas Graham discovered *by experiment* a simple relationship known as Graham's law. This law states that *the rates of diffusion of different gases are inversely proportional to their molecular weights.* This relationship follows directly from Eq. (1A.11). It seems plausible that the rate of diffusion of a gas through a small orifice should be proportional to the speed with which its molecules move; this is borne out by detailed investigation. At any specified temperature, all the terms in Eq. (1A.11) are constant except m and $\overline{\mathbf{v}^2}$, and hence $m\overline{\mathbf{v}^2}$ must be constant. As a result the mean-square velocity $\overline{\mathbf{v}^2}$ is inversely proportional to m, the mass of an individual molecule of the gas:

$$\overline{\mathbf{v}^2} = \text{const}/m.$$

Since MW, the molecular weight, is simply N_0 times m, it follows that $\overline{\mathbf{v}^2} = \text{const}/MW$. Then the root-mean-square velocity $(\overline{\mathbf{v}^2})^{1/2}$ is inversely proportional to the square root of the molecular weight, or

$$(\overline{\mathbf{v}^2})^{1/2} = \text{const}/MW^{1/2}.$$

The root mean square velocity is about eight percent greater than the absolute magnitude of the average velocity $|\overline{\mathbf{v}}|$. But since the ratio $(\overline{\mathbf{v}^2})^{1/2}/|\mathbf{v}|$ is a constant, it also follows that

$$|\overline{\mathbf{v}}| = \text{const}/MW^{1/2}. \tag{1A.13}$$

Equation (1A.13), which was derived from the ideas of the kinetic molecular theory of gases, is consistent with Graham's experimentally based law of diffusion, which may be expressed as

$$r_1/r_2 = \sqrt{MW_2/MW_1}, \tag{1A.14}$$

where r represents the rate of diffusion. The experimentally measured values of r may be expressed in any convenient units, since the units will cancel in the ratio r_1/r_2. Incidentally, we may use Graham's law as expressed in

Eq. (1A.14) to calculate *relative* molecular weights. For instance, it is found experimentally that the ratio r_{H_2}/r_{O_2} is 4:1. We may then calculate that M_{O_2}/M_{H_2} will be 16:1. If we have evaluated the molecular weight of *any* gas by other means, we can use this experimentally determined relative diffusion ratio r_1/r_2 to calculate the molecular weights of other gases.

1A.3 THE ABSOLUTE VELOCITIES OF MOLECULES

We may use Eq. (1A.12) to calculate the root-mean-square speed $(\overline{v^2})^{1/2}$ and thence the average speed of the molecules of any gas at a specific temperature.

Table 1A.1

Root-mean-square speeds of molecules at 298°K

Molecule	cm/sec	Molecule	cm/sec
Argon	4.31×10^4	Hydrogen	1.93×10^5
Carbon dioxide	4.11×10^4	Oxygen	4.84×10^4
Chlorine	3.27×10^4	Water	6.43×10^4
Helium	1.37×10^5	Xenon	2.38×10^4

To make this calculation we must use the value of Boltzmann's constant which is expressed in cgs units and the value of the molecular weight MW. (Recall $m = MW/N_0$.) Some root-mean-square speeds are given in Table 1A.1. The average speed of the hydrogen molecule at room temperature turns out to be just about one mile per second.

1A.4 COMMENTS ON WHAT WE HAVE DONE

In the last few pages, we have given a brief account of an argument that had profound consequences. First, it shows how the chemists' conception of a molecule can be rendered physical and as a result can lead to an explanation of the pressure exerted by a gas on its container that is at least plausible. The idea defended by Dalton with purely chemical reasoning turns out to be richer than was originally realized. This is an important characteristic of a successful scientific theory. The "model" suggests explanations of more phenomena than it was originally introduced to explain. Moreover, this new argument leads to the suggestion that the effects which we feel as temperature and measure by using thermometers in which some substance expands can be related to objects in motion. This is a profound *unification* of different areas of our experience.

Now there is reason to worry about a few things. As is easily shown experimentally, the equation $PV = nRT$ is only a first approximation to the behavior of gases. Careful measurements show that each gas deviates from this behavior in its own way and that the ideal equation represents only the low-pressure limit for all gases. Does this limitation represent a fundamental difficulty for our theory? Let's consider some slight sophistications in the treatment to show that the kinetic theory can be extended to deal more realistically with the behavior of actual gases as the results of more precise measurements of P, V, and T are considered.

1A.5 REAL AND IDEAL GASES: THE VAN DER WAALS EQUATION

We have seen that the ideal gas equation ($PV = RT$ for 1 mole of gas) as derived from the kinetic theory of gases is based on the assumption that the molecules have no attraction for each other and, being regarded as point centers of mass, have no finite volume. Of course in a real gas neither of these assumptions is strictly appropriate.

The first successful attempt to modify the ideal gas equation, in order to correct for the volume and attraction of molecules, is attributed to J. D. van der Waals (1873), and the resulting *equation of state* is known as the van der Waals equation. This equation is given as

$$(P + a/V^2)(V - b) = RT. \tag{1A.15}$$

Here P, V, and T are the measured pressure, volume, and temperature. The constants a and b are chosen to give the best agreement with experiment for each gas, but the equation may be derived (we won't do it) from an elaborated kinetic theory. The term a/V^2 is a measure of the attractive force of the gas molecules and has the dimensions of pressure. Because of the attractive intermolecular forces, the measured pressure P is less than the ideal pressure and hence the correction term a/V^2 must be added. To make allowance for the finite volume of the molecules of gas, the ideal volume is obtained by subtracting the term b from the measured volume V. This term b has the dimensions of volume and is a function of the molecular diameter d. Table 1A.2 gives some van der Waals diameters. Note that we have here a first possible way to estimate molecular sizes.

Table 1A.2

**Some diameters of gas molecules inferred
from the van der Waals equation**

Molecule	Diameter, cm
H_2	2.20×10^{-8}
O_2	2.80×10^{-8}
N_2	3.00×10^{-8}
I_2	4.40×10^{-8}

Tables of van der Waals constants for various real gases are found in standard texts* of physical chemistry and in the *Handbook of Chemistry and Physics*. At high temperatures and low pressures, the terms a/V^2 and b approach being negligibly small *compared to P and V* and hence the van der Waals equation approaches the ideal gas equation ($PV = RT$). Other, more sophisitcated treatments of the problem of nonideality of real gases have been developed. There are parallel developments of kinetic theory for each stage of sophistication in the treatment of PVT data.

---

* See for example F. Daniels and R. A. Alberty, *Physical Chemistry*, Third Edition (John Wiley and Sons, Inc., 1966), p. 19.

CHAPTER 2

Valence, Molecular Structure, and Organic Chemistry

In Chapter 1 we traced the development of the idea that substances are ultimately composed of atoms of the elements and that certain groupings of atoms form molecules, which are the smallest units of compounds. In this chapter we shall extend molecular theory considerably. But it is important to begin with a warning. So far, our determination of the constitution of a molecule has depended upon our ability to measure the vapor density of the compound. As a result, the scope of our measurements has been limited to substances reasonably easily studied in the gas phase. There are many substances that are very difficult to vaporize, and we cannot be sure that the compounds that we do vaporize are exactly the same in the gas phase as they are in the liquid or solid. It will, in fact, emerge that the idea of a molecule has its limits and that certain situations will require the introduction of other structural units (e.g., ions, see Chapter 5), but for the moment let's ask some more questions about molecular substances and molecules.

2.1 THE IDEA OF VALENCE

What is the next question? In Chapter 1, we developed a theory which describes a molecule in terms of the numbers and types of atoms that it contains. Is there anything we can say about the relationships among these atoms? Do they, for example, form something like a planetary system: the biggest one at (or near) the center with the others orbiting about that center? Or, is there a definite *structure* to a molecule with each atom occupying a more or less fixed position in relation to the others? Again, to try to answer such questions, it will be necessary to see what regularities can be discovered in the constitution and behavior of molecules and then to try to develop an explanation of the regularities.

The most obvious regularity in molecular constitution is a simple numerical one. Recognizing it is a necessary prelude to considering molecular structure. It was foreshadowed in the discussion in Chapter 1. The laws of definite and multiple proportions, which played an important part in the development of the Dalton-Avogadro theory, depend for their usefulness on the fact that most molecules studied are relatively simple. One might not have noticed the laws if it were true that the compounds of hydrogen and chlorine were $H_{145}Cl_{250}$ and $H_{148}Cl_{251}$. In such a case, there would seem to be a continuous range of compositions. Instead, the discovery and explanation of the laws show that at least very many molecules have simple formulas involving small numbers of atoms. This suggests that atomic combining capacity is easily saturated; that an atom may unite with only a few others. By inspection of formulas we can try to find the rules governing atomic combining capacity.

Such rules of chemical combination must take into account not only the simple formulas that do occur, but also those that do not.

Why do H and Cl always form HCl, but not HCl_3 or H_5Cl_2 or some other reasonably simple combination? Why do hydrogen and oxygen atoms form either H_2O or H_2O_2, the compound hydrogen peroxide, but not HO_2 or H_3O_2? Or considering carbon and oxygen, how is it that we find CO or CO_2 but never C_2O? A crude first approach to answers to these questions, which was discussed in Dalton's time, is to picture the atoms as having something like "hooks" that serve to couple them together. To explain the existence of HCl and H_2O, we assert that *hydrogen has only one hook* and that the chlorine atom has one hook but that the oxygen atom has two hooks. With this scheme we can also obtain a picture for the hydrogen peroxide molecule H_2O_2 by suggesting that oxygen with its two hooks has one hook attached to another oxygen and then one hook attached on each end of this chain to a hydrogen. On this basis we see that there would be no way to construct such a molecule as H_3O_2. Nor, for that matter, would there be a molecule HO_2. HO_2 cannot form since hydrogen has used its one hook when it has bound the first oxygen. If oxygen has two hooks, then the formula CO_2 suggests that carbon may have as many as four hooks, so as to fully engage both of the oxygen atoms.

Even the most naive who have employed this idea of hooks to express the simple numerical relationships of the combining powers of elements do not by any stretch of the imagination really believe that the mechanism of linking atoms together in molecules really involves physical hooks. The model is too crude to be taken seriously. This is simply a convenient way to express a *numerical rule*. We can either improve the situation or perhaps just cloud the issue by inventing a *name* for the numerical property of the atom *which is expressed as a number related to its ability to combine with other elements*. The name we give this number is the *valence* of the element. The valence of an element is to be regarded as a number characterizing its chemical combining power. Hydrogen is regarded *a priori* as having a valence of 1. We say that it is monovalent. The simplest way to find the valence of an element is to find the molecular formula of some substance, if one exists, that involves only this element and hydrogen. For example, the formula H_2O for water indicates that the valence of oxygen is 2 (divalent), the formula of NH_3 for ammonia indicates that the valence of nitrogen is 3 (trivalent), and the formula CH_4 for methane indicates that the valence of carbon is 4 (tetravalent). We also recognize, in some cases, groups of atoms that exist unchanged in a number of different compounds. A good example is the sulfate group, SO_4, which appears in the sodium salt Na_2SO_4, in the calcium salt $CaSO_4$, in the barium salt $BaSO_4$, and in the compound H_2SO_4, which indicates that the valence of this unit (formerly simply called a "radical" but in modern theory recognized as a charged unit or ion—see Chapter 5) which persists through reactions is 2. We say that the sulfate group is divalent. Now, a number of elements do not combine with hydrogen. One can often determine the valence of these elements by finding out how much hydrogen they will displace during a reaction. Thus the reaction of calcium with water at high temperatures to produce hydrogen gas and calcium oxide shows that the valence of calcium is 2 because calcium replaces two hydrogens from H_2O to form CaO plus one molecule of H_2.

Problem 2.1   Given the formulas HCl, NaCl, $CaCl_2$, $AlCl_3$, $SnCl_4$, and PCl_5, find possible valence numbers of sodium, calcium, aluminum, tin, and phosphorus. Predict the formula for expected oxides (or compounds with oxygen) of aluminum and phosphorus.

At the conclusion of this discussion, one warning must be entered: It is not true that the valence number of an element is always the same. Witness, for example, the compounds CO and CO_2 and the compounds $FeCl_2$ and $FeCl_3$. However, the idea of valence is a useful one because it is true that for almost all elements the number of possible valences is limited. To answer the underlying question of *why* atoms have specific valences, more is necessary than an accurate knowledge of molecular formulas or even of the molecular structures, an issue which we shall take up next. The key must be found in the nature of the forces which bind atoms together to form molecules, and these questions of chemistry have had to wait for the developments of atomic physics which have taken place in our own century. However, it is fair to say that the understanding of atomic structure and the development of the physics appropriate to understanding the forces that hold atoms together in molecules has benefited greatly from the evolution in the nineteenth century of the theory of chemical valence. The development of this valence theory goes far beyond the mere listing of some characteristic numbers for each element. We shall see that valence numbers may even mean different things for different types of compounds.

2.2 ORGANIC CHEMISTRY: THE STRUCTURE OF MOLECULES

We turn now to organic chemistry, that is, to the study of the compounds of carbon especially in combination with the elements hydrogen, nitrogen, phosphorus, oxygen, sulfur, and the halogens (fluorine, chlorine, bromine, and iodine). The reason for interest at this point in the compounds of carbon, or organic chemistry, is that it is in this area that the next major development in the theory of valence took place. Up to now, we have talked only about the numbers and types of atoms that are present within a molecule; but the crude language of hooks that we used in the introduction of the theory of valence rather suggests that one might talk about *structures of molecules*, the array of the atoms within a molecule. We urgently need such a theory to make sense out of the formulas of organic compounds. The notions applied to formulas in Problem 2.1 can lead to confusion with even the simplest organic substances. Let us try to apply the rule that hydrogen is monovalent to the calculation of the valence of carbon in some compounds of these two elements. The simplest is CH_4, which yields a valence 4 for carbon, but another is C_2H_6, which suggests a valence of 3 for carbon. A third is the compound C_2H_4, which yields a valence of 2 for carbon. Another is C_2H_2, which might suggest a valence of 1 for carbon. Still another is C_4H_{10}, which could imply a valence of 2.5 for carbon! Things would be worse if we considered still other examples that would yield all sorts of values, both integral and nonintegral, for the valence of carbon. The famous German chemist Wöhler wrote in a letter to Berzelius in 1835,

Organic chemistry just now is enough to drive one mad. It gives me an impression of a primeval tropical forest full of the most remarkable things, a monstrous and boundless thicket, with no way to escape, into which one may well dread to enter.

In 1859 August Kekulé introduced two postulates which enormously clarified the situation. These are:

1. Carbon always has a valence of **four.**

2. Carbon atoms **may link with each other.**

Using these two postulates we can make quite good sense of the formulas given above. This is shown in Fig. 2.1. (Note especially the *multiple* carbon-carbon links in the diagrams for C_2H_4 and C_2H_2. This is necessary to maintain the valence of 4 for carbon.)

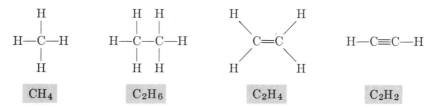

Fig. 2.1. Formulas of organic compounds using Kekulé's postulates.

Problem 2.2   Using Kekulé's postulates, write reasonable formulas for the following molecules after the fashion of Fig. 2.1: C_3H_8, C_4H_{10}, C_3H_6, C_3H_4.

Problem 2.3   Are the solutions to Problem 2.2 unique, or are there other ways to draw the formulas that are still consistent with Kekulé's postulates?

2.3 ISOMERISM

One of the most interesting puzzles of organic chemistry arises from the existence of isomers. Isomers are *different* compounds with the same molecular formula (in the Dalton-Avogadro sense of Chapter 1). There are many examples. Potable alcohol and a relative of anaesthetic ether both have the molecular formula C_2H_6O and yet these are truly *different* substances with different physical and chemical properties.

First let us apply the Kekulé postulates to the problem of explanation of the existence of *two* different hydrocarbons with the formula C_4H_{10}. We can construct this formula on the basis of two different ways of linking together the four carbons as shown in Figs. 2.2 and 2.3.

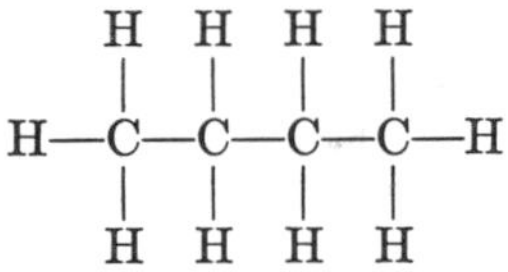

Fig. 2.2. Butane. 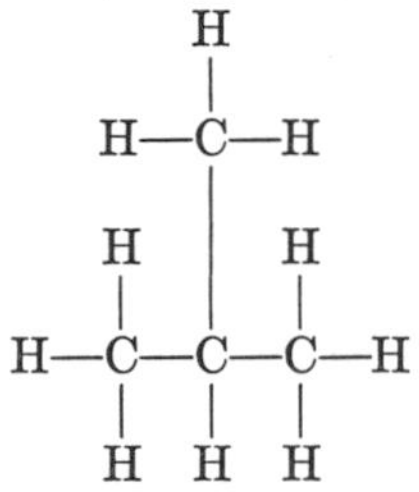   **Fig. 2.3.** *Isobutane.*

Figures 2.2 and 2.3 suggest that the difference between the two substances resides in the fact that one has a carbon that is linked to three other carbons and only one hydrogen, whereas the other has carbons linked to one or two carbons and three or two hydrogens respectively.* Experience reveals *only* two substances with the formula C_4H_{10}. We cannot call Fig. 2.4 a different compound.

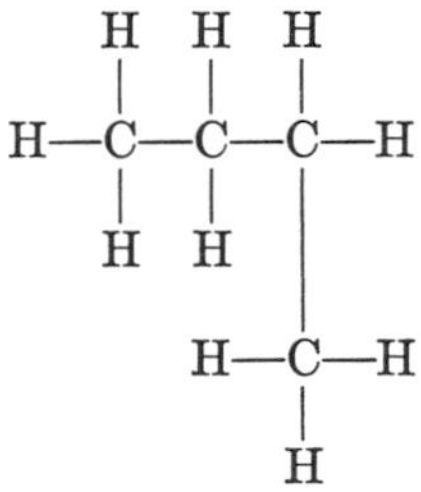

Fig. 2.4. Proposed but not observed butane isomer.

It has the same *linkage* as in Fig. 2.2 with the *chain bent.* Thus, if Kekulé's postulates are to help us, we must assume that they give *only* the order of linking and *not* the positions of the atoms in space. Our drawings tell us only which atoms are "hooked" to which. Bending them around is of no significance.

---

* Bending the 4-carbon chain of Fig. 2.2 to produce Fig. 2.4 does not result in a truly different structural formula.

Problem 2.4 Which of the following pairs represent genuine isomers and which irrelevant "chain bending" that is of no experimental significance?

a)

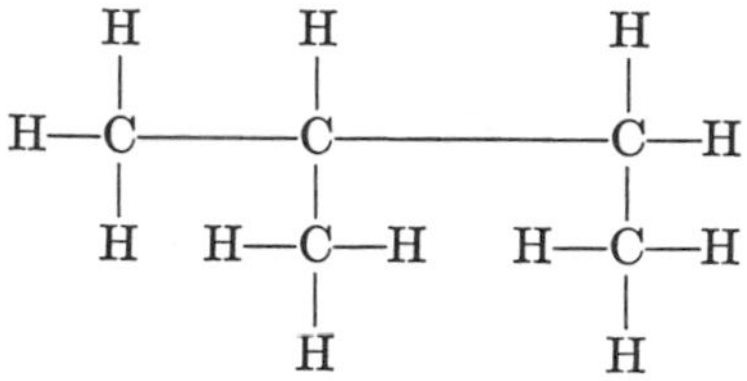

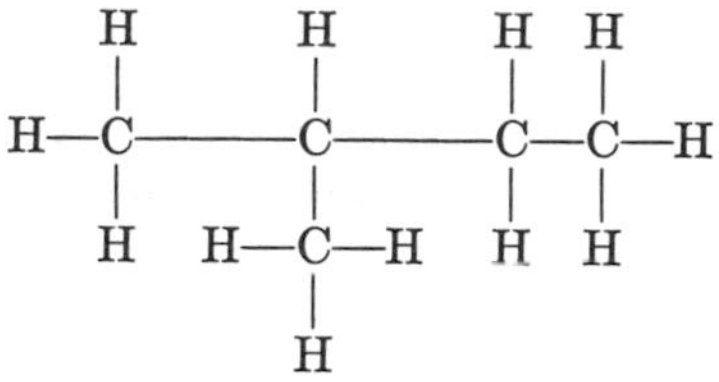

b)

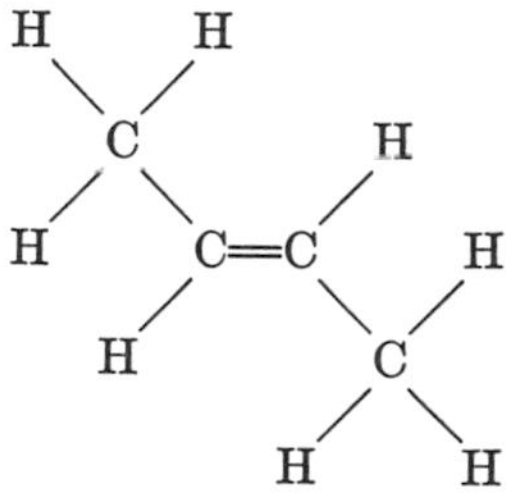

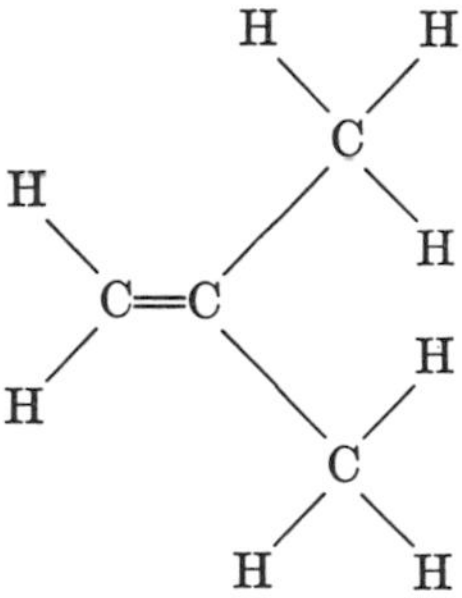

c)

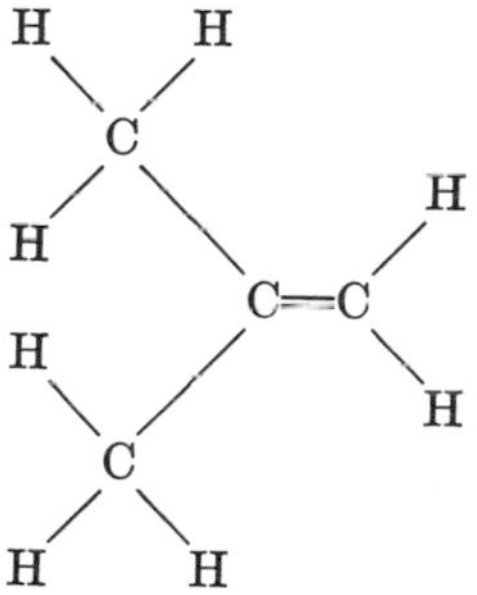

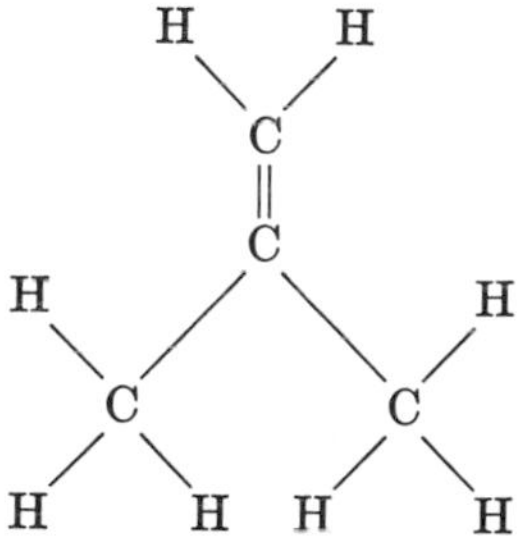

Problem 2.5 Draw the isomers of C_5H_{12} and C_6H_{14} expected from Kekulé's postulates. (Experiment reveals three and five respectively.) Don't be fooled by "chain" bending similar to that shown in Fig. 2.4.

Problem 2.6 Predict the number of "isomers" of C_7H_{16}. (Note that nineteenth century chemists were hard pressed indeed to check this prediction, since the isomers were not easy to separate, but it may be relatively easily done with modern gas chromatography.)

2.4 REACTIONS AND STRUCTURES

The theory of molecular structure that we are discussing would have only limited value if it were not possible to extend it to the explanation of reactions. Next we shall examine the reactions of a pair of isomers and attempt to show how reactions are understood in such a way that we can tell which structure belongs to which substance.

In this discussion, we shall introduce some terminology. We shall call the line representing a valence link between two atoms a *"chemical bond"* or simply *"bond,"* and we shall call a formula which displays the order of linking of atoms in a molecule a *"structural formula."*

Consider two isomeric C_4H_8 compounds. Assigning a valence of four to each carbon requires some double bonding. Three possible structures (where, for simplicity, the H's attached to a C are now written next to it in a "condensed" formula) are the following:

$$H_2C\!\!=\!\!CH—CH_2—CH_3 \qquad\qquad CH_3—CH\!\!=\!\!CH—CH_3$$

$$\text{(I)} \qquad\qquad\qquad\qquad \text{(II)}$$

$$\begin{array}{c} CH_2 \\ \| \\ H_3C—C—CH_3 \end{array}$$

$$\text{(III)}$$

Problem 2.7 Rewrite the structural formulas in Problem 2.5 using the simplification of writing H's attached to the C's following or preceding them on a line. That is, convert the awkward "spiderweb" formulas to "condensed" formulas.

It turns out that C_4H_8 compounds react with O_3 (ozone) breaking up into fragments with oxygen incorporated in the fragments. It is plausible to suppose that the point of attack in this reaction is at the *double bond*, since O_3 does not degrade (break down) C_4H_{10} compounds which presumably have no $C\!\!=\!\!C$ bonds.

With the initial hypothesis that O_3 reacts at double bonds and the normally accepted postulate that oxygen has a valence of *two*, let's attempt to analyze reactions of two substances having the molecular formula C_4H_8. We shall call these substances A and B. By experiment it is found that substance A reacts with ozone to give CH_2O and C_3H_8O. Substance B reacts with O_3 to give C_2H_4O *only*. Nearly two moles of this product per mole of C_4H_8 used may be recovered. We write abbreviated and unbalanced equations:

$$A \quad (C_4H_8) \xrightarrow{O_3} CH_2O + C_3H_6O,$$

$$B \quad (C_4H_8) \xrightarrow{O_3} 2C_2H_4O.$$

The compound CH_2O can have only one Kekulé formula, that is:

$$\begin{array}{c} O \\ \| \\ H—C—H, \end{array}$$

where we postulate that H, O, and C are respectively mono-, di-, and tetravalent. Looking at the possible structures I and III, we can easily imagine how $CH_2\!\!=\!\!O$

is produced by attack of O_3. We assume that the carbon-carbon double bond is split, leaving two valence units on each carbon to be satisfied. Each of them could then be taken up by a divalent oxygen as is illustrated in Fig. 2.5 using structural formula I. (Note that we are *not* saying anything about how this happens!)

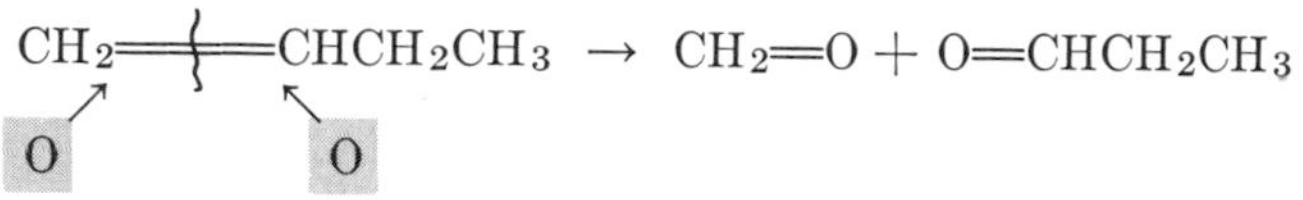

$$CH_2\!\!=\!\!\!\!\!\fbox{}\!\!\!\!\!=CHCH_2CH_3 \;\rightarrow\; CH_2\!\!=\!\!O + O\!\!=\!\!CHCH_2CH_3$$

$$O \qquad\qquad O$$

Fig. 2.5. The effect of ozone on a double bond.

According to Fig. 2.5, the other product should be C_3H_6O, and this is observed experimentally. With this picture in mind it follows easily that the reaction of ozone with compound B is the break-up of formula II, which has the same group on both sides of the double bond. Formula II should give $CH_3CH\!\!=\!\!O + O\!\!=\!\!CH\!\!-\!\!CH_3$.

We can now associate one of the Kekulé structures with B (formula II), but we do not yet know whether A is I or III (both should give CH_2O and C_3H_6O). A useful reaction that is observed for C_2H_4O is "oxidation" with $KMnO_4$ under mild conditions of low temperature and low concentration of inorganic acids:

$$C_2H_4O \xrightarrow{\;KMnO_4\;} C_2H_4O_2 \quad acetic \; acid.$$

Perhaps a better reagent for this reaction would be Ag_2O instead of $KMnO_4$. Writing Kekulé formulas for the acids on the assumption that the reaction took place on the carbon that already had an O attached (again $KMnO_4$ or Ag_2O doesn't react with C_4H_{10}!), we get IV:

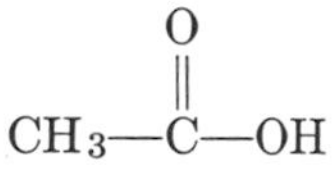

$$\begin{array}{c} O \\ \| \\ CH_3\!\!-\!\!C\!\!-\!\!OH \end{array}$$

$$(IV)$$

It appears that the reaction might be *represented* as the "insertion" of an oxygen atom between C and H:

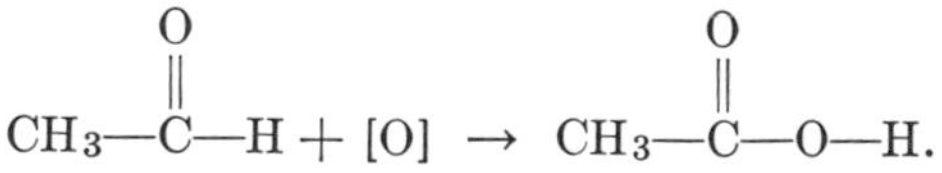

$$\begin{array}{c} O \\ \| \\ CH_3\!\!-\!\!C\!\!-\!\!H \end{array} + [O] \;\rightarrow\; \begin{array}{c} O \\ \| \\ CH_3\!\!-\!\!C\!\!-\!\!O\!\!-\!\!H. \end{array}$$

We can show by experiment that the product of the $KMnO_4$ or Ag_2O oxidation is an *acid*. Acids show a convenient characteristic reaction with NaOH to produce a water-soluble crystalline material (e.g., $NaC_2H_3O_2$) and water (HOH).

Looking at the products of "ozonolysis" of A, we could have, in addition to $CH_2{=}O$,

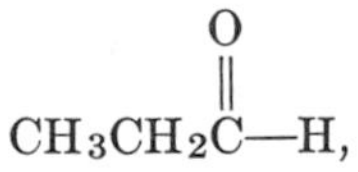

$$CH_3CH_2\overset{\overset{\displaystyle O}{\|}}{C}{-}H,$$

if A is I, or

$$CH_3{-}\overset{\overset{\displaystyle O}{\|}}{C}{-}CH_3,$$

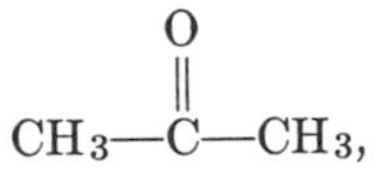

if A is III. The former compound is like C_2H_4O in that it has the group

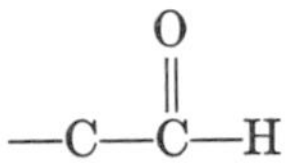

$${-}\overset{}{C}{-}\overset{\overset{\displaystyle O}{\|}}{C}{-}H$$

available for "insertion" of an oxygen atom. It should react with Ag_2O to give an acid

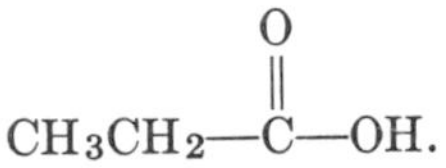

$$CH_3CH_2{-}\overset{\overset{\displaystyle O}{\|}}{C}{-}OH.$$

The latter may not react in this way, since it has only the

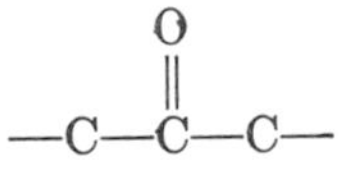

$${-}\overset{}{C}{-}\overset{\overset{\displaystyle O}{\|}}{C}{-}\overset{}{C}{-}$$

structure and no place for the "insertion" of an oxygen. If it is found experimentally that the C_3H_6O produced does *not* react with Ag_2O (or $KMnO_4$ under mild conditions), we conclude provisionally that III represents compound A. We would become more confident if we found an isomer of C_3H_6O that did produce an acid.

We have now not only accounted for the existence of isomers but also assigned structures to each. But, note that *several assumptions* were employed along the way. They are plausible enough, but cannot be viewed as secure unless the structures assigned will also account for *all of the other patterns of reactions of the substances involved*. Ultimately, the scheme is to be checked by its consistent ability to assign *one* structure to each *compound* such that *all* reactions of that compound are rationalized. Much more data is needed to be secure in assignments. Problem 2.8 represents one possible beginning.

Problem 2.8 The C_4H_8 isomers will add HBr to form C_4H_9Br compounds. If the above assignments are correct, predict the number of C_4H_9Br isomers that should be derivable from compounds A and B. (It isn't always easy to produce in the laboratory all of the expected compounds.)

Problem 2.9 Write at least three Kekulé structures consistent with the formula $C_4H_{10}O$. Recall that the valence of O is usually 2.

Let's consider another example which includes some atoms other than C and H. There are several isomers of the molecular formula $C_4H_{10}O$. Two of the isomers are the familiar "anaesthetic ether" used in medicine, which boils at 38°C, and a relative of potable alcohol, which boils at 83°C. In this case, it's interesting to dig right in to the reactions before writing any structural formulas. Both compounds react with HI in solution. "Ether" gives two moles of one product containing carbon which has the molecular formula C_2H_5I. The "alcohol" gives one mole of one organic product, C_4H_9I. We write:

$$C_4H_{10}O \text{ (ether)} \quad \xrightarrow[\substack{\text{(in boiling} \\ \text{solution)}}]{\text{HI}} \quad 2C_2H_5I,$$

$$C_4H_{10}O \text{ (alcohol)} \quad \xrightarrow[\substack{\text{(at room} \\ \text{temperature)}}]{\text{HI}} \quad C_4H_9I.$$

In both cases, the oxygen has been eliminated in the reaction and a compound called an "alkyl iodide" has been formed. But, removal of the oxygen from the "ether" has resulted in fragmentation of the carbon skeleton. Fragmentation was not the fate imposed on the "alcohol." Perhaps the structural difference lies in the fact that the oxygen is in the middle of the "ether" carbon chain, but at the end of the "alcohol" carbon chain.

A partial structural formulation of the reactions might be:

$$\text{ether:} \quad C_2H_5\text{—}O\text{—}C_2H_5 + 2HI \;\rightarrow\; 2C_2H_5I + H_2O,$$

and

$$\text{alcohol:} \quad C_4H_9OH + HI \;\rightarrow\; C_4H_9I + H_2O.$$

This suggests that oxygen "removal" will break the "ether" into two parts but will take along only a stray hydrogen in the alcohol case.

There is only one finished set of structural formulas possible for the "ether" if we accept the above argument:

$$CH_3CH_2\text{—}O\text{—}CH_2CH_3 + 2HI \;\rightarrow\; 2CH_3CH_2I + H_2O.$$

Several Kekulé structures are possible for C_4H_9OH. One consistent formulation is:

$$
\begin{array}{ccc}
CH_3 & & CH_3 \\
| & & | \\
CH_3\overset{\,}{C}OH + HI & \rightarrow & CH_3\text{—}C\text{—}I + H_2O. \\
| & & | \\
CH_3 & & CH_3
\end{array}
$$

But,

$$CH_3CH_2CH_2CH_2OH, \quad CH_3\overset{\displaystyle CH_3}{\underset{|}{CH}}-CH_2OH, \quad \text{and} \quad CH_3\overset{\displaystyle OH}{\underset{|}{CH}}CH_2CH_3$$

would do just as well. To distinguish among these isomers other reactions must be explored. The HI reaction has simply substituted —I for the —OH group. It provides no obvious distinction among the various arrangements of the carbon chain to which the —OH group is attached.

A useful reaction in this context is again oxidation with potassium permanganate ($KMnO_4$) solutions. If we had samples of all four of the isomeric C_4H_9OH alcohols, we would find that two were oxidized to compounds, $C_4H_8O_2$, showing acid reaction similar to acetic acid (see above), a third was oxidized to C_4H_8O but not beyond (reminiscent of

$$CH_3-\overset{\displaystyle O}{\overset{||}{C}}-CH_3$$

in our first problem), and the last was not affected by $KMnO_4$ under *mild* conditions. Structurally, these results are understood as follows:

1. $CH_3CH_2CH_2CH_2OH \xrightarrow{KMnO_4} CH_3CH_2CH_2COOH$

and

$$CH_3-\overset{\displaystyle CH_3}{\underset{|}{CH}}-CH_2OH \xrightarrow{KMnO_4} CH_3-\overset{}{\underset{\underset{\displaystyle CH_3}{|}}{CH}}-COOH,$$

2. $CH_3-\overset{\displaystyle OH}{\underset{|}{CH}}-CH_2CH_3 \xrightarrow{KMnO_4} CH_3-\overset{\displaystyle O}{\overset{||}{C}}-CH_2CH_3$ (Any further oxidation would have to be at the expense of C—C bonds.),

3. $CH_3-\overset{\displaystyle CH_3}{\underset{\underset{\displaystyle CH_3}{|}}{\overset{|}{C}}}-OH \xrightarrow{KMnO_4}$ no reaction   (no C—H bond to attack).

Finding that the alcohol boiling at 83°C does not react with $KMnO_4$ under mild conditions, we could assign it the structure on the right. One interesting way to check this assignment would be to carry out a "dehydration" reaction. In the presence of concentrated sulfuric acid

$$CH_3-\overset{\displaystyle OH}{\underset{\underset{\displaystyle CH_3}{|}}{\overset{|}{C}}}-CH_3$$

(H_2SO_4), alcohols lose water (H_2O) according to the reaction:

$$C_4H_9OH \xrightarrow{H_2SO_4} C_4H_8 + H_2O.$$

C_4H_8 must be one of the compounds with a carbon-carbon double bond if it is to have a satisfactory Kekulé structure. If we dehydrate

$$
\begin{array}{c}
OH \\
| \\
CH_3\!-\!\overset{\displaystyle}{\underset{\displaystyle |}{C}}\!-\!CH_3 \\
CH_3
\end{array}
$$

we might expect:

$$
\begin{array}{c}
HO \quad H \\
| \quad\;\; | \\
CH_3\!-\!\overset{}{\underset{|}{C}}\!-\!CH_2 \xrightarrow{H_2SO_4} \quad CH_3\!-\!\overset{}{\underset{|}{C}}\!=\!CH_2 + H_2O \\
CH_3 \qquad\qquad\qquad CH_3
\end{array}
$$

so that the C_4H_8 obtained is identical to substance A (structure III) of our first problem.

Remember, we are making many assumptions and reasoning plausibly. To be sure the structures are right we will watch to see if they also account for the *other reactions* of the *compounds* in question.

Problem 2.10   Write the *structural* formulas for the molecules involved in the following three paths of reaction by using reasoning analogous to that used in the preceding paragraphs. If more than one structure is reasonable, give all of them.

a) $C_2H_6O + 2HBr \rightarrow 2CH_3Br + H_2O$

b) $C_2H_6O + HBr \rightarrow C_2H_5Br + H_2O$

    $\Big\downarrow_{H_2SO_4}$ $C_2H_4 \xrightarrow{Br_2} C_2H_4Br_2$

c) $C_3H_8O \xrightarrow{KMnO_4} C_3H_6O$ (won't go on to acid)

    $\Big\downarrow_{HI}$ $C_3H_7I + H_2O$

        $\Big\downarrow_{NaOH}$ C_3H_6 (This new reaction is removing HI from an alkyl iodide. Try a C=C structure for C_3H_6.).

Hint: If you can't see your way into a problem, write out all Kekulé structures for the simpler molecules involved and work backwards.

2.5 FUNCTIONAL GROUPS

We must now examine the character of one key assumption made above. In several places we assumed that a reaction did not take place in the part of the molecule which resembles simple hydrocarbons like $CH_3CH_2CH_2CH_3$ (because these are relatively unreactive), but rather occurred where an atom other than C or H was substituted into the molecule or where a double bond occurred. This is an example of the very useful idea that organic molecules may be (on paper) resolved into unreactive hydrocarbon parts and *functional groups*, the latter being the *usual sites of reaction*. The functional groups in the first problem solved were the *double bond* (alkene) function

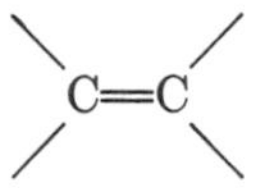

$$\diagdown_{C=C}\diagup$$

and the *aldehyde* function

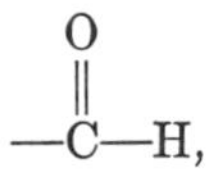

$$\overset{\displaystyle O}{\underset{\displaystyle }{\overset{\|}{-C-H,}}}$$

the less oxidizable *ketone* function

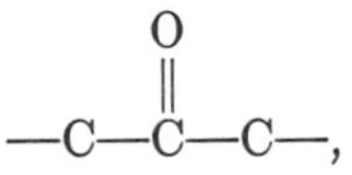

$$\overset{\displaystyle O}{\overset{\|}{-C-C-C-,}}$$

and the *acid* function

$$\overset{\displaystyle O}{\overset{\|}{-C-OH.}}$$

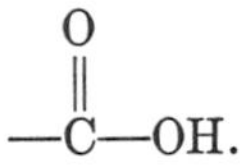

In the second example we introduced the alcohol function

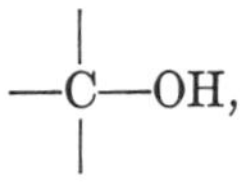

$$\overset{\displaystyle |}{\underset{\displaystyle |}{-C-OH,}}$$

the ether function

$$\overset{\displaystyle |}{\underset{\displaystyle |}{-C-}}O\overset{\displaystyle |}{\underset{\displaystyle |}{-C-}},$$

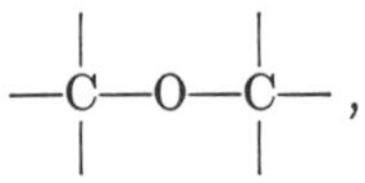

and the alkyl iodide

$$\overset{\displaystyle |}{\underset{\displaystyle |}{-C-I.}}$$

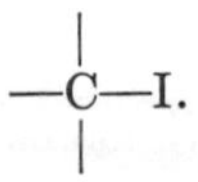

A little inspection of the arguments reveals that we also *assumed that a given functional group will react in more or less the same way whatever the structure of the rest of the molecule happens to be.* This idea proves to be a very useful one in organic chemistry although there are certainly many exceptional cases in which neighboring functional groups influence each other's reactivity.

Problem 2.11 Identify the functional groups present in each of the following structures. (For example,

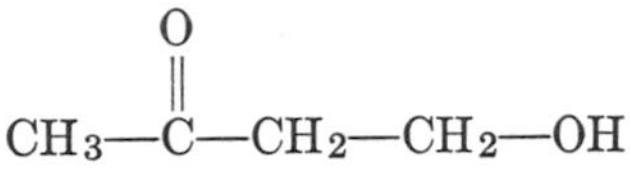

$$\text{CH}_3-\overset{\displaystyle O}{\overset{\|}{\text{C}}}-\text{CH}_2-\text{CH}_2-\text{OH}$$

contains the *ketone*,

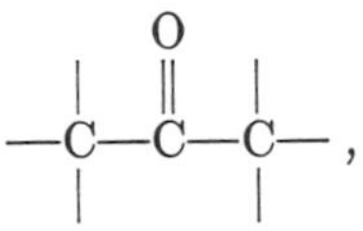

$$-\overset{\displaystyle |}{\underset{|}{\text{C}}}-\overset{\displaystyle O}{\overset{\|}{\text{C}}}-\overset{\displaystyle |}{\underset{|}{\text{C}}}-,$$

and *alcohol*,

$$-\overset{|}{\underset{|}{\text{C}}}-\text{O}-\text{H},$$

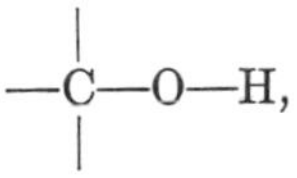

functions.) You may wish to consult Table 2.1 and the appendix to this chapter.

a) 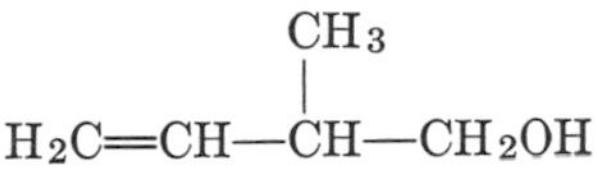 $\text{H}_2\text{C}=\text{CH}-\overset{\displaystyle \text{CH}_3}{\overset{|}{\text{CH}}}-\text{CH}_2\text{OH}$

b) $\text{CH}_3-\text{O}-\text{CH}_2\text{CH}_3$

c) 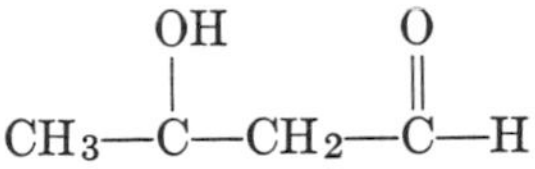 $\text{CH}_3-\overset{\displaystyle \text{OH}}{\overset{|}{\text{C}}}-\text{CH}_2-\overset{\displaystyle O}{\overset{\|}{\text{C}}}-\text{H}$

d) $\overset{\displaystyle \text{CH}_3}{\underset{\displaystyle \text{CH}_3}{>}}\text{C}=\text{CH}-\text{CH}_2-\text{I}$

e) $\text{CH}_3-\overset{\displaystyle O}{\overset{\|}{\text{C}}}-\text{CH}_2-\overset{\displaystyle O}{\overset{\|}{\text{C}}}-\text{CH}_3$

f) $\text{CH}_3-\text{CH}_2-\overset{\displaystyle \text{CH}_3}{\overset{\cdot|}{\text{CH}}}-\overset{\displaystyle O}{\overset{\|}{\text{C}}}\text{OH}$

g) $\text{CH}_3-\overset{\displaystyle O}{\overset{\|}{\text{C}}}-\text{CH}_2-\text{CH}_2-\text{OH}$

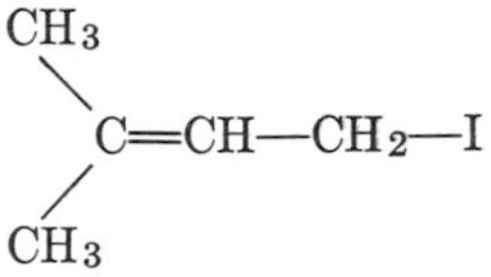

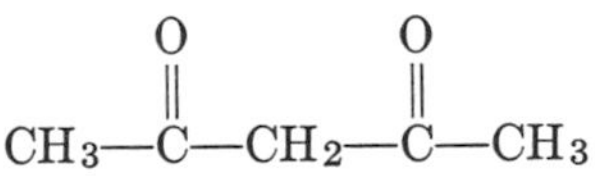

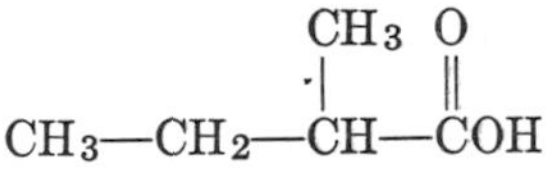

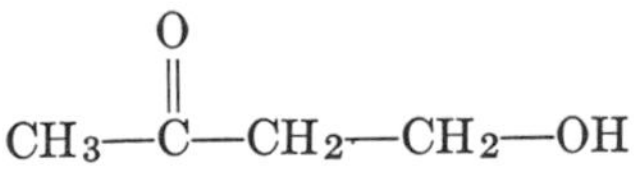

h) $\text{CH}_3-\overset{\displaystyle \text{C}=\text{O}}{\overset{\displaystyle |}{\overset{\text{HO}}{\underset{|}{\text{C}}}}}-\text{CH}_2-\text{CH}_2-\text{I}$

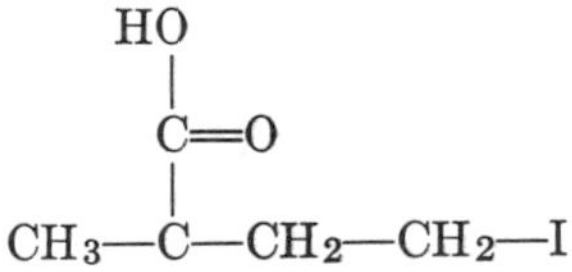

Note: Some of these examples with several functional groups would be unstable and quite reactive.

Table 2.1
Some functional groups and characteristic reactions

ALKENES

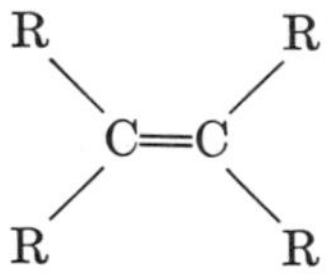

$$\underset{R}{\overset{R}{\diagdown}}C=C\underset{R}{\overset{R}{\diagup}}$$

Ozonolysis

$$\underset{R}{\overset{R}{\diagdown}}C=C\underset{R}{\overset{R}{\diagup}} \xrightarrow{O_3} \underset{R}{\overset{R}{\diagdown}}C=O \;+\; O=C\underset{R}{\overset{R}{\diagup}}$$

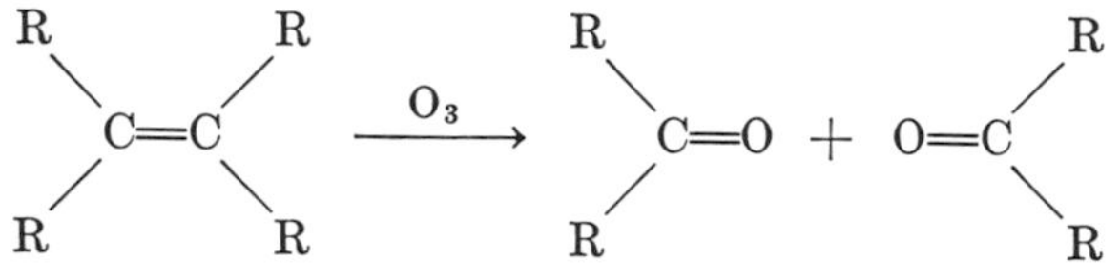

alkene   ketone (or aldehyde if one R = H)

ALCOHOLS   R—OH

Substitution

$$R—OH + HI \longrightarrow R—I + H_2O$$

alcohol   alkyl iodide

Oxidation

I) $R—CH_2OH \xrightarrow{KMnO_4} \underset{\text{carboxylic acid}}{R—\overset{\displaystyle O}{\overset{\|}{C}}—OH}$

primary alcohol

II) $\underset{H}{\overset{R}{\overset{|}{R—\underset{|}{C}—OH}}} \xrightarrow{KMnO_4} \underset{\text{ketone}}{R—\overset{\displaystyle O}{\overset{\|}{C}}—R}$

secondary alcohol

Dehydration $R—CH_2—CH_2OH \xrightarrow{H_2SO_4} R—CH=CH_2 + H_2O$

alcohol   alkene

ALDEHYDE $R—\overset{\displaystyle O}{\overset{\|}{C}}—H$ (distinguish from ketone $R—\overset{\displaystyle O}{\overset{\|}{C}}—R$)

Oxidation

$$R—\overset{\displaystyle O}{\overset{\|}{C}}—H \xrightarrow{KMnO_4} R—\overset{\displaystyle O}{\overset{\|}{C}}—OH$$

aldehyde   carboxylic acid

ETHER

Cleavage

$$R—O—R + 2HI \longrightarrow 2RI + H_2O$$

Armed with the idea that particular atomic and bond arrangements give characteristic reactions whenever they occur in a molecule, we can organize our information about reactions of organic molecules by simply cataloging the reactions of each functional group as they are encountered. The examples of Section 2.4 suggest the brief catalog shown in Table 2.1. In that table the notation R for either an H atom or a group like CH_3— or CH_3CH_2— or a large hydrocarbon fragment is introduced to emphasize that the reactions of the functional groups are presumed to be general reactions, despite the rest of the molecule.

With a good catalog (such as Table 2.1) derived from analysis of relatively simple cases such as those in Section 2.4 it becomes possible to find *structures* of even quite complicated molecules. The strategy is to find conditions under which the molecules react then to decide what functional groups are responsible for the reactions and finally to put together the jigsaw puzzle of the parts discovered to form a picture of the whole molecule. Of course, the process can also be reversed. A particular large structure can be synthesized out of small molecular parts by choosing reactions of functional groups that will result in linking the parts together in the right order to give the desired structure. Customarily, an organic chemist confronted with a complicated problem will work both ways. He will first carry out reactions breaking down his original substance until he can guess what the structure is. Then he will attempt to develop a series of reactions that assemble the structure he has guessed and he will see whether or not the resulting substance is identical to the original unknown.

2.6 DETERMINATION OF STRUCTURES OF NATURAL PRODUCTS

One of the most impressive accomplishments of structural organic chemistry is the mastery of the structures of many products isolated from plant and animal tissue, and the synthesis of these natural structures and variations on them in the laboratory. This achievement has made available invaluable tools to the practice of medicine and set the stage for the joint enterprise of chemists and biologists: the study of the molecular basis of the life process. Similarly, it has provided a new and deeper understanding of the materials used in technology and the opportunity to create new synthetic technology.

Perhaps it is valuable to cite one or two of the more sophisticated accomplishments in this field in order to show what can be achieved. Cholesterol is an interesting example. It is a substance widely distributed in the human body, especially in nerve and brain tissue. It has received notoriety in recent years for its connection with ailments of the circulatory system, particularly hardening of the arteries. Cholesterol was recognized as a distinct chemical substance as early as 1812, but all aspects of its structure were not settled until about 1955. It may be represented as shown on the next page.

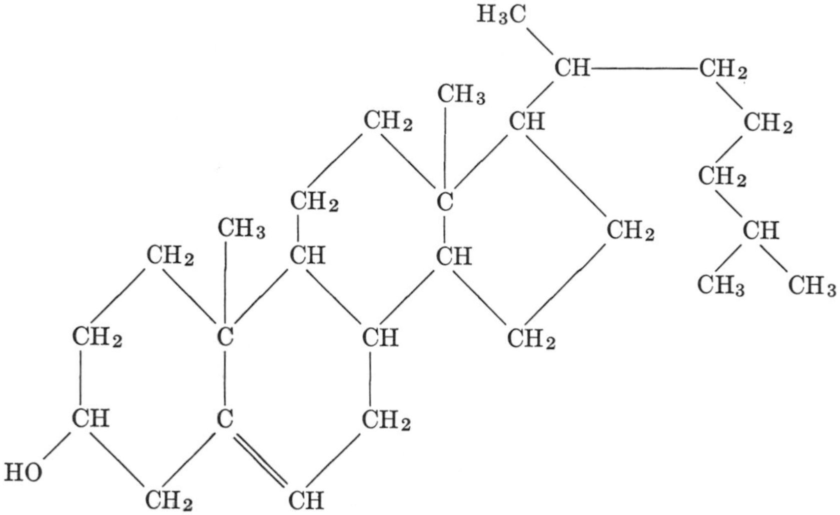

Cholesterol

Another well-known story is associated with the widely effective antibiotic compound, penicillin. Its clinical value was first realized at the beginning of the Second World War (1941), and intensive work during the war led to the isolation of pure crystalline samples from the mold cultures in which it is produced. Actually, four very closely related substances with very similar structures were isolated. The structure of one of them is shown below.

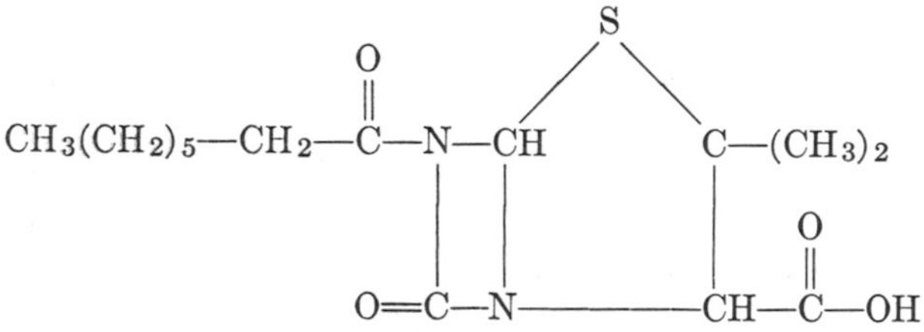

The laboratory synthesis of one of the penicillin structures was finally accomplished in 1957.

Let us undertake the elucidation of the structure of one very simple natural product, which may be accomplished using the information about functional groups and their reactions derived in Section 2.4 and collected in Table 2.1 with only one or two additions. The substance is *geraniol*, which is a perfume ingredient isolated from geraniums. (It is responsible for the flower's characteristic fragrance, but in high concentrations encountered in working with the pure material, the smell is by no means obviously pleasant!)

The methods of Chapter 1 lead to assigning the formula $C_{10}H_{18}O$ to the molecule. One of the simplest reactions it will undergo is addition of Br_2 (see

Problem 2.8): $C_{10}H_{18}O + 2Br_2 \rightarrow C_{10}H_{18}OBr_4$.

This is a characteristic reaction of *alkenes* and suggests the presence of *two* double bonds in the molecule as shown below.

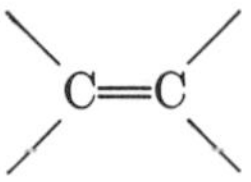

$$\underset{/}{\overset{\backslash}{C}}=\underset{\backslash}{\overset{/}{C}}$$

Another simple reaction is an oxidation to produce a product having the properties of an *acid*:

$$C_{10}\underline{H_{18}}O \xrightarrow{\text{oxidation}} C_{10}\underline{H_{16}}O_2.$$

This should suggest the behavior of the alcohol group, $-CH_2OH$. So we write, at least tentatively, the partial structure:

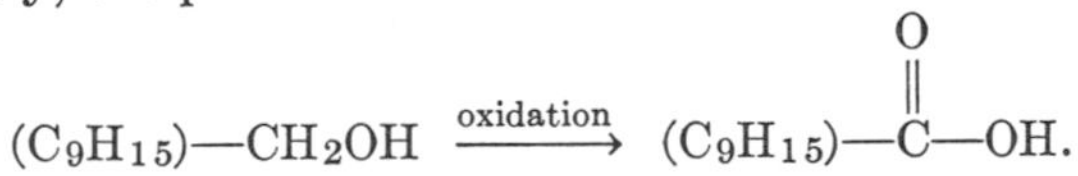

$$(C_9H_{15})-CH_2OH \xrightarrow{\text{oxidation}} (C_9H_{15})-\overset{\overset{\textstyle O}{\|}}{C}-OH.$$

The hypothesis that double bonds are present suggests the use of the ozonolysis reaction to break the molecule down into simpler fragments. This experiment has the result that *geraniol* breaks down into *three* fragments (as it should if two double bonds are present). The reaction is:

$$C_{10}H_{18}O \xrightarrow{O_3} \underset{A}{C_2H_4O_2} + \underset{B}{C_5H_8O_2} + \underset{C}{C_3H_6O}.$$

Fragment C is the easiest to assign. It may be only

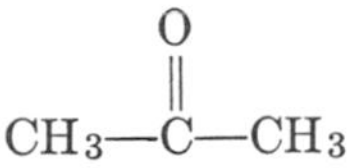

$$CH_3-\overset{\overset{\textstyle O}{\|}}{C}-CH_3$$

(the ketone called acetone) or

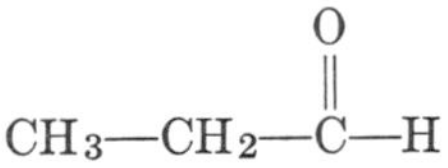

$$CH_3-CH_2-\overset{\overset{\textstyle O}{\|}}{C}-H$$

(the aldehyde called propionaldehyde). If $KMnO_4$ will oxidize it, it is the aldehyde. If not, it is the ketone. The observed failure of oxidation shows that it is the ketone.

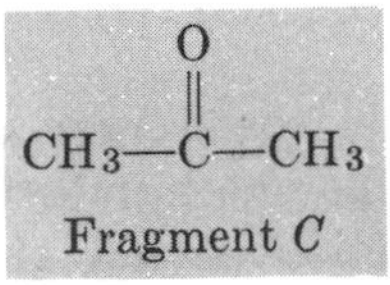

$$CH_3-\overset{\overset{\textstyle O}{\|}}{C}-CH_3$$
Fragment C

Fragment A is not an acid despite its two oxygens. Therefore it probably has *two* functional groups. Oxidation of A leads to the remarkable formula $C_2H_2O_4$. This *is* an acid and can *only* be:

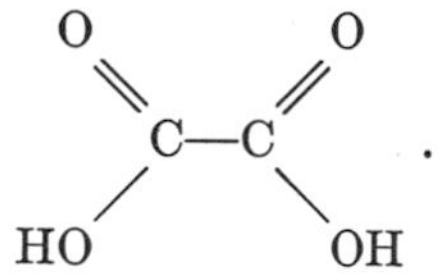

$$\underset{HO}{\overset{O}{\diagdown}}C-C\underset{OH}{\overset{O}{\diagup}} \; .$$

The two ends of A, then, were either aldehyde or alcohol so that they would oxidize to acids. Note that the alcohol group, $-CH_2OH$, identified at the outset must be on one fragment and that the structure,

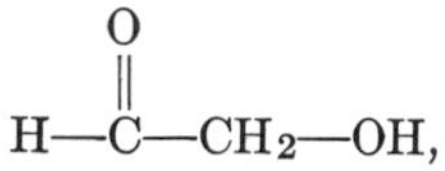

$$\overset{O}{\underset{\|}{H-C-CH_2-OH,}}$$

agrees with the $C_2H_4O_2$ formula and the reactions of A.

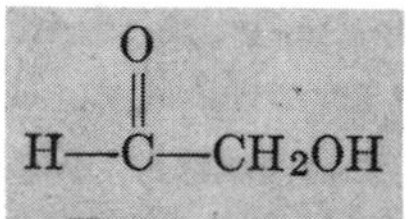

$$\overset{O}{\underset{\|}{H-C-CH_2OH}}$$
Fragment A

Fragment B, $C_5H_8O_2$, remains. It is *not* an acid either, so the two oxygens presumably represent two functional groups. They must be either aldehyde or ketone functions generated in the ozonolysis of the double bonds, since we have already located the alcohol function of the original molecule (which contained only one oxygen). The addition of two oxygens, of course, means that two double bonds were broken. This must be the *middle* fragment of the original molecule. It turns out that oxidation of B gives $C_4H_8O_3$, which is an acid. Clearly, *only* one functional group was oxidized so $C_4H_8O_2$ (B) must have one aldehyde and one ketone group. (The aldehyde was oxidized to an acid; the ketone was not.) The structure is, in fact:

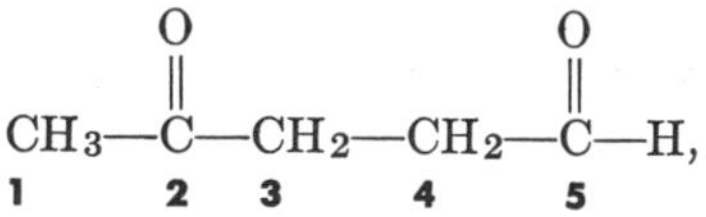

$$\underset{1}{CH_3}-\overset{O}{\underset{2}{\overset{\|}{C}}}-\underset{3}{CH_2}-\underset{4}{CH_2}-\overset{O}{\underset{5}{\overset{\|}{C}}}-H,$$

although it is not yet clear that the oxygen of the ketone is located on the carbon numbered 2 instead of the one numbered 3 or 4. The position of this oxygen is established by a reaction outside the short list in Table 2.1. The ketone may be reduced to an alcohol by the reaction:

$$
\underset{\text{CH}_3-\overset{\displaystyle\text{O}}{\overset{\|}{\text{C}}}-\text{CH}_2}{} \quad\xrightarrow{\text{H}_2}\quad \underset{\text{CH}_3-\overset{\displaystyle\text{OH}}{\underset{\displaystyle\text{H}}{\overset{|}{\underset{|}{\text{C}}}}}-\text{CH}_2}{} \quad.
$$

The alcohol may then be subjected to dehydration to produce a *double bond*, and the double bond is then cleaved (e.g., by ozone). The size of the fragment thus obtained will show how close to the end the ketone oxygen was situated. (For completeness, note that the dehydration can give two isomers with distinct ozonolysis fragmentation patterns. One is most helpful to examine at this point.)

$$
\text{CH}_3-\overset{\displaystyle\text{O}}{\overset{\|}{\text{C}}}-\text{CH}_2-\text{CH}_2-\overset{\displaystyle\text{O}}{\overset{\|}{\text{C}}}-\text{H}
$$

Fragment *B*

Now all that remains is to find how the fragments join together. Fragment *B* received two oxygens in the ozonolysis, so it must be the middle fragment with the two double bonds located at the position where the oxygens attached. We can identify the ends of the original double bonds by locating all the oxygens introduced in ozonolysis:

$$
\underset{\text{CH}_3}{\overset{\text{CH}_3}{\diagdown}}\text{C}\!=\!\!=\!\text{O} \qquad \text{O}\!=\!\!=\!\text{CH}-\text{CH}_2\text{CH}_2-\underset{\text{CH}_3}{\overset{\text{CH}_3}{|}}\text{C}\!=\!\!=\!\text{O} \qquad \text{O}\!=\!\!=\!\text{CH}-\text{CH}_2\text{OH}.
$$

$$\quad\quad C \qquad\qquad\qquad\qquad\qquad B \qquad\qquad\qquad\qquad A$$

Taking the oxygens out and *reforming* the original double bonds reconstructs the molecule. But there is *one remaining ambiguity*. Fragment *B* could be turned around and reattached in a way opposite to that shown in the sketch. How do we know that the ketonic end of *B* joins fragment *A*? This is discovered by a *mild* (low-temperature) oxidation of the original molecule with KMnO₄. The oxidation leads to conversion of the *alcohol* group to an acid, then the breaking of *only one* double bond as follows:

$$
\text{C}_9\text{H}_{15}-\text{CH}_2\text{OH} \;\rightarrow\; \text{C}_9\text{H}_{15}-\text{COOH} \;\rightarrow\; \text{C}_8\text{H}_{14}\text{O} + \text{HOOC}-\text{COOH}.
$$

Fragment *A only* has been removed. When $\text{C}_8\text{H}_{14}\text{O}$ turns out to be a *ketone*, it follows that *A* was attached to what became the ketonic end of *B*. When *C* was left undisturbed, we didn't produce any aldehydic end. The structure of *geraniol* is then established as:

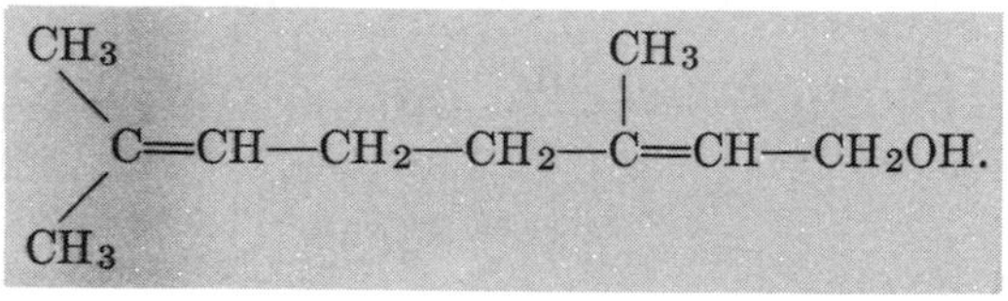

$$\underset{\underset{CH_3}{\diagdown}}{\overset{\overset{CH_3}{\diagup}}{}} C = CH - CH_2 - CH_2 - \underset{\overset{|}{CH_3}}{C} = CH - CH_2OH.$$

Problem 2.12   Write possible structural formulas for **A**, **B**, **C**, and **D** below.

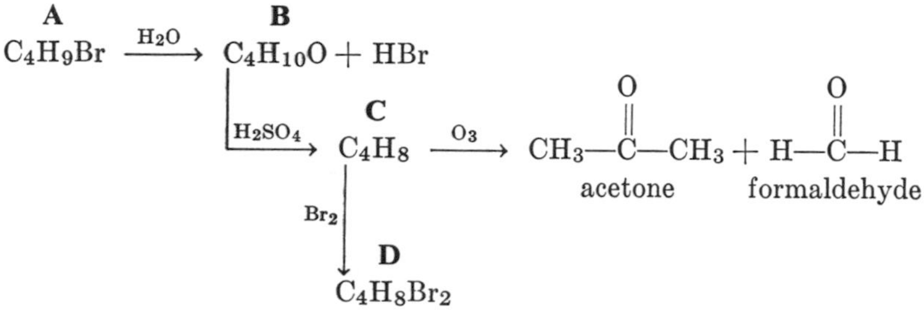

$$\overset{\textbf{A}}{C_4H_9Br} \xrightarrow{H_2O} \overset{\textbf{B}}{C_4H_{10}O} + HBr$$

$$\downarrow H_2SO_4$$

$$\overset{\textbf{C}}{C_4H_8} \xrightarrow{O_3} CH_3 - \overset{\overset{O}{\|}}{C} - CH_3 + H - \overset{\overset{O}{\|}}{C} - H$$
$$\text{acetone} \qquad \text{formaldehyde}$$

$$\downarrow Br_2$$

$$\overset{\textbf{D}}{C_4H_8Br_2}$$

Problem 2.13   Compound A, called isoprene, is the basic structural fragment of a number of substances derived from plants. It may be isolated from pyrolysis of rubber. Its formula is C_5H_8. One mole of A adds two moles of Br_2. Ozonolysis of A yields CH_2O and one-half as many moles of a compound, B, with the formula $C_3H_4O_2$. Compound B is not an acid but may be oxidized to C. Compound C has the formula $C_3H_4O_3$ and is an acid. Compounds B and C show some characteristic reactions of ketones. Write structures for C, B, and A.

2.7 THE BENZENE PROBLEM

Every theory seems to have its limits. The most obvious limitation of the spectacularly successful theory of structure diagrams and functional-group reactivity patterns is encountered in the attempt to understand the molecule C_6H_6 (benzene) and its derivatives. Kekulé suggested the following diagram for C_6H_6 on the basis of his postulate of a valence of four for carbon.

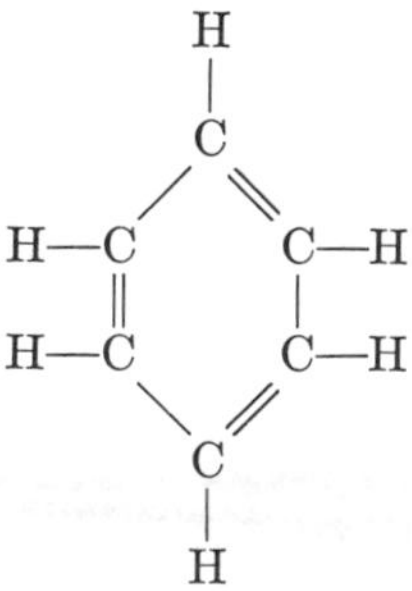

This is usually abbreviated

,

where each corner represents a carbon and is understood to have one hydrogen attached. Looking at this structure one might expect to find the following isomers if two of the hydrogens were *replaced* by bromine atoms.

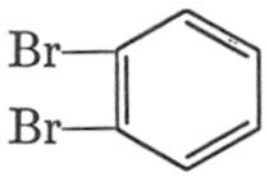

$$\text{Br—}\bigcirc\text{—Br} \quad \text{and} \quad \text{Br—}\bigcirc\text{—Br}$$

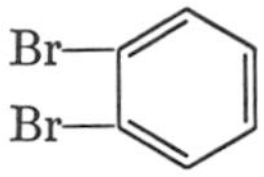

In one case the two Br's are separated by a carbon-carbon double bond

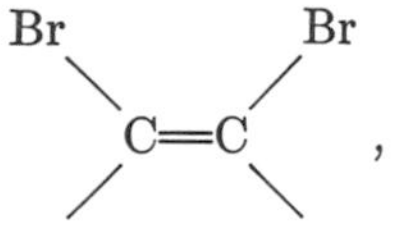

$$\overset{\text{Br}}{\diagdown}\text{C}=\text{C}\overset{\text{Br}}{\diagup} \quad,$$

in the other case, by a carbon-carbon single bond

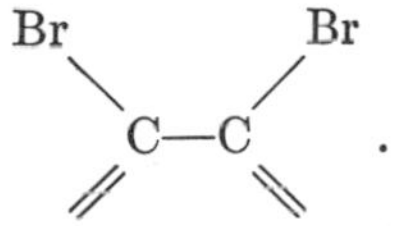

$$\overset{\text{Br}}{\diagdown}\text{C}-\text{C}\overset{\text{Br}}{\diagup} \quad.$$

No such isomers can be found.

Kekulé suggested it might be only a matter of the experimental difficulty of separating the very similar isomers. But the difficulty is more profound. Benzene does not react like a compound containing the normal double bond. Compare the following:

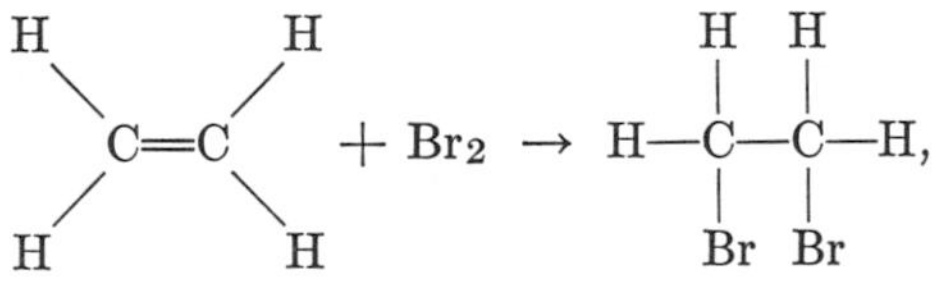

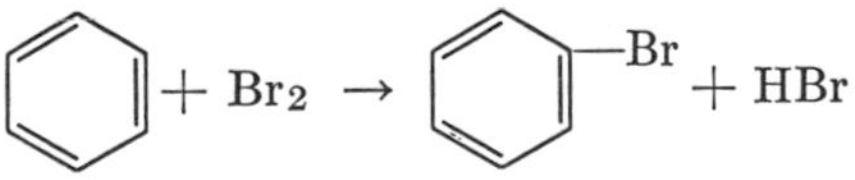

$$\overset{\text{H}}{\underset{\text{H}}{\diagdown}}\text{C}=\text{C}\overset{\text{H}}{\underset{\text{H}}{\diagup}} + \text{Br}_2 \rightarrow \text{H—}\overset{\overset{\text{H H}}{|\;|}}{\underset{\underset{\text{Br Br}}{|\;|}}{\text{C—C}}}\text{—H,}$$

$$\bigcirc + \text{Br}_2 \rightarrow \bigcirc\text{—Br} + \text{HBr}$$

$$(\text{C}_6\text{H}_6 + \text{Br}_2 \rightarrow \text{C}_6\text{H}_5\text{Br} + \text{HBr}).$$

Bromine does not add to the double bond in benzene; it replaces a hydrogen instead. A double bond in benzene is clearly *less reactive* or *more stable* than a normal double bond.

There are, thus, two problems. In the *six-membered ring* with *alternating* double and single bonds, the molecule doesn't seem to be able to "tell" which are the single and which are the double bonds, *and* the double bonds seem to have acquired extra resistance to reaction. Modern chemists have been forced to take the position that a benzene molecule in some sense reflects both arrangements of double bonds:

and that the best representation is to picture the molecule as intermediate between these two arrangements with *six equivalent* carbon-carbon bonds that are approximately one and one-half bonds. In order to stay within the rules of drawing formulas derived from Kekulé's postulates, the situation is represented

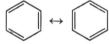

where the double-headed arrow implies that both structures enter into the best description of the molecule. It is a *rule* that whenever a molecule must be so represented to account for its behavior, it is in fact more stable than either of the individual structures would imply. This extension of structural-formula theory is called the *theory of resonance.*

The introduction of the theory of resonance adds an additional challenge we shall face when we try to develop a theory of the origin of the interactions between atoms that lead to the bonds we have drawn in our pictures of molecules.

Problem 2.14 One of the "dibromobenzenes" ($C_6H_4Br_2$) was diagrammed above. If all six carbon-carbon bonds in benzene are equivalent, draw the structures of the three *observed* isomers of dibromobenzene.

2.8 FUNCTIONAL GROUPS AND INFRARED SPECTROSCOPY

The concepts of functional group and structure have proved to be very fruitful. Not only is the theory encompassing these ideas able to account for the patterns of chemical reactions of organic molecules, but it also correlates many aspects of the *physical* behavior of organic molecules. We shall explore one example which is also a very useful tool in modern efforts at structure determination.

White light, passed through a prism, is resolved into constituent colors. Since the development of the wave theory of light (a development beginning contemporaneously with the development of Daltonian atomic theory), the resolution has been regarded as separating light into different wavelengths.

Use of instruments other than the eye as detectors reveals that light of wavelengths both longer and shorter than visible light exists. We shall be interested in the light of longer wavelengths lying beyond the red limit of the visible "spectrum" and called *infrared*. If infrared light of varying wavelength is directed onto a sample of a given thickness and the percent of that light transmitted by the sample is recorded, it is observed that a given substance absorbs infrared light strongly at a *limited number of rather well-defined wavelengths*. It is possible to "fingerprint" (so to speak) any compound by observing the wavelengths at which infrared is strongly absorbed. The process is called infrared spectroscopy, and the record an infrared spectrum. The regions of strong absorption are called *absorption bands*. A typical spectrum (that of allylalcohol, $CH_2=CH-CH_2-OH$) is shown in Fig. 2.6. The wavelengths are measured in microns (1 micron, symbolized by the Greek letter μ, $= 10^{-6}$ m).

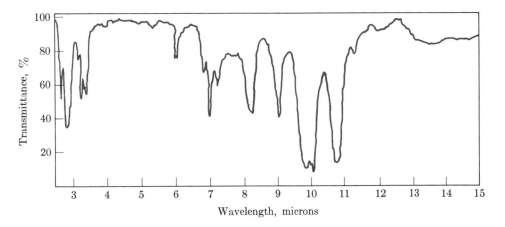

Fig. 2.6. Infrared spectrum of allylalcohol ($CH_2=CH-CH_2-OH$). Note the evidence for the presence of —OH bonds in region around 3 microns and the C=C bond in the 6-micron region.

The interaction of light with matter has played a very important role in bringing the thinking of physicists and chemists together in the twentieth century. It is important to note that the interaction of infrared radiation with matter has turned out to be connected with the vibration of molecules. Absorption of the energy of infrared light excites molecular vibration. However, we shall be content for now with an obvious and very important empirical correlation. Observation of a number of infrared spectra reveals that particular *functional groups* lead to infrared absorption at *particular wavelengths*. Alcohols have infrared absorption maxima (transmittance minima) from 2.7 to $3.2\,\mu$ (microns). Ketones display strong absorption bands from 5.8 to $5.9\,\mu$, alkenes absorb between 5.95 and $6.15\,\mu$, and the basic C—H of the skeleton of pure hydrocarbons is associated with absorption between 3.0 and $3.5\,\mu$. Thus it appears that functional groups not only have characteristic patterns of reaction

Table 2.2

Functional group	Wavelength range, μ
Alcohol, ROH	2.7 to 3.2 (strong)
Aldehyde, R—C(=O)—H	3.5 to 3.7 and 5.7 to 5.9 (very strong absorption)
Ketone, R—C(=O)—R	5.8 to 5.9 (very strong absorption)
Alkene, $R_2C{=}CR_2$	5.95 to 6.15
Alkyl groups, —CH$_2$—	3.0 to 3.5 (weak)

but also characteristic interactions with infrared light. A summary of a few wavelengths at which functional groups absorb is shown in Table 2.2.

Of course, it is possible to turn the discovery of a connection between infrared absorption and functional groups around. Infrared spectroscopy has been, since the late 1940's, one of the most powerful tools for the determination of the functional groups present in new substances.

Problem 2.15   A compound C_3H_6O has strong infrared absorption bands in the 2.7- to 3.2-μ region and in the 6.0- to 6.2-μ region. What is its structure?

Problem 2.16   A compound $C_4H_{10}O$ has no infrared absorption band in the 3-μ region. How would it be expected to react with HI?

2.9 THE CHEMICAL BOND—A LAST WORD

In this chapter, we have encountered two key ideas that have played a central role in the development of a systematic theory of chemical combination. The first is easy to phrase. It is the notion of *valence*. The concept of *valence* suggests that each atom has a characteristic combining power which may be expressed by a small integer. (Recall Section 2.1.) This idea, although easily explained, turns out to be a bit slippery in application. Ultimately, we shall find it desirable to generate more than one set of valence numbers based on different points of view. (See especially Chapter 6.)

The other idea is the related one of a *chemical bond*. This notion has become crucial in modern chemistry with each new area of experimentation reinforcing its usefulness. However, the idea is much more difficult to summarize in a few words because it is essentially a name for a set of rules of manipulation of lines drawn between atoms when writing structural formulas for molecules as we have done in this chapter. Our present concern is to try to say a few words about the significance of these lines that have been drawn.

About the best that can be said is that the significance of these bond lines is *implicit* in the discussion earlier in this chapter. The *circumstances* that lead to our use of them, the way that they function to interpret reaction patterns give the meaning of these symbols. The intent of these last few remarks is to justify the *absence* in these pages of a concise definition of an idea as central to our subsequent discussion as the idea of *a chemical bond*.

Actually, the circumstance we are talking around here is not uncommon in science. Einstein has remarked, "If you want to find out anything from theoretical physicists about the methods they use, I advise you to stick closely to one principle: don't listen to their words, fix your attention on their deeds." The distinguished philosopher of science, Stephen Toulmin, has observed that concepts like the chemical bond idea represent not so much objects, which, like elephants, may be unfamiliar but nonetheless something we might expect to meet someday, *but* instead represent a name for a set of techniques of reasoning by means of which we deal with a class of phenomena. The chemical bond is defined by the way in which manipulation of bond lines in structure diagrams allows us to explain what happens in chemical reactions. This whole chapter has been the first stage of "definition" of chemical bond. *Take another look at it from that point of view!*

READINGS AND EXTENSIONS

Good organic chemistry textbooks are too numerous to list here. One, however, which discusses many topics in the developmental spirit is A. H. Corwin and M. Bursey, *Elements of Organic Chemistry*, Addison-Wesley, Reading, Mass., 1966. A good paperback which extends not only the ideas of this chapter but also those of Chapter 3 and Appendix 1 to give a modern mechanistic account of organic reactions is Werner Hertz, *The Shape of Carbon Compounds*, W. A. Benjamin, New York, 1962.

The tangled history of the problem of organic structure is recreated in a collection of the original papers commented on and edited by O. T. Benfey, *Classics in the Theory of Chemical Combination*, Dover Publications, New York, 1963. Benfey has also written a history of the problem called *From Vital Forces to Structural Formulas* (Houghton Mifflin, Boston, 1965). These books had a strong influence on the formulation of the attitude we took in preparing Chapters 2 and 3 of this text.

The nomenclature of organic chemistry is an important technical accomplishment for the future chemist or biochemist. It is also a matter well adapted to programmed instruction. See J. G. Traynham's workbook, *Organic Nomenclature: A Programmed Introduction*, Prentice-Hall, Englewood Cliffs, N. J., 1966.

A Small Catalog of Some Functional Groups and Some of their Reactions

Using reaction schemes like the ones illustrated in Section 2.4, we can assign structures to a variety of simple organic molecules and learn the characteristic reactions of the various functional groups. It will not be useful to repeat the details of the arguments for each case. In the next few pages we summarize some important results. We give nomenclature and some characteristic reactions for a number of important types of functional groups in terms of structural formulas. *It may be assumed that these were derived in the way we have illustrated.* Once we *know* characteristic reactions of *functional groups*, our analysis of organic molecules can proceed to a higher stage of sophistication. The reactions of complicated molecules reveal the presence of functional groups, often making it possible to write a fairly securely established Kekulé type structural formula. The following catalog should be useful in solving structural problems or planning the synthesis of a particular structure.

2A.1 ALKANES

The family of compounds called alkanes consists of the pure hydrocarbons with no functional groups in the sense of that term employed in Section 2.4. They display relatively few reactions and those tend to be somewhat nonspecific. We list examples of three. (The equations show only products and are not balanced. This practice will be common in listing organic reactions below.)

1. Oxidation (combustion)

$$C_2H_6 + \text{excess } O_2 \rightarrow CO_2 + H_2O,$$

2. Halogenation $\quad CH_4 + Cl_2 \xrightarrow{\text{light}} CH_3Cl + HCl,$

$$CH_3Cl + Cl_2 \xrightarrow{\text{light}} CH_2Cl_2 + HCl \ldots,$$

3. Cracking $\quad CH_3CH_2CH_2CH_3 \xrightarrow[\text{catalyst}]{\text{heat}} CH_3CH_3 + CH_2{=}CH_2.$

2A.2 ALKENES

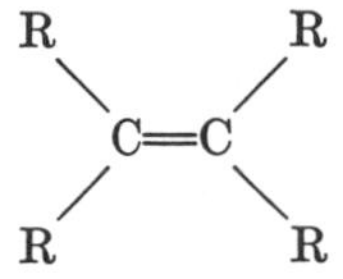

Alkenes are those molecules with a carbon-carbon double bond which may be symbolized generally as shown on the right, where R is a general symbol for a general substituent group. Names are derived from those of the corresponding alkanes by replacing the *ane* ending by *ene*.

$$\begin{array}{c}R\\ \diagdown\\ \end{array}\hspace{-0.5em}C\hspace{-0.3em}=\hspace{-0.3em}C\hspace{-0.5em}\begin{array}{c}R\\ \diagup\\ \end{array}$$

When there are different possible locations for the double bond, its position may be designated by prefixing the number of the lower numbered carbon involved in the double bond. The double bond function is richly reactive. The four typical reactions given represent a modest start. Note that (2) and (3) are closely related. Both involve adding to each side of the double bond to produce a single bond.

1. Ozonolysis

$$CH_2=CH—CH_2—CH_3 + O_3 \rightarrow CH_2=O + O=CHCH_2CH_3$$

  1-butene   formaldehyde   propanal (propionaldehyde)

$$\overset{\displaystyle CH_3}{\underset{\displaystyle |}{CH_3—CH_2—C}}=CH_2 + O_3 \rightarrow CH_3CH_2\overset{\displaystyle CH_3}{\underset{\displaystyle |}{C}}=O + CH_2=O$$

   2-methyl-1-butene   butanone   formaldehyde

2. Halogenation

$$CH_2=CH_2 + Cl_2 \rightarrow ClCH_2CH_2Cl$$

   ethene (ethylene)   1,3 dichloroethane

$$CH_3—CH=CH—CH_3 + Br_2 \rightarrow CH_3—\overset{\displaystyle Br}{\underset{\displaystyle |}{CH}}—\overset{\displaystyle Br}{\underset{\displaystyle |}{CH}}—CH_3$$

     2-butene   2,3 dibromobutane

3. HX Addition (X = Cl, Br, I, OH)

$$CH_3—CH=CH_2 + HX \rightarrow CH_3—\overset{\displaystyle X}{\underset{\displaystyle |}{CH}}—CH_3$$

(or, under special circumstances, $CH_3CH_2CH_2X$)

4. $KMnO_4$ oxidation

$$CH_3—CH=CH_2 \xrightarrow{KMnO_4} CH_3—\overset{\displaystyle O}{\overset{\displaystyle \|}{C}}—OH + H—\overset{\displaystyle O}{\overset{\displaystyle \|}{C}}—OH*$$

           acetic acid   formic acid

The last reaction, no doubt, goes by way of aldehydes.

---

* This is a simplification. Formic acid is unstable and goes on to $H_2O + CO$.

The HX addition reactions in (3) require various conditions. For example, HX = HOH (water) is run in the presence of H_2SO_4 and actually is better written as:

$$CH{=}CH_2 + H_2SO_4 \longrightarrow CH_3{-}CH_2SO_4H \xrightarrow{\text{HOH}} CH_3CH_2OH + H_2SO_4.$$

Under appropriate conditions, reaction 3 may be run in the opposite direction to *produce* alkenes.

2A.3 ALDEHYDES

Aldehydes may be represented in general as

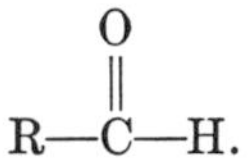

$$\overset{\displaystyle O}{\underset{\displaystyle }{R{-}\overset{\|}{C}{-}H.}}$$

There must be an H bonded to the carbon, which is double bonded to oxygen (called a carbonyl group). Names are generated by dropping *e* from the alkane name and adding *al*. For historical reasons, $H_2C{=}O$ is usually called "formaldehyde" and $CH_3CH{=}O$ "acetaldehyde." Two of the many aldehyde reactions are:

1. $KMnO_4$ oxidation

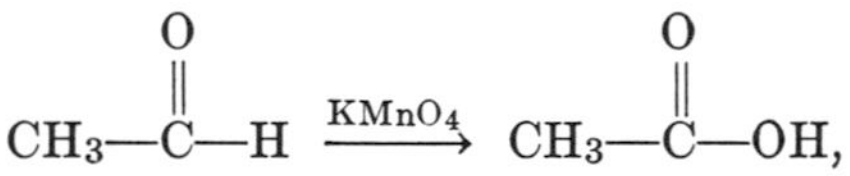

$$CH_3{-}\overset{\displaystyle O}{\overset{\|}{C}}{-}H \xrightarrow{\text{KMnO}_4} CH_3{-}\overset{\displaystyle O}{\overset{\|}{C}}{-}OH,$$

$$\text{ethanal or} \qquad\qquad \text{acetic acid}$$
$$\text{acetaldehyde}$$

2. the Grignard reaction

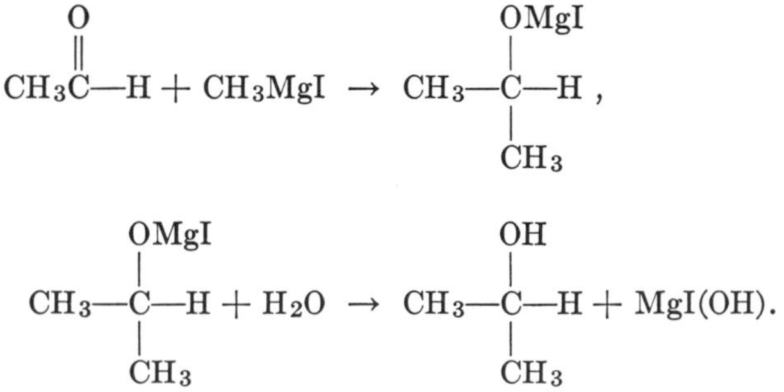

$$CH_3\overset{\displaystyle O}{\overset{\|}{C}}{-}H + CH_3MgI \;\rightarrow\; CH_3{-}\underset{\underset{\displaystyle CH_3}{|}}{\overset{\overset{\displaystyle OMgI}{|}}{C}}{-}H\,,$$

$$CH_3{-}\underset{\underset{\displaystyle CH_3}{|}}{\overset{\overset{\displaystyle OMgI}{|}}{C}}{-}H + H_2O \;\rightarrow\; CH_3{-}\underset{\underset{\displaystyle CH_3}{|}}{\overset{\overset{\displaystyle OH}{|}}{C}}{-}H + MgI(OH).$$

$$\text{2-propanol}$$

The Grignard reaction is widely used, since it increases the number of carbons in a molecule. The compound CH_3MgI (or in general RMgI) is prepared by reacting CH_3I (or RI) with magnesium filings in scrupulously dry ether.

2A.4 KETONES

The general formula for ketones may be written

$$R-\overset{\overset{\displaystyle O}{\|}}{C}-R.$$

They differ from aldehydes in that they do not have an H attached to the carbonyl group. To name a ketone, drop the final e from the name of the corresponding alkane and add *one*. (The simplest ketone,

$$CH_3-\overset{\overset{\displaystyle O}{\|}}{C}-CH_3,$$

is more commonly called acetone than the systematic name, propanone.)

Ketones react similarly to aldehydes. Both, for example, undergo the Grignard reaction, but an important difference in reactivity is that ketones are not oxidized by $KMnO_4$ under conditions which lead to oxidizing aldehydes to acids:

1. $CH_3-\overset{\overset{\displaystyle O}{\|}}{C}-CH_3 \xrightarrow{KMnO_4}$ no reaction,
    propanone
    (acetone)

2. $CH_3CH_2-\overset{\overset{\displaystyle O}{\|}}{C}-CH_3 + CH_3MgI \longrightarrow CH_3CH_2\overset{\overset{\displaystyle OMgI}{|}}{\underset{\underset{\displaystyle CH_3}{|}}{C}}-CH_3 \xrightarrow{H_2O}$
    butanone

$$CH_3CH_2-\overset{\overset{\displaystyle OH}{|}}{\underset{\underset{\displaystyle CH_3}{|}}{C}}-CH_3 + MgI(OH).$$
2-methyl-2-butanol

2A.5 ALCOHOLS

The general formula for alcohols is $R-OH$. Names are formed by dropping the final e from the name of the alkane and adding *ol*. Alcohols are flexible in reactivity and may be used in reactions leading to a variety of other functional groups.

1. Oxidation by $KMnO_4$

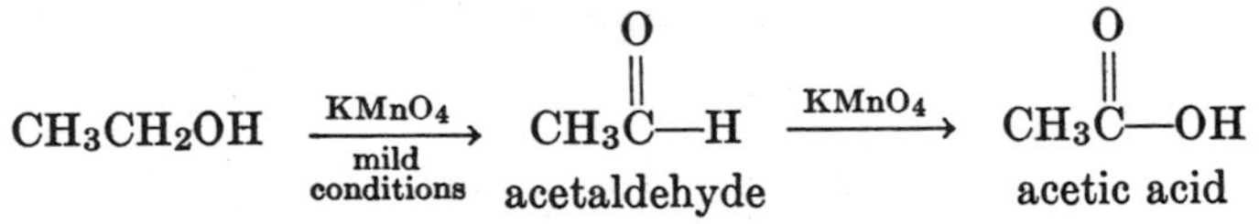

$$CH_3CH_2OH \xrightarrow[\text{conditions}]{\substack{KMnO_4 \\ \text{mild}}} \underset{\text{acetaldehyde}}{CH_3\overset{\displaystyle O}{\overset{\|}{C}}-H} \xrightarrow{KMnO_4} \underset{\text{acetic acid}}{CH_3\overset{\displaystyle O}{\overset{\|}{C}}-OH}$$

Great care must be exercised to stop this reaction at the aldehyde stage. The reaction is not useful for preparation of aldehydes. Note that an alcohol

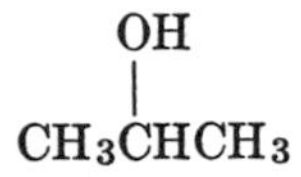

$$\overset{\displaystyle OH}{\underset{}{CH_3\overset{|}{C}HCH_3}}$$

would yield only a *ketone*

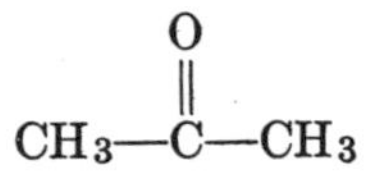

$$CH_3-\overset{\displaystyle O}{\overset{\|}{C}}-CH_3$$

and that

$$CH_3-\overset{\displaystyle OH}{\underset{\displaystyle CH_3}{\overset{|}{\underset{|}{C}}}}-CH_3$$

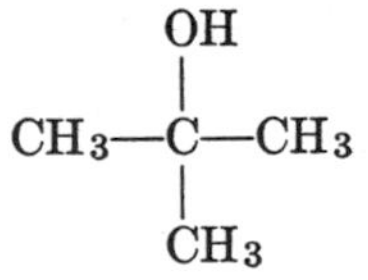

will not be oxidized (forming a group

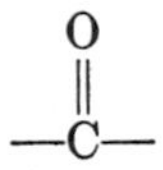

$$-\overset{\displaystyle O}{\overset{\|}{C}}-$$

would require five bonds to C.)

2. Ether formation

$$CH_3OH + Na \rightarrow CH_3ONa + \tfrac{1}{2}H_2,$$

$$\underset{\text{dimethyl ether}}{CH_3ONa + CH_3Cl \rightarrow CH_3OCH_3 + NaCl.}$$

Ethers have the functional group R—O—R. (Note the ether cleavage reaction:

$$CH_3CH_2-O-CH_2CH_3 + HI \rightarrow 2CH_3CH_2I + H_2O.)$$

3. Dehydration (to alkenes)

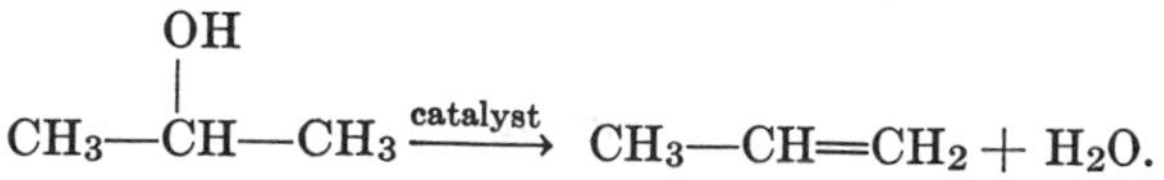

$$
\begin{array}{c}
\text{OH} \\
| \\
CH_3-CH-CH_3
\end{array}
\xrightarrow{\text{catalyst}}
CH_3-CH=CH_2 + H_2O.
$$

2-propanol   propene

4. HX substitution (X = Cl, Br, I)

$$
CH_3CH_2OH + HX \rightarrow CH_3CH_2X + H_2O.
$$

5. Ester formation

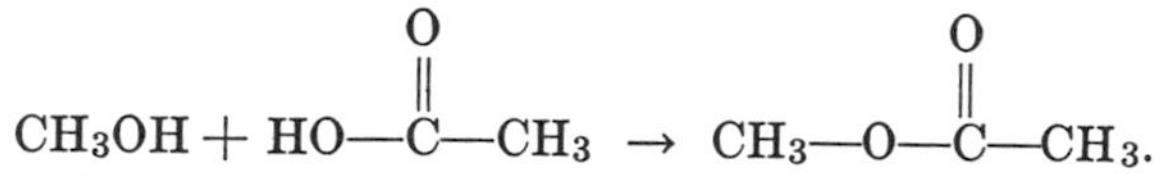

$$
\begin{array}{ccc}
 & \text{O} & & & \text{O} \\
 & \| & & & \| \\
CH_3OH + HO-C-CH_3 & \rightarrow & CH_3-O-C-CH_3.
\end{array}
$$

methanol   acetic acid   methyl acetate

The ester functional group is, in general,

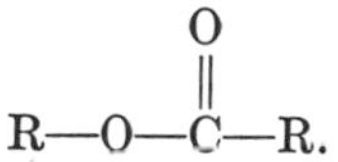

$$
\begin{array}{c}
\text{O} \\
\| \\
R-O-C-R.
\end{array}
$$

2A.6 CARBOXYLIC ACIDS

These compounds have the general formula

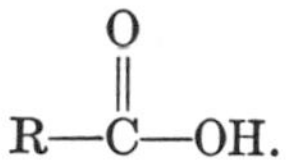

$$
\begin{array}{c}
\text{O} \\
\| \\
R-C-OH.
\end{array}
$$

The designation "acid" connects them to a larger family of substances about which we shall have much more to say in Chapter 5. For the present we may note that the idea of a category of compounds called acids which react with a category called bases to produce so-called salts and water is a very old one antedating all of atomic-theory-based chemistry. The observational characteristics which led to classing substances together as acids include their sour taste and their ability to change the colors of certain naturally occurring dyes. Lavoisier (1789) was persuaded that a structural interpretation of this acidity could be given; he suggested (incorrectly) that the key feature was the presence of oxygen in the molecule. (The name oxygen is derived from Greek roots meaning acid-producing.) Liebig (1838) contended more accurately that the key feature was the presence of a reactive hydrogen atom. In the "carboxylic acids" of organic chemistry it is the H of the

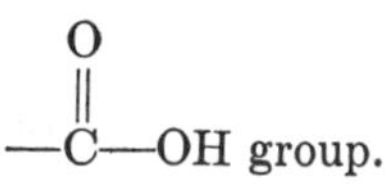

$$
\begin{array}{c}
\text{O} \\
\| \\
-C-OH
\end{array}
$$
group.

We have noted above the formation of some acids by oxidation and the reaction of acids with alcohols to form esters.   Here it seems appropriate to add the reaction of an acid with a typical base and the characteristic reaction with reactive metals (These reactions are not limited to carboxylic acids but are general to acids of all kinds.):

$$
\text{1. } \underset{\text{acetic acid}}{CH_3\overset{\overset{\displaystyle O}{\|}}{-}C-OH} + NaOH \text{ (a base)} \rightarrow \underset{\text{sodium acetate}}{Na-O-\overset{\overset{\displaystyle O}{\|}}{C}-CH_3^*} + H_2O,
$$

$$
\text{2. } \underset{\text{formic acid}}{H-\overset{\overset{\displaystyle O}{\|}}{C}-OH} + Na \rightarrow \underset{\text{sodium formate}}{Na-O-\overset{\overset{\displaystyle O}{\|}}{C}-H^*} + \tfrac{1}{2}H_2.
$$

Seminar: The Benzene Structure Problem

In this chapter we argued for representation of benzene by the structural formula

but were forced to admit that we had to draw two such diagrams

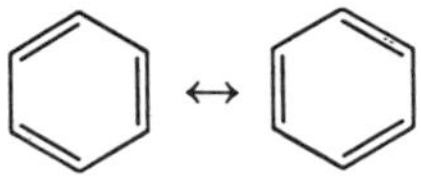

because benzene doesn't behave like an ordinary *alkene*. Thus we are forced to a special modification of structure theory called resonance. Or are we?

Before we accept such an arcane elaboration of structure theory as the idea of resonance, we should carefully examine other ways to represent C_6H_6. Over the years, ingenious chemists have offered alternatives to the Kekulé

---

* The role of sodium in these compounds is misleadingly represented by the line from Na to O suggesting a bond just like the others. See Chapter 5.

formula. Some of the major proposals include:

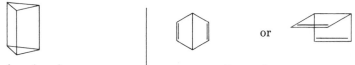

Ladenburg's prism   Dewar benzene

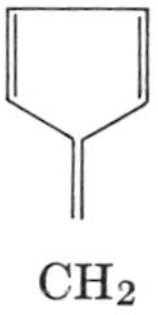

CH₂

A substance now called fulvene

Meyer's centric formula

(The arrows imply a valence from each carbon directed toward the center and interacting with each other to stabilize the system.)

The absence of isomers of

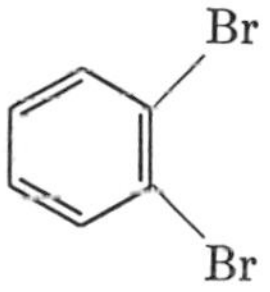

led us to argue that the location of the three double bonds in the ring was indefinite and that all six carbons and all six carbon-carbon bonds were equivalent. Clearly, the number of isomers can be very useful for testing possible structures. There are six hydrogens, and all six may be replaced by, for example, Br ($C_6H_6 \rightarrow C_6Br_6$). The number of isomers at each stage of substitution is given in the following table.

Substitution	mono	di	tri	tetra	penta	hexa
Isomers of substituted benzene	1	3	3	3	1	1

Does this evidence alone eliminate some possibilities for the C_6H_6 structure?

A similar line of argument can be developed from the following evidence. Benzene will undergo addition reactions under special circumstances. One can add H_2 to benzene in the presence of a finely divided Ni catalyst, as shown in Fig. 2S.1. This reaction does not automatically exclude a formula such as the Ladenburg prism because the "special conditions" might be those that

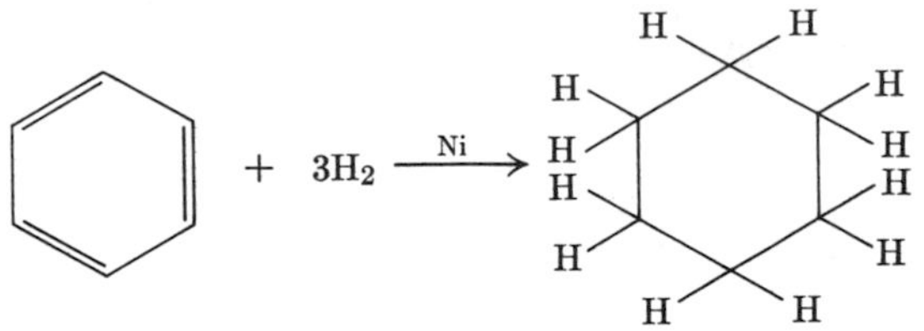

Figure 2S.1.

C_6H_{12}, cyclohexane

would break down the C—C single bond to add H at the ends:

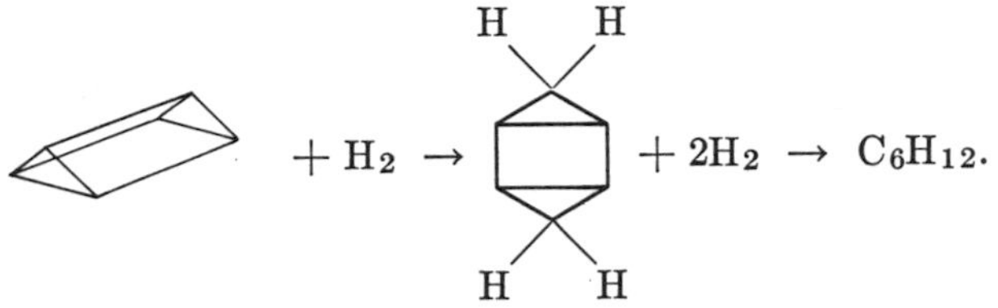

$+ H_2 \rightarrow$ $+ 2H_2 \rightarrow C_6H_{12}$.

But, there is an interesting isomer-count experiment. The three isomers of xylene (dimethylbenzene) may be hydrogenated to give three dimethylcyclohexanes. It is found that each one gives one and that there is *no way* for a *given* dimethylcyclohexane to arise from two different xylenes. This has been cited as evidence against the Ladenburg prism. Can you explain why?

In addition to isomer counts, the other important aspect of benzene chemistry is the dissimilarity of benzene reactions and those of simple alkenes. An interesting comparison is between benzene and hexatriene,

$$CH_2\!\!=\!\!CH\!-\!CH\!\!=\!\!CH\!-\!CH\!\!=\!\!CH_2.$$

The following table may suggest the comparative "unreactivity" of benzene.

Reagent	Benzene product	Hexatriene product
HNO_3/H_2SO_4	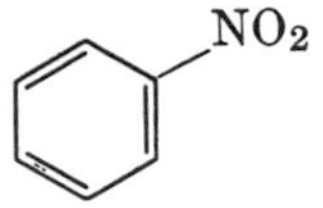	*Instantaneous* polymerization and oxidation
Br_2	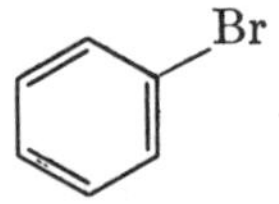 $+ HBr$ (in presence of a catalyst)	$BrCH_2\!-\!CH\!\!=\!\!CH\!-\!CH\!\!=\!\!CH\!-\!CH_2Br$ (in $CHCl_3$ solution below 13°C)
H_2 with PtO_2 catalyst in acetic acid solution	C_6H_{12} (cyclohexane) after 25 hr at 25°C	C_6H_{14} (hexane) after 1 to 2 hr at 25°C

Giving due consideration to all these facets of benzene chemistry *organize* an argument leading to assignment of the structure of benzene. Your emphasis should be on the interpretation of *the experiments cited.* Do these observations settle the issue? *What other experiments might clarify matters?*

Useful reading

Corwin, A. H., and M. Bursey, *Elements of Organic Chemistry*, Addison-Wesley, Reading, Mass., 1966, Chapter 5.

Fieser, L., and M. Fieser, *Organic Chemistry*, 3rd ed., D. C. Heath, Boston, 1956, Chapter 20.

Morrison, R. T., and R. N. Boyd, *Organic Chemistry*, 2nd ed., Allyn & Bacon, Boston, 1966, Chapter 10.

CHAPTER 3

Stereochemistry — Molecular Geometry

3.1 INTRODUCTION

In Chapter 2 we saw evidence that what was called the "structure" of a molecule did not, in fact, describe the arrangement of atoms in space but only the *order* in which they are *connected*. For example the structure of dichloromethane is

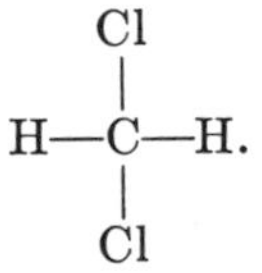

$$
\begin{array}{c}
\text{Cl} \\
| \\
\text{H--C--H.} \\
| \\
\text{Cl}
\end{array}
$$

If this "structure" on the plane of the page represented the actual spatial geometry of the molecule, the compound would be reasonably expected to have an isomer:

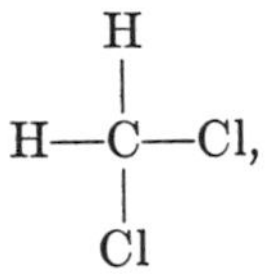

$$
\begin{array}{c}
\text{H} \\
| \\
\text{H--C--Cl,} \\
| \\
\text{Cl}
\end{array}
$$

which, of course, it does not. There is only one substance with the formula CH_2Cl_2. This result is equivalent to the claim in Chapter 2 that "bending" the representation of a formula doesn't make any difference and, for example, that the two structures

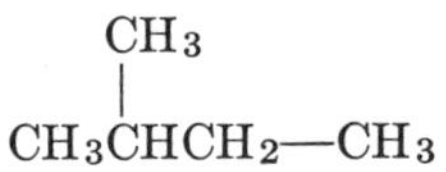

$$
\begin{array}{cc}
\begin{array}{c} \text{CH}_3 \\ | \\ \text{CH}_3\text{CHCH}_2\text{--CH}_3 \end{array} & \quad\text{and}\quad \begin{array}{c} \text{CH}_3\ \ \text{CH}_3 \\ |\quad\ \ | \\ \text{CH}_3\text{CH--CH}_2 \end{array}
\end{array}
$$

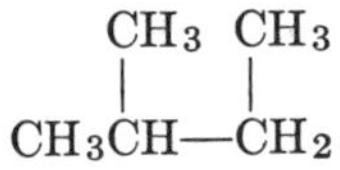

are equivalent. Thus we may conclude one thing about the geometry of organic molecules. They are probably *not* arrangements of atoms *in a plane* as diagrams shown on a plane page tend to suggest.

The obvious next question to investigate is the actual spatial arrangement of atoms in molecules, but how are we to proceed? Before we can meaningfully ask such questions, we must consider what is experimentally accessible that might depend upon the spatial arrangements within the molecule. Historically, the relevant experimental results came from a most unexpected quarter. The subject of the spatial arrangement of atoms in a molecule is called *stereochemistry*, and the first definitive lines of relevant evidence came from the study of the behavior of *polarized light*.

3.2 POLARIZED LIGHT AND OPTICAL ACTIVITY

At the end of the year 1808, the French physicist Malus announced the discovery that light reflected from both opaque and transparent objects acquired extraordinary properties. Malus designated the change *polarization*. The phenomenon can be produced in various ways. One of the earliest discovered is passage of a beam of light through a rhomboid of crystalline calcium carbonate. A much more recent method is probably familiar to most readers. We shall use this familiar case to explain what is meant by polarization. Polaroid filters are readily available and they are used in many common products (e.g., sunglasses). If one mounts a pair of Polaroid filters in such a way that a beam of light passes through both, the surprising phenomenon reported by Malus may be observed. If the second filter is oriented properly with respect to the first, very little light passes through the filters at all. If the second filter is now rotated by 90°, light passes readily. By rotating the filter, one may cause the light to be either transmitted or blocked. The effect is diagrammed in Fig. 3.1.

The light behaves as though, when polarized, it is confined to a plane defined by the first filter (often called the polarizer), and if the second filter (often called the analyzer) is oriented so that it is perpendicular to that plane, the light cannot pass. This idea can perhaps be clarified by an analogy which has more than a purely incidental relationship to the situation under discussion since light is a wave phenomenon. Suppose there is a rope strung through a closely spaced picket fence. If we wave the end of the rope in the direction of

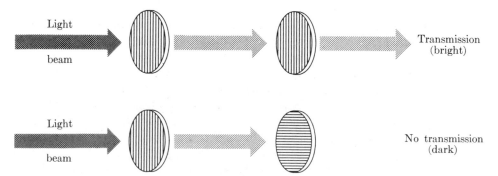

Fig. 3.1. Polarization of light.

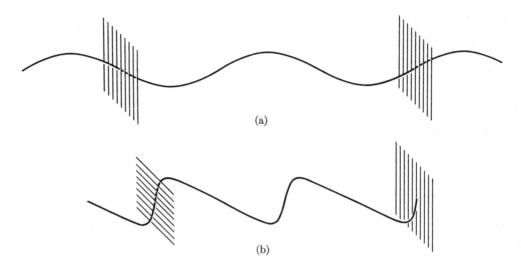

(a)

(b)

Fig. 3.2. Waves through "filters."

the openings, the waving will pass through unimpeded (see Fig. 3.2a). In that plane, the picket fence has no effect on the passage of the disturbance. If a first fence defines a plane in which a disturbance may pass, a second fence must be oriented the same way if the disturbance is to pass it also. Now suppose we attempt to wave the rope in the plane perpendicular to the openings in the fence. In this case the rope hits the pickets and the disturbance is impeded (see Fig. 3.2b).

We shall not attempt to justify our analogy in detail because the main point is that passing light through the polarizer *defines a plane and that plane may be detected by orienting the analyzer to pass or block the beam that has been polarized* by the polarizer. It is not necessary to understand how the phenomenon of polarization arises in order to follow the relationship to the problem of the shape of molecules. In fact, when the discoveries which we shall review in the rest of this chapter were made, there was no detailed theory of the mechanism of polarization.

Malus died in 1812 at the age of 37. But pursuit of the phenomenon of polarization was pressed actively by two other French physicists, D. F. Arago and J. B. Biot. It was soon discovered that a quartz crystal cut parallel to its axis has an interesting effect on polarized light passed through normal to the surface. The cut quartz crystal *rotates* the plane of polarization. That is, when the quartz crystal is placed between polarizer and analyzer arranged for minimum transmission of light, the analyzer must be rotated to a *new angle* in order to restore minimum transmission. Appropriate samples of a number of other minerals also turned out to rotate the plane of polarized light *and in 1815 Biot was able to show that a number of naturally occurring organic compounds, either in the form of liquids or in solution, rotated the plane of polarized light.* The instru-

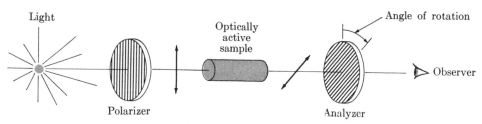

Fig. 3.3. Polarimeter (schematic).

ment used in these experiments (a polarimeter) is shown in Fig. 3.3. Between a polarizer and an analyzer a tube containing the liquid or solution to be studied may be placed. The analyzer is mounted on a ring calibrated in degrees of arc. First, in the absence of the sample, the analyzer is set to minimum light transmittance, and the angle is recorded. Then with the tube containing the sample in the light path, the analyzer is reset to minimum transmittance and the new angle recorded. The difference is the angle by which the plane of polarization is rotated by the sample. Not surprisingly, it emerges that the extent of rotation of the plane of polarization depends on the thickness of the crystal or the concentration and thickness of the solution placed in the light path as well as the wavelength of light. Thus it is important to compare the rotation caused by different substances under the same conditions. (All our remaining discussion will concern the rotation of polarized yellow light from a sodium lamp.)

The phenomenon of rotation of the plane of polarized light is known as *optical activity*. A substance which rotates the plane of polarized light is said to be *optically active*.

3.3 THE EXPLANATION OF OPTICAL ACTIVITY IN CRYSTALS

Biot had noticed that some cut quartz crystals rotated the plane of polarized light to the left and others rotated the plane to the right. The problem lay fallow for a few years before the mineralogist Häuy noticed that some specimens of quartz crystals exist in two hemihedral forms each characterized by the presence of a set of faces arranged in either a right-handed or left-handed fashion and constituting just *half* of the faces required to give a *symmetrical* crystal.

In 1820, Sir John Herschel communicated to the Royal Society of London the opinion that the hemihedry of quartz crystals was connected with their optical activity. This view was easily supported by experiment. The crystals which rotate the plane of polarized light one way are "right-handed." Let's pause for a moment to examine the nature of the relationship between the two hemihedral forms of quartz crystals. Figure 3.4 indicates this relationship. It is identical to the relationship of right and left hands. Hands are very similar but they are not identical. It is not possible (even neglecting small defects) to superimpose a left hand on a right one. It is this fact which makes it possible

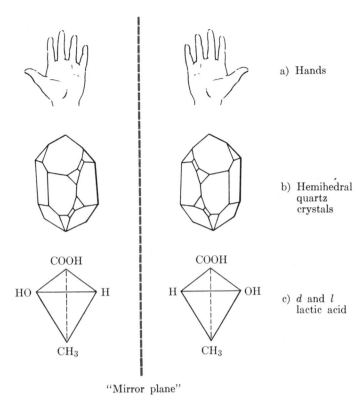

a) Hands

b) Hemihedral quartz crystals

c) d and l lactic acid

"Mirror plane"

Fig. 3.4. Right- and left-handedness: the mirror relationship for nonequivalent structures.

to tell the difference between left and right hands. However, if we construct the *mirror image* of a right hand, it is superposable on a left hand. The *difference* between *the similar structures* of two hands is the *difference of a reflection in a mirror*. This is the relationship of the hemihedral crystals cut from quartz. The mirror image forms are distinct and *not superposable, not equivalent*. (The condition of a nonequivalent mirror image is not possible for geometrical structures which contain a "plane of symmetry," that is, a geometrical structure which may be reflected through some plane to yield a configuration identical to the original.) Herschel's suggestion is that it is the *symmetry property* associated with a *mirror image distinguishable from the original* structure which is responsible for the ability of crystals to rotate the plane of polarized light.

3.4 THE CONTRIBUTIONS OF PASTEUR

Tartaric acid is now known to have the structure

$$HOOC—CH—CH—COOH.$$
$$\qquad\quad OH \quad OH$$

It is present in grapes and is isolated as a by-product in wine making. Shortly after his graduation from the Ecole Normale in Paris, Louis Pasteur undertook to repeat some crystallographic observations published (1841) by de la Provostaye on the crystals of salts of tartaric acid. The experiments were undertaken in the spirit of an exercise to improve Pasteur's competence as a crystallographer, but in the course of his experiments Pasteur discovered one fact which had gone unnoticed by his predecessor. The *tartrate salts*, like quartz, would, under appropriate conditions, exhibit hemihedry. That is, it was possible to isolate crystals which were not symmetrical on reflection in a mirror, but had non-equivalent mirror images. Moreover, all the hemihedral tartrates Pasteur obtained were of the *same handedness*. He was, of course, interested in this hemihedry and its connection to optical activity. Biot had observed that tartrates *in solution* can rotate the plane of polarized light, and Pasteur realized the possibility of a correlation between crystalline hemihedry and rotation in solution analogous to the suggestion of Herschel concerning quartz crystals. Unfortunately, there was some reason for pessimism. In 1844, Mitscherlich had reported that the sodium ammonium salt of racemic acid (which was an obvious isomer* of tartaric acid) was optically inactive (did not rotate the plane of polarized light) and was crystallographically identical to the corresponding salt of tartaric acid. But, Pasteur carefully reexamined this conflicting case hoping to find that Mitscherlich had overlooked hemihedry in the $NaNH_4$ tartrate. He observed that the salt of tartaric acid examined by Mitscherlich did in fact, as all of the other tartrate salts had, display hemihedry, but he also made the startling discovery that the *salt of racemic acid was hemihedral, with the important difference that crystals of both the right- and left-handed forms were present.* Pasteur carefully separated (by hand!) a collection of crystals hemihedral to the right and another hemihedral to the left. When dissolved in water, the former rotated the plane of polarized light to the *right exactly as is observed with the naturally occurring tartaric acid salt.* The latter, in solution, rotated the plane of polarized light a corresponding amount *to the left.* Further examination showed that the "right-handed" crystals sorted out from the careful crystallization of sodium ammonium racemate were no longer different from the natural tartrate in physical properties. These were not crystals of an isomer of the tartrate but were plainly the tartrate itself. Moreover, the "left-handed" hemihedral crystals were like the salt of the naturally occurring tartrate in all respects save the *one*, that they caused rotation of the plane of polarized light to the *left.* The material differing in overall physical properties (the salt of racemic acid) was a *mixture of the two tartrates* which differ only with respect to the direction of rotation of the plane of polarized light.

In this experiment of 1844 Pasteur had accomplished the first *resolution* of an optically inactive material into optically active components and established

---

* Actually, the term *isomer* came into use following the discovery by Berzelius in 1831 that the two acids isolated in fermentation of grapes, tartaric and racemic acids (Latin *racemus*, bunch of grapes), had identical atomic composition but different physical properties.

the relationship of three isomeric materials: the right rotating tartaric acid of nature called d-tartaric acid (d for dextrorotatory), the material differing from it *only* in that it rotates polarized light to the left called l-tartaric acid (l for levorotatory), and the substance which sometimes crystallizes in a symmetrical form—racemic acid—which turns out to be an equal mixture of d- and l-tartaric acids.

Pasteur's discovery was referred to Biot for review before presentation to the Academy of Sciences. The discoverer of the activity of organic compounds was skeptical and sufficiently aware of the importance of Pasteur's claim that he required the younger worker to repeat the resolution. In Biot's laboratory Pasteur crystallized and separated the hemihedral crystals of d- and l-tartaric acid salts from a sample of racemic acid supplied by Biot. Biot prepared solutions of the d- and l-forms and examined first the one Pasteur declared should be levorotatory. To quote Pasteur's own account of the incident:

He first placed in the apparatus the more interesting solution which ought to deviate to the left. Without even making a measurement, he saw by the appearance of the tints of the two images . . . in the analyser that there was a strong deviation to the left. Then, very visibly affected, the illustrious old man took me by the arm and said:—'My dear Child, I have loved science so much throughout my life that this makes my heart throb.' You will pardon me . . . these personal recollections which have never been effaced from my mind.*

It was only much later that it was realized that Pasteur's success depended on a critical temperature factor. Above 28°C the sodium ammonium salt of racemic acid crystallizes from solution not as two separate types of crystal hemihedral, respectively left- and right-handed, but as a single symmetrical crystal containing the mixture of d- and l-tartaric acids.

The key point of the tartaric acid researches is that Pasteur had established a correlation between *hemihedry in crystals* and *rotation of the plane of polarized light by the same substance when dissolved in solution.* With the background of Herschel's idea that nonequivalent mirror-image crystals were required to give rotation, it was natural to suggest that the same symmetry requirement applied in solution. But, in solution, it must be the *molecules* of the compound themselves that have left- and right-handed forms. The atoms in the molecules must be so arranged that the mirror image forms are not superposable. *Thus optical activity appears to be diagnostic of the spatial arrangements of atoms in molecules.* (We should perhaps note that the existence of right- and left-handed molecules may in many cases explain the fact that they pack in right- and left-handed crystals. Thus we are one step closer to an overall explanation of optical activity.) Again, let us let Pasteur speak on this point:

We know, on the one hand, that the molecular structures of the two tartaric acids are asymmetric and on the other rigorously the same with the sole difference of showing asymmetry

---

* From *Researches of Molecular Asymmetry*, 1858 (Alembic Club Reprint No. 4, 1899, Simpkin, Marshall, Hamilton, Kent & Co., Ltd. Reissue Edition, E. & S. Livingstone, Ltd., Edinburgh, 1960).

in opposite senses. Are the atoms of the right acid grouped on the spirals of a dextrogyrate helix, or placed on the points of an irregular tetrahedron, or disposed according to some particular asymmetric grouping or other? We cannot answer these questions. But it cannot be a subject of doubt that there exists an arrangement of the atoms in an asymmetric order, having a non-superposable mirror image.*

Pasteur was unable to give an explicit answer in these remarks of 1860 to our big question. But he had established the basis for further study. What was needed was a knowledge of more "structures" (in the sense of Chapter 2) of optically active substances. After all, in 1860 the ideas of Cannizzaro and Kekulé were very new, and the structural formulas of few organic molecules were well understood.

3.5 RESOLUTION

A fair number of optically active compounds derived mainly from plant sources have been mentioned above, and living systems do provide many more. However, many substances either natural or prepared in the laboratory resemble racemic acid in that they occur as equal mixtures of d- and l-forms which must be resolved before they can be shown to be optically active. Unfortunately, only very few spontaneously resolve themselves on crystallization as the forms of the salt of tartaric acid do. In most cases, a *symmetrical* crystal forms that is still an *equal mixture* of d- and l-isomers. Pasteur was also the discoverer of the most generally applicable method for resolving such mixtures. It depends on using some optically active substance which will react (reversibly!) with the substance to be resolved. Suppose that we have a compound A existing as a d,l-mixture. If this reacts with a substance B which we have, for example, as the d-isomer, then the possibilities are:

$$d,l\ A + dB \left\{ \begin{array}{l} \nearrow dA\text{-}dB \\ \searrow lA\text{-}dB \end{array} \right. .$$

One AB adduct is d for both components. The other is l for A and d for B. The adducts usually turn out to have differing physical properties, e.g., solubilities, melting points, etc. By ordinary methods (see Chapter 1), they may be separated to give separate samples dA-dB and lA-dB. The reasons that adducts have differing physical properties may emerge from examination of Fig. 3.5, which shows both gloves and molecular models. The important thing to notice is that dA-lB differs from dA-dB in *more ways* than just a mirror reflection. There are actually different distances between similar "atoms" or "groups" in these two isomers. Note that, after all, the mirror image of dA-dB would

---

* From *Researches of Molecular Asymmetry*, 1858 (Alembic Club Reprint No. 4, 1899, Simpkin, Marshall, Hamilton, Kent & Co., Ltd. Reissue Edition, E. & S. Livingstone, Ltd., Edinburgh, 1960).

Models

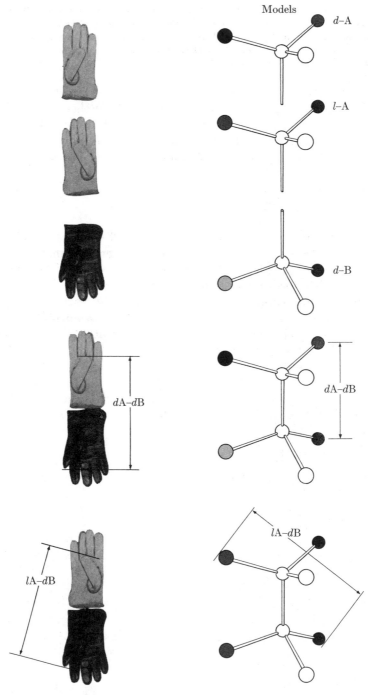

Figure 3.5.

be *lA-lB*, *not dA-lB*, which is of interest to us for the "resolution" process under discussion. If the original reaction combining *A* and *B* can be reversed, it is possible to obtain separate samples of *dA* and *lA* (and even recover the *dB* to be used again later).

3.6 TERMINOLOGY

We shall require some special terminology to distinguish the kinds of isomerism discussed in this chapter from those of Chapter 2. We call a pair of isomers differing only with respect to their ability to rotate the plane of polarized light (mirror image forms) *enantiomers* (*dA* and *lA* are enantiomers). When compounds involve more than one source of optical activity, as in our adduct *AB*, *dA-dB* is the enantiomer (and mirror image) of *lA-lB*. It is not the mirror image of *dA-lB*, but is clearly an isomer. The *dA-lB* is called a *diastereomer* of *dA-dB*. When we have a sample which is not optically active because it is an equal mixture of *d* and *l* forms, it is known as a *racemic* mixture after the first recognized case. For purposes of the discussion in this book, we shall limit the term *structural isomers* to the sense of the term used in Chapter 2 and describe these further kinds of isomerism as *optical isomerism* or *stereoisomerism*. (This last term is a little more inclusive than "optical isomerism." We have yet to encounter a final case which it includes.)

3.7 VAN'T HOFF AND LE BEL—TETRAHEDRAL CARBON

Given Pasteur's theory that optical rotation arises from asymmetry in the molecules themselves, we might begin to look at the structures of molecules that exhibit optical rotation to see if we find any regular structural feature.

In 1874 two young chemists, van't Hoff and Le Bel, quite independently proposed a specification of the feature in structural formulas required for the occurrence of optical activity among common organic compounds. This they did by examining the then known structures of compounds that could be resolved into enantiomers and looking for a common feature. The number of available instances was not large. All of the examples cited in Le Bel's paper are shown in Fig. 3.6. Both noted that in each of these structures there was at least one carbon (marked * in the figure) which is connected to four different groups. (This arrangement has come to be called an asymmetric center.)

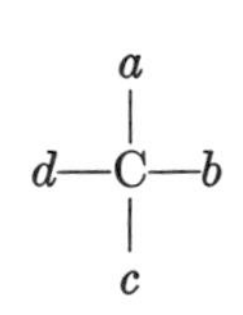

$$
\begin{array}{c}
a \\
| \\
d—\overset{\displaystyle}{\underset{\displaystyle}{C}}—b \\
| \\
c
\end{array}
$$

No known optically active structure lacked this feature. If two of the groups

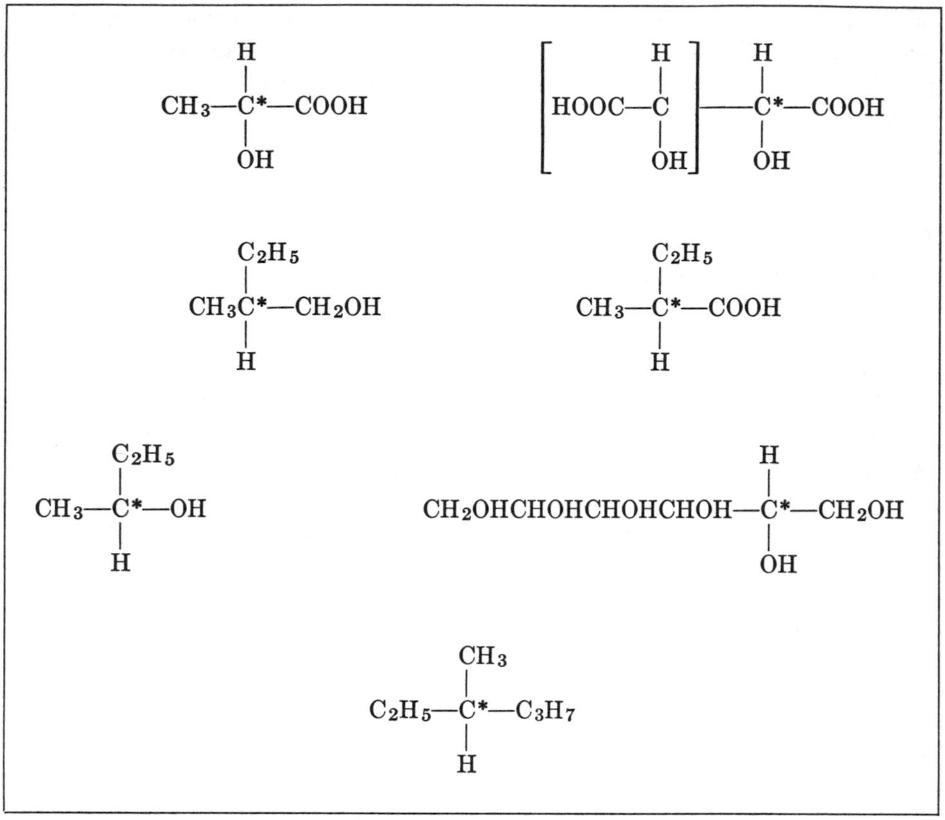

$$CH_3-\overset{\overset{\displaystyle H}{|}}{\underset{\underset{\displaystyle OH}{|}}{C^*}}-COOH \qquad \left[HOOC-\overset{\overset{\displaystyle H}{|}}{\underset{\underset{\displaystyle OH}{|}}{C}} \right]-\overset{\overset{\displaystyle H}{|}}{\underset{\underset{\displaystyle OH}{|}}{C^*}}-COOH$$

$$CH_3\overset{\overset{\displaystyle C_2H_5}{|}}{\underset{\underset{\displaystyle H}{|}}{C^*}}-CH_2OH \qquad CH_3-\overset{\overset{\displaystyle C_2H_5}{|}}{\underset{\underset{\displaystyle H}{|}}{C^*}}-COOH$$

$$CH_3-\overset{\overset{\displaystyle C_2H_5}{|}}{\underset{\underset{\displaystyle H}{|}}{C^*}}-OH \qquad CH_2OHCHOHCHOHCHOH-\overset{\overset{\displaystyle H}{|}}{\underset{\underset{\displaystyle OH}{|}}{C^*}}-CH_2OH$$

$$C_2H_5-\overset{\overset{\displaystyle CH_3}{|}}{\underset{\underset{\displaystyle H}{|}}{C^*}}-C_3H_7$$

Fig. 3.6. Le Bel's examples of structures leading to the existence of optical activity and some of their derivatives. The last example is from van't Hoff's paper. Asymmetric centers are marked with an asterisk.

attached to the key carbon are made identical, the activity disappears. As Le Bel noted,

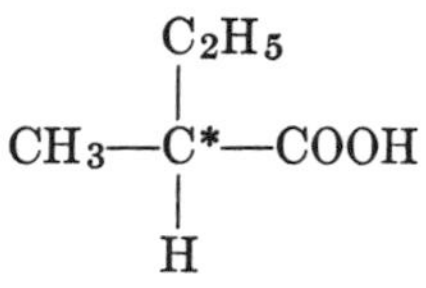

$$CH_3-\overset{\overset{\displaystyle C_2H_5}{|}}{\underset{\underset{\displaystyle H}{|}}{C^*}}-COOH$$

is active, whereas

$$CH_3-\overset{\overset{\displaystyle C_2H_5}{|}}{\underset{\underset{\displaystyle H}{|}}{C}}-CH_3$$

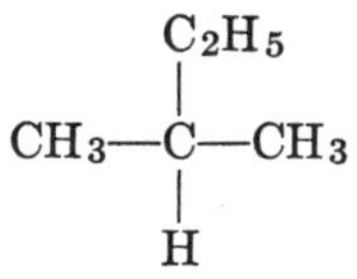

is not.

The question is, then, how can four groups be arranged around a given center so that if *all four* are different, the structure has a *nonequivalent* mirror image, but if any two are the same, the mirror image of the structure is superposable on, and identical to, the original? To begin with, it is obvious that a planar arrangement will not do. Consider C(*abcd*) as shown below.

$$a$$
$$|$$
$$c—C—b$$
$$|$$
$$d$$

If all the groups lie in the plane of the page, a mirror held over the page would give a reflection of the structure which could merely be dropped back onto the page and would superpose upon the original. The mirror image of any planar structure will be equivalent to the structure because *the plane of the structure is a plane of symmetry.* On the other hand, as van't Hoff and Le Bel both observed, if the four valence bonds of carbon were directed toward the corners of a tetrahedron, then the four groups *a*, *b*, *c*, and *d* would occupy the corners of a tetrahedron and the mirror image of such a structure would be nonequivalent as shown in Fig. 3.7 (and for the particular case of lactic acid in Fig. 3.4c).

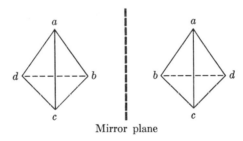

Mirror plane

Fig. 3.7. A tetrahedral disposition of four different groups and its mirror image.

To see that the two forms in Fig. 3.7 are different, note that moving the form on the right over the form on the left without any rotation would accomplish superposition of *a* and *c*, but would leave *b* on *d* and *d* on *b*. To superpose *b* and *d* correctly, one might rotate the form on the right by 180° before moving it over on the form on the left. In that case, *b* and *d* would superpose correctly, but *c* and *a* would be mismatched. Examining the various possible ways to turn one form, one can convince oneself that *there is no way to superpose the form on the right on the form on the left.* Figure 3.8 gives a pattern for constructing a tetrahedron. The reader is strongly urged to construct a tetrahedral *a,b,c,d* type compound and its mirror image and test these conclusions for himself.

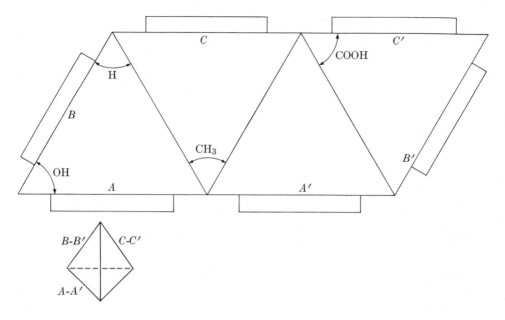

Fig. 3.8. Pattern for tetrahedron construction. Copy on stiff paper. Fold along the lines and attach *A* to *A'*, *B* to *B'*, and *C* to *C'* with gluing tabs or cellophane tape. The pattern is marked to show one enantiomer of lactic acid, assuming that the asymmetric carbon is in the center of the tetrahedron. To construct the other enantiomer (mirror image of this form) exchange the H and OH labels on the pattern.

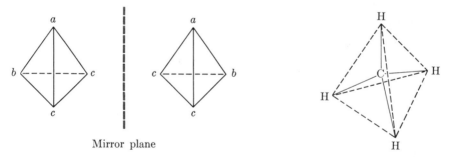

Fig. 3.9. A tetrahedral disposition of four groups, two of which are the same, and its mirror image.

Fig. 3.10. The geometry of methane.

    The tetrahedral distribution of four groups produces a structure with a nonequivalent mirror image. Thus a tetrahedral C(*abcd*) should show optical activity. It remains to show that once two groups are made the same, the mirror-image forms become equivalent. Consider the case where group *d* is replaced by a second group *c*. Figure 3.9 shows C(*abcc*) and its mirror image. Note in this case that after a rotation of 120° the right-hand form can be superposed on the left. Thus the mirror image is not a distinct form. The tetrahedral distribution of four groups bound to a carbon in the center thus exactly satisfies the rules noted by van't Hoff and Le Bel for the occurrence of optical activity.

Thus they concluded that the tetrahedron was the geometrical distribution of bonding about carbon. As van't Hoff explicitly suggested, methane has the geometry shown in Fig. 3.10. For most purposes, it is more convenient to have a diagram in which it is easy to include the carbon in the middle. A useful convention defines ordinary bond lines (—) as lying in the plane of the page, dashed bonds (---) as projecting behind the page and heavy triangular bonds (◀) as projecting forward from the plane of the page. Using this convention we may diagram a tetrahedral distribution as below.

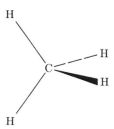

It turns out that a tetrahedral distribution is equivalent to having two of the groups in *one plane* and the other two in the *plane perpendicular* to the first. Using diagrams in this form, we may show a pair of enantiomers as below.

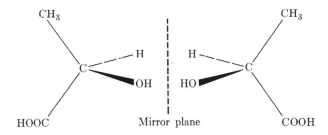

It is easy to visualize the geometry of an extended carbon chain.

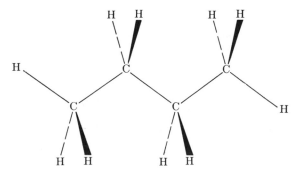

The example is, of course, butane.

3.8 GEOMETRICAL ISOMERISM—VAN'T HOFF'S PREDICTION

Van't Hoff observed that a molecule containing a carbon-carbon double bond (alkene) might be represented by joining two corners of each tetrahedron as shown.

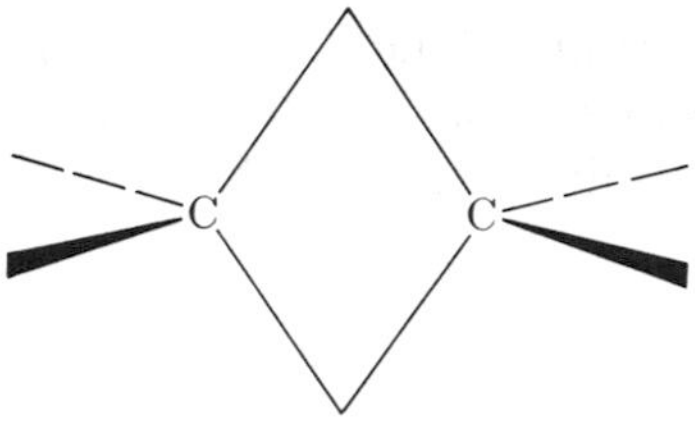

The four groups attached around the carbon-carbon double bond would all lie in the same plane.

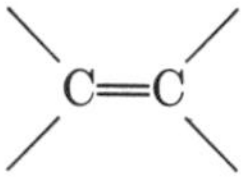

We have noted that the nonexistence of isomers of CH_2Cl_2 indicates the non-planarity of bonding about a single carbon; on the other hand, the planar arrangement of four groups about a double bond should lead to two isomers for formulas like CHCl=CHCl. These are:

H   H   Cl   H

C=C   and   C=C

Cl   Cl   H   Cl

cis   *trans*

The form with like atoms adjacent is called *cis*, and the form with like atoms opposite is called *trans*. These isomers will *not* rotate the plane of polarized light. All atoms are in a plane, and there is therefore a plane of mirror symmetry. Recall that enantiomers differ *only* in their ability to rotate the plane of polarized light. Examination of tetrahedral models will suggest why this is so. Any atom around the tetrahedron is equally distant from all others. The spatial distances between like atoms are the same in both *d*- and *l*-forms. In the isomers of CHCl=CHCl, it is clear that the distance separating the two hydrogens or the two chlorines in the *cis* form differs from the distance in the *trans* form. We might expect that these isomers would differ in physical properties. This prediction of van't Hoff is readily verified. There are two substances with the structure CHCl=CHCl. One melts at −80.5°C and boils at 60.1°C. The other melts at −50°C and boils at 48.4°C. This form of isomerism is considered a type of *stereoisomerism* and is called *geometrical* isomerism.

3.9 MOLECULES CONTAINING MORE THAN ONE ASYMMETRIC CENTER

Pasteur's prime example of an optically active compound, tartaric acid, actually contains two carbons with four different groups attached:

$$HOOC—\overset{\overset{\displaystyle H}{|}}{\underset{\underset{\displaystyle OH}{|}}{C^*}}—\overset{\overset{\displaystyle H}{|}}{\underset{\underset{\displaystyle OH}{|}}{C^*}}—COOH.$$

The naive expectation is that there should be a *d*- and *l*-arrangement at each carbon and, as a result, *four isomers* (*d,d′ l,d′ d,l′ l,l′*, where letters without primes indicate the first carbon and letters with primes, the second). Generalizing, we can say that if there are *n* carbons in a molecule, which are asymmetric centers, there should be 2^n isomers. However, tartaric acid represents an important exception to this simple picture. The two asymmetric centers are exactly the same. If the arrangement about both carbons is *d*, the plane of polarized light is rotated to the right. If the arrangement about both carbons is *l*, the plane of polarized light is rotated to the left. These are the *d*- and *l*-tartaric acids of Pasteur's work. If, however, the arrangement about one carbon is *d* and about the other is *l*, the fact that the two centers are identical leads to a cancellation of rotation and an optically inactive substance. This isomer is called a *meso* form (Fig. 3.11).

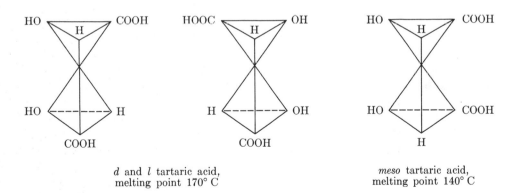

d and *l* tartaric acid,
melting point 170° C

meso tartaric acid,
melting point 140° C

Fig. 3.11. The three isomers of tartaric acid.

3.10 X-RAY DIFFRACTION

A modern method for studying structures of solid materials provides quite independent evidence for the ideas of molecular geometry developed above. This method is outlined here because it is usually regarded as providing the *"ultimate"* test of theories of molecular geometry. It depends on ideas of wave diffraction, which are discussed in Appendix 3 and which will be used as crucial notions in a very different context in Chapter 8.

Infrared radiation is a lightlike phenomenon associated with waves that have a wavelength longer than the wavelengths of visible light. In 1895, W. K. Röntgen succeeded in producing an emission with the wavelike properties of light but with wavelengths *very much* shorter than those associated with visible light. These have come to be known as x-rays. The experimental circumstances that lead to the conclusion that light, infrared, or x-rays have wavelike properties are those associated with interference and diffraction. The basic circumstance is shown in Fig. A3.28*. In the figure, a series of wave crests originating from a light source are represented by lines approaching a pair of *slits*. As light passes through the slits, the slits function as though they were new sources for the wave disturbance: new circular wavelets originate from each slit. (This can be observed easily with ripples on water. We shall wish to explore the matter in detail in Chapter 8.) Beyond the slits the waves from each "interfere." If they meet "in phase" (i.e., either crest to crest or trough to trough), they reinforce. If they meet crest to trough, they cancel out and the wave disturbance is damped. Thus, at particular angles, the light viewed coming through the slit will appear bright or weak. At a bright point, the light coming from the farther slit must have traveled just *one wavelength* farther than the light coming from the nearer slit in order for reinforcement (or constructive interference) to occur. As a result, the angles at which bright images *are observed are determined by the spacing between the slits and the wavelength of the light.* If one of the quantities is known, the other may be determined!

So-called "diffraction patterns" (such as the simple one described) are produced whenever light or other wave disturbances pass through or are reflected from a *periodic lattice* that has a regularly repeating arrangement of points that interact with the incident wave. The simplest example might be a row of posts in a harbor producing a diffraction pattern in incoming waves. An only slightly less familiar one is the use of a so-called diffraction grating to resolve a beam of white light into the colors of the spectrum. For observable effects, the *slit size must be similar to the wavelength of the wave disturbance.* A solid crystal turns out to be a satisfactory lattice for diffraction of x-rays. This implies that the spacing of the lattice points (by inference, the atoms!) is comparable to the wavelength of x-rays. If those wavelengths can be determined, then the *spacings* may be *found*. If the pattern is found, the arrangement (and spacings) of atoms in the crystal may be found and from such relationships one may check inferences about molecular geometry.

The first quantitative connection between the wavelength λ and the lattice spacing in a crystal d was obtained in 1913 by W. L. Bragg from consideration of the consequences of *reflecting* x-rays from the planes defined by the successive layers in a crystal. The wave description of reflection from a surface treats each

---

* Figure A3.28 is in Appendix 3 at the end of this book.

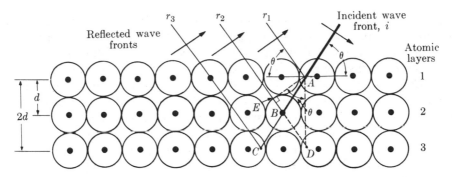

Fig. 3.12. Bragg reflection of an incident wave front i from successive parallel atomic layers (1, 2, 3, etc.) in a crystal. A small fraction of the incident energy is reflected at each layer to form parallel reflected wave fronts (r_1, r_2, r_3, etc.) all traveling in the same direction. The reflected wave fronts lag behind one another by a distance $AE = 2d \cos \theta$. In a sinusoidal wave train the reflected waves will interfere constructively if the distance AE is an integral number of wavelengths. The decrease in width of line i is intended to imply successive decreases of incident wave intensity with reflections occurring at each atomic layer. (From A. B. Arons, *Development of Concepts of Physics*, Addison-Wesley, Reading, Mass., 1965.)

point where an incoming wave front strikes as a *source* for a circular wave and asks the question: "In what direction will the reflected wave which forms from the superposition of the adjacent circular waves be going?" The construction of the reflected wave front is discussed in Appendix 3 and shown in Fig. A3.18. The result is the familiar one that the reflected wave front travels away from the surface at an angle *equal* to the angle of the incident wave front.

If we now consider reflection from successive planes in a crystal, we must (with Bragg) add another condition for the observation of a strong reflection. The waves reflected from the *second* and *third* layers must meet those reflected from the *first* and *second* layers in phase (crest-to-crest or trough-to-trough). Otherwise, there will be destructive interference and no strong reflection. This added condition will exclude strong reflection at some angles. The condition will be satisfied only if the waves reflected from lower layers travel an *integral* number of wavelengths farther than those reflected from the upper layers. The paths of wave fronts from lower layers must be $n\lambda$ times longer than paths of fronts reflected from the upper layer where n is an integer (1, 2, 3, . . .). We need now to relate this to the lattice spacing in the crystal.

The problem is set out in Fig. 3.12, which shows a series of plane waves r_1, r_2, r_3 arising from successive layers as the incident wave i is reflected partially from layers 1, 2, and 3. As the figure suggests, the resolution of the problem after a little trigonometry is that strong reflections occur if

$$n\lambda = 2d \cos \theta^*,$$

---

* Many books quote this relation in terms of the *sin* of the angle made by the x-ray "ray" (direction in which the wave front is moving). The result is clearly equivalent since the "ray" direction is perpendicular to the wave front.

where d is the vertical distance or spacing between the rows of atoms that are thought of as forming the reflecting planes. (Note that more than one kind of atom in the lattice will complicate the analysis since the different atoms may have different reflecting abilities or, as they are now called, "scattering factors.")

Unfortunately, the analysis is only this simple when the lattice can be described with a *single* spacing parameter, d. In general, the arrangement of atoms in a crystal will involve a variety of distances, and the diffraction pattern and its analysis become quite complicated, the treatment requiring quite sophisticated mathematical tools. Let us, however, record two important results here. First, it does appear that a carbon atom has *four* neighbors, which may be regarded as members of the *same molecule* and which are disposed tetrahedrally confirming the results from optical activity. Second, the lattice spacing *may* be interpreted as the sum of the radii of the two adjacent atoms giving an insight into the question of atomic size. In fact, we have here one way to resolve an important problem from the discussion in Chapter 1. If the dimensions of an atom are known (radii turn out to be of the order of 10^{-10} m or 10^{-8} cm), the number in a macroscopic sample may be calculated. Thus *Avogadro's number*, the number of atoms per mole, becomes accessible. The modern result is 6.02×10^{23} atoms. Of course, relative atomic weights in grams divided by 6.02×10^{23} also provide the weights of individual atoms. Thus an atom of hydrogen weighs approximately 2×10^{-24} g.

3.11 BOND LENGTH

It is interesting that the use of x-rays to determine the *spacings* between atomic centers in crystals also offers us another enrichment of our idea of bond, which was introduced in Chapter 2. It turns out that there is a *nearly constant* value for the carbon-carbon spacing in those circumstances which are represented by a single bond in structural formulas. The same is true for carbon-carbon double bonds. In general it appears that one may quote a characteristic distance which will (to a good approximation) give the separation of atoms involved in a particular kind of bond. This distance is called the *bond length*. Bond lengths are

Table 3.1

Bond lengths

Bond	Molecule in which found	Distance, Å
C—C	Diamond	1.54
C—C	C_2H_6	1.54
C—C	C_2H_5OH	1.55
C=C	$H_2C\!=\!CH_2$	1.34
C≡C	$H\!-\!C\!\equiv\!C\!-\!H$	1.20

most conveniently quoted in angstrom units ($1 \text{ Å} = 10^{-8}$ cm) because bond lengths usually fall in the region of one to two angstroms or between one and two hundred-millionths of a centimeter. Bond lengths for some organic compounds are shown in Table 3.1.

SUMMARY PROBLEMS

Problem 3.1 Draw the structure of the simplest (a) alkane, (b) alcohol, (c) acid that would be resolvable into optically active enantiomers.

Problem 3.2 Mark the carbons that are centers of optical activity (asymmetric centers) in the following structures.

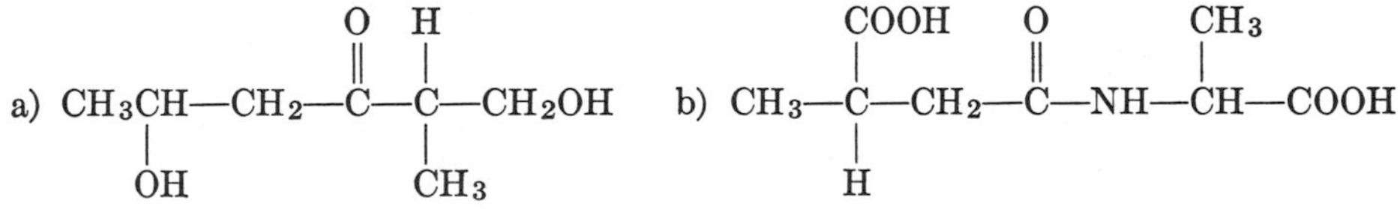

a)
$$
\begin{array}{c}
\quad\quad\quad\quad\quad\overset{O}{\underset{}{\|}}\ \ \overset{H}{\underset{}{|}} \\
CH_3CH-CH_2-C-C-CH_2OH \\
\quad|\quad\quad\quad\quad\ |\ \\
\quad OH\quad\quad\quad\quad CH_3
\end{array}
$$
   b)
$$
\begin{array}{c}
COOH\quad\ O\quad\quad CH_3 \\
\ \ \ |\quad\quad\ \|\quad\quad\quad| \\
CH_3-C-CH_2-C-NH-CH-COOH \\
\quad\quad| \\
\quad\quad H
\end{array}
$$

Problem 3.3 Indicate the types of stereoisomerism possible in the following compounds (geometrical or optical or both) if any.

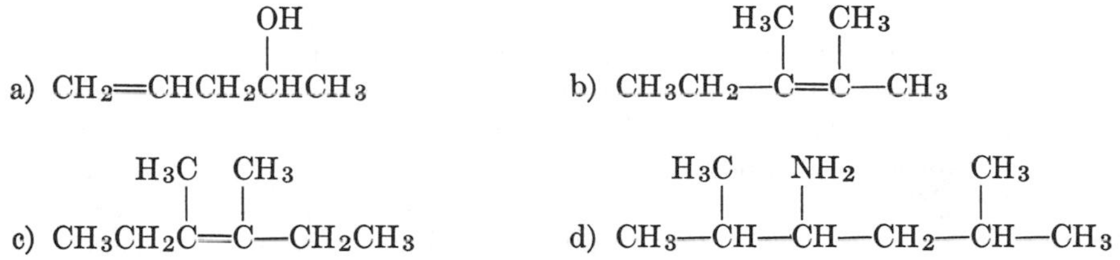

a)
$$
\begin{array}{c}
\quad\quad\quad OH \\
\quad\quad\quad\ | \\
CH_2=CHCH_2CHCH_3
\end{array}
$$
   b)
$$
\begin{array}{c}
H_3C\ \ CH_3 \\
\ \ |\quad\ | \\
CH_3CH_2-C=C-CH_3
\end{array}
$$

c)
$$
\begin{array}{c}
H_3C\ \ CH_3 \\
\ \ |\quad\ | \\
CH_3CH_2C=C-CH_2CH_3
\end{array}
$$
   d)
$$
\begin{array}{c}
H_3C\ \ NH_2\quad\quad CH_3 \\
\ \ |\quad\ |\quad\quad\quad\ | \\
CH_3-CH-CH-CH_2-CH-CH_3
\end{array}
$$

Problem 3.4 Two acids are known, one with a melting point of 72°C and one with a melting point of 15°C. Develop a series of reactions that would suggest that both have the structure $CH_3CH=CHCOOH$. Illustrate the relationship between the isomers.

Problem 3.5 Historically, the first examples of geometrical isomerism elucidated were maleic and fumaric acids, which have the structure:

$$HOOC-CH=CH-COOH.$$

a) Draw diagrams of the *cis* and *trans* isomers.
b) Both isomers will undergo a dehydration reaction (loss of water) to form an "anhydride" with the structure:

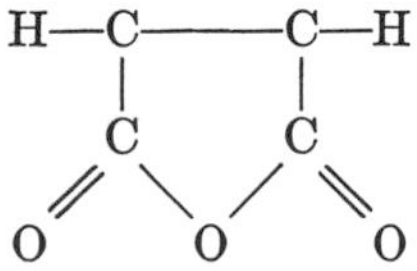

$$
\begin{array}{c}
H-C\text{——}C-H \\
\ \ |\quad\quad\ | \\
\ \ C\quad\quad\ C \\
\ /\!/\ \backslash\ /\ \backslash\!\backslash \\
O\quad\ O\quad\ O
\end{array}
$$

Maleic acid heated to 160°C yields the anhydride. Fumaric acid must be heated to a temperature well above 200°C to accomplish this. When fumaric acid is heated to 200° in a sealed tube to prevent escape of water it yields maleic acid. Which acid is probably *cis*? Support your conclusion.

Problem 3.6 Van't Hoff predicted optical activity (verified some 60 years later) in certain compounds of the type:

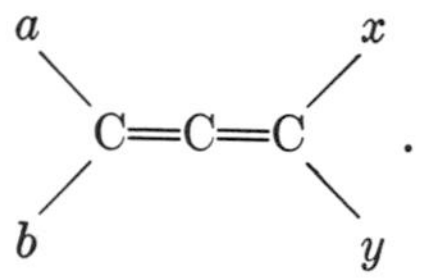

$$
\begin{array}{c}
a \diagdown \qquad \diagup x \\
\qquad C{=}C{=}C \\
b \diagup \qquad \diagdown y
\end{array}
$$

Show that this is a reasonable consequence of the postulate of tetrahedral carbon despite the lack of an apparent "asymmetric center."

Problem 3.7 A compound $C_4H_{10}O$ strongly absorbs infrared radiation in the 3-μ region and may be resolved into optically active forms. What is its structure? How would you expect it to react with $KMnO_4$?

READINGS AND EXTENSIONS

The subject of stereochemistry is obviously the completion of the developments begun in Chapter 2. All of the books mentioned there are relevant to this chapter too. Especially, it should be pointed out that the papers of van't Hoff and Le Bel are reprinted in the collection edited by Benfey. They are entirely accessible to the modern student reader. If you look into these papers, note the "scientifically" conservative style in which these two very young chemists introduced their revolutionary ideas. If you would be convinced that feelings actually did run high and that the "style" represents carefully suppressed passion, read the scathing attack on van't Hoff written by H. Kolbe (a distinguished older German organic chemist) and reprinted in translation in G. W. Wheland, *Advanced Organic Chemistry*, 2nd ed., John Wiley and Sons, New York, 1949, p. 132.

The story of the impact of x-ray crystallography on modern chemistry has recently been told by one of the founders, Sir Lawrence Bragg. The article appears in *Scientific American*, Volume 219, Number 1, page 58 (July 1968). This article is accessible and authoritative and is warmly recommended.

Seminar: Ring Conformations

We have seen that the normal C—C—C angle in an alkane is 109°, the tetra-hedral angle

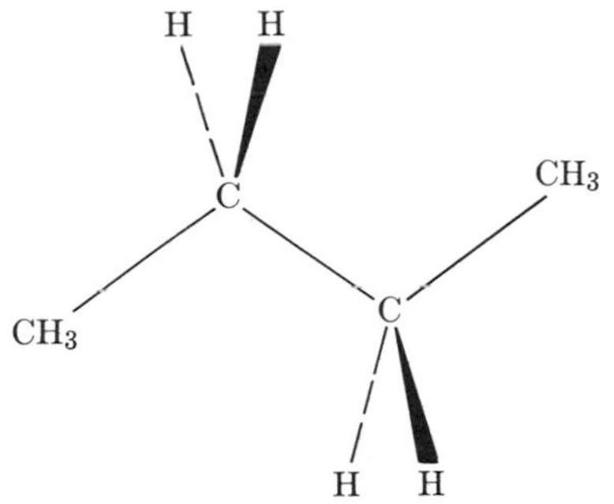

This angle may be adopted in any "straight-chain" compound, but it is not clear that it is always possible in alkanes which have cyclic structures like the following.

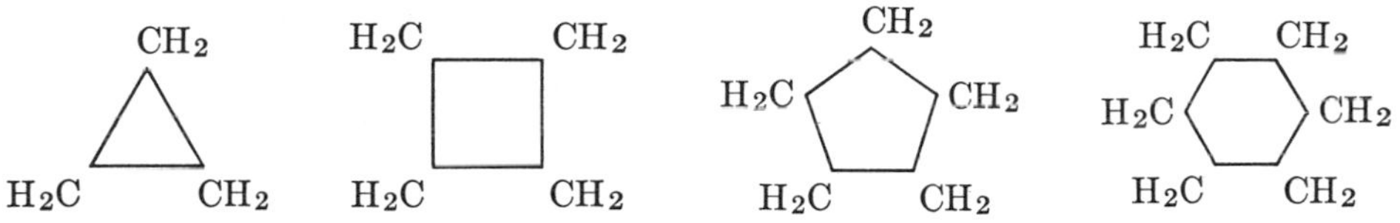

What would be the internal angle

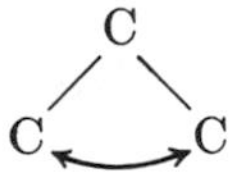

in each of the rings shown if the rings were flat on the page? For the first three, C_3H_6, C_4H_8, and C_5H_{10}, can you imagine any distortions of the rings out of the plane of the page that would result in bringing all of the internal C—C—C angles closer to 109°?

There are two possible structures for C_6H_{12}, cyclohexane, which permit the C—C—C angle of 109°. They are shown below.

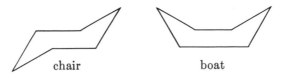

chair   boat

As early as 1890, Sachse suggested that the C_3 to C_5 rings were planar but that C_6 and higher rings were not. His argument was based on observation of the number of calories of heat released on combustion of the "cyclo" alkanes in oxygen. The data follow.

Heat of combustion, thousands of calories per mole	C_3H_6	C_4H_8	C_5H_{10}	C_6H_{12}	C_7H_{14}
	506	663	797	939	1103
Heat per CH_2 unit	169	166	159	157	158

For "straight-chain" CH_2 the heat of combustion is 157 calories.

Can you reconstruct his argument?

Against Sachse's ideas stood the fact that only one form of C_6H_{12} was known, whereas he had anticipated separable isomers of the chair and boat forms. In 1918, Mohr revived Sachse's proposal with the additional suggestion that the problem might be no more than that the boat and chair forms were too easily interconvertible for them to be experimentally separable. Mohr also suggested that interconversions might occur less readily in *decalin:*

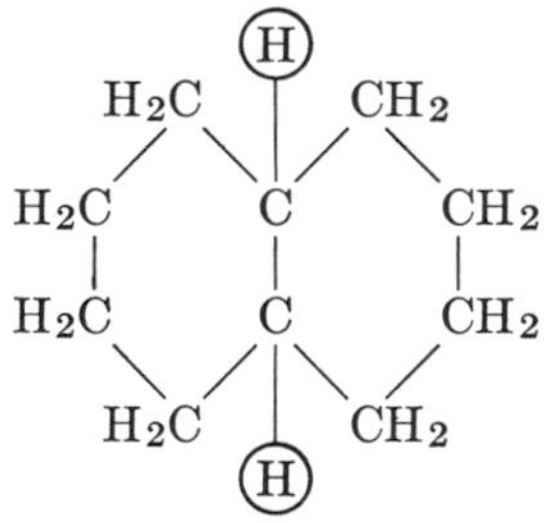

If the two rings in decalin are planar, how many isomers should exist? If the two rings may adopt either the *boat* or *chair* conformations, how many isomers should exist? In particular, how many distinct ways may the two hydrogens circled in the diagram of decalin be disposed with respect to each other?

In 1925, W. Hückel isolated two distinct isomers of decalin with the following characteristics:

	Melting point, °C	*Boiling point,* °C	*Density, g/ml*
Decalin "A"	−43.3	194	0.895
Decalin "B"	−31.5	185	0.870

What do you suppose the relationship between these two isomers is?

What now seems the most plausible description of C_6H_{12}? How confident can you be? Does the evidence answer such questions as which of the conformations (boat or chair) is more stable and whether or not cyclohexane is normally a mixture of the two?

Useful reading

Fieser, L., and M. Fieser, *Organic Chemistry*, 3rd ed., D. C. Heath, Boston, 1956, Chapter 12.
Morrison, R. T., and R. N. Boyd, *Organic Chemistry*, 2nd ed., Allyn & Bacon, Boston, 1966, Chapter 9.
Roberts, J. D., and M. Caserio, *Basic Principles of Organic Chemistry*, W. A. Benjamin, New York, 1964, Chapter 4.

CHAPTER 4

Solutions of Nonelectrolytes

4.1 INTRODUCTION

Up to this point we have dealt primarily with compounds which have molecules as the fundamental structural units. The determination of molecular weights has depended on the fact that these compounds could be studied in the gas phase. A large class of common substances is not easily studied in the gas phase. To study these substances experimentally, it will be necessary to find alternatives to the complex of ideas surrounding Avogadro's hypothesis for gases. The alternatives arise from the regularities which may be discovered in the behavior of solutions. From a simple experimental point of view, solutions may be classified as nonelectrolyte solutions (those that do not conduct electric current) and electrolyte solutions (those that do conduct electric current).* It will be expedient to consider nonelectrolyte solutions first. Most of the compounds we have studied so far fall into this class.

4.2 CERTAIN PROPERTIES OF PURE LIQUIDS

In this chapter we shall consider the effect of dissolved solutes on the vapor pressures, boiling points, and freezing points of liquids. Before proceeding to the solutions, it is imperative to have a clear grasp of these phenomena as they apply to the pure liquids themselves.

Vapor Pressure of a Liquid

If we isolate a pure liquid in an enclosed space, it is found that the liquid will evaporate in whole or in part and that the vapor will exert a pressure on the walls of the container. This process can be visualized in either one of the following experiments.

---

* It will be the point of Chapter 5 on electrolytic solutions to show that the notion of molecules is not entirely relevant to the discussion of the structures of electrolytes.

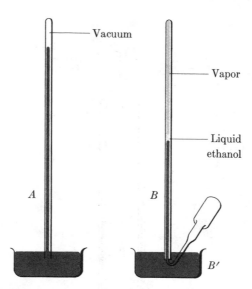

Fig. 4.1. Demonstration of the vapor pressure exerted by a liquid.

1. Suppose we set up two simple barometer tubes as represented in Fig. 4.1 keeping tube A for reference. The space above the mercury in A is a vacuum except for traces of moisture or air which may not have been completely displaced in setting up the tubes. Now by means of a medicine dropper with a curved tip we may insert a drop of a liquid substance, say ethanol, into the open end of tube B which dips under the mercury in cup B'. The ethanol, being much less dense than mercury, rises to the top of the column. We note that the mercury column in B is greatly depressed below the level of that in A, and that some liquid ethanol is floating on top of the column. Addition of another drop of ethanol will have no further marked effect in depressing the column. The difference in height of the mercury columns A and B, with a slight correction for the hydrostatic pressure of the liquid ethanol on top of the mercury, now represents the pressure exerted by the vapor over the liquid. We say this is the vapor pressure of the liquid at the temperature of the experiment.

2. A somewhat more precise but yet rather simple experiment can be conducted in the all-glass apparatus of Fig. 4.2. The only critical dimension in this apparatus is the one from the stopcock C to the bottom of the manometer B, which should be somewhat in excess of 760 mm. (Why?) A few milliliters of the liquid ethanol is poured into the bulb A, which is of convenient capacity, say 25 ml. Bulb A is attached by means of the ground-glass joint J.

By opening the stopcocks D and C and keeping bulb A immersed in the low-temperature bath, the air is drawn out of the system. Stopcock C is now closed. If the bulb A is allowed to warm up to room temperature and a part of the liquid ethanol is evaporated and drawn out of the system as a vapor into the

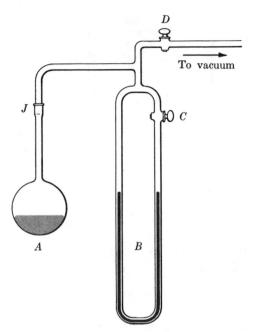

Fig. 4.2. A simple apparatus to measure the vapor pressure of a liquid.

vacuum pump, the ethanol remaining behind is thus freed of any dissolved air. Stopcock D is now closed, and we are ready to measure the vapor pressure of ethanol at any temperature *less* than that of the room by surrounding A with a water bath held at the desired temperature.

Problem 4.1 What readings in the apparatus of Fig. 4.2 would be required to find the vapor pressure of the liquid at each successive temperature? Explain in a few words.

Problem 4.2 Explain why the apparatus as described could not be used to measure the vapor pressure at some temperature above that of the room. Suggest how the experimental arrangement could be modified to make such a measurement possible.

It should be noted that in either of the experimental procedures cited above, it is necessary that some liquid phase shall remain so that equilibrium is maintained between the liquid and its vapor. The vapor pressure registered would be independent of the amount of excess liquid phase.

Vapor-Pressure Curves

A plot of vapor pressure against temperature gives the vapor-pressure curve for a liquid. Several vapor-pressure curves are shown in Fig. 4.3. It will be noted that each liquid has its own characteristic vapor-pressure curve and that

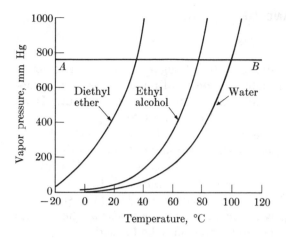

Fig. 4.3. Vapor-pressure curves.

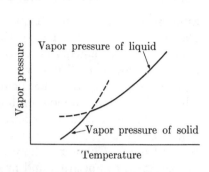

Fig. 4.4. Relation of the freezing point of a substance to the vapor pressure curves of its solid and liquid phases.

at any given temperature there is one number which represents the vapor pressure for any given liquid, depending of course on the units in which it is expressed.

The Normal Boiling Point of a Liquid

The boiling point of a liquid is very crudely defined as the temperature at which the liquid boils. The phenomenon of boiling involves the formation of bubbles of vapor, which are formed against the pressure of the prevailing atmosphere plus the small additional hydrostatic pressure of the liquid above the bubble. It is apparent that the temperature at which the vapor pressure is sufficient to produce vapor bubbles, i.e., at which the liquid boils, will vary with the atmospheric pressure and even slightly with the depth of the liquid above the region of bubble formation.

The normal boiling point is defined as that temperature at which the vapor pressure equals the normal atmospheric pressure at sea level, i.e., 76 cm of mercury. The normal boiling points of the liquids cited in Fig. 4.3 can be read from the intersections of the horizontal line AB at 76 cm with the vapor-pressure curves.

The Freezing Point of a Pure Liquid

When a liquid is slowly cooled, there is a point at which the first crystals of the solid phase appear. This gives us the approximate freezing point. The freezing point can be found quite accurately if due attention is given the phenomenon of supercooling, which is often involved. The freezing point may be defined as the temperature at which the vapor-pressure curves for the solid and liquid phases intersect (see Fig. 4.4), i.e., the temperature at which the vapor pressures of the solid and liquid phases are equal.

4.3 SOLUTIONS—VOCABULARY AND DEFINITIONS

Solutions as Homogeneous Mixtures

Suppose that we dissolve a quarter-teaspoonful of sugar or of salt in separate glasses of water and stir each thoroughly. The resulting mixtures are homogeneous, that is, equally sweet or equally salty to the taste throughout. If we evaporate equal small unit volumes, say 1 ml of sugar solution, we shall find equal residual weights of the sugar. The same would be true for the salt solution. Thus we say that we have uniform composition throughout the solutions, and we say that each solution is homogeneous. This homogeneity or uniformity of composition and properties is also a characteristic of pure substances. As we pointed out in Chapter 1, however, there is a difference. The pure substance has constant composition and properties for all samples, whereas a whole series of homogeneous mixtures, say of sugar and water, may be formed.

Pertinent Definitions

In discussing solutions it is essential to establish a relevant vocabulary. For the examples cited in the preceding paragraph, we refer to the water as the *solvent* and to the sugar or salt as the *solute*. This designation may however become a rather arbitrary one, especially when both the solute and the solvent are liquids* For instance, ethanol and water are miscible in all proportions. It would be natural when considering a solution of 1 g of ethanol in 10 g of water to designate these two substances as solute and solvent respectively. It is equally reasonable to refer to 1 g of water as the solute when it is dissolved in 10 g of ethanol as solvent. What shall we do in the case of solutions containing 5 g of each? Obviously any distinction becomes an arbitrary one, and we may equally well call either component the solvent.

The *concentration* expresses the amount of solute in a specified amount of solvent or solution. The terms dilute and concentrated are qualitative only, indicating that the concentration is low or high. The term saturated solution, however, has a definite and quantitative meaning; this concept is discussed in Section 4.8.

Methods of Expressing Concentration

It is convenient to express concentration of the solute in one of several different ways. We may designate grams of solute per 100 g of solvent or grams per liter of solution. However, because we are frequently concerned with the

---

* For our present purposes we shall be concerned with solutions in which the solvent is in the liquid state while the solute may be either solid, liquid, or gaseous. However, the solvent or solute may represent any one of the three states of matter. The solutions of the two metals, called alloys, would be solid solutions in which both solute and solvent are solids. Air may be considered a solution of its various gaseous components.

number of *molecules* and therefore with the number of *moles* in a solution, it is often desirable to designate concentrations in terms of moles.

1. A *molar* solution contains one mole of solute per liter of solution. Thus we say that a solution containing 0.15 moles of solute per 500 ml of solution has a molarity of 0.3 moles per liter. This solution is 0.30 molar or 0.3 *M*. When the solute is molecular in form, the mole of course designates N_0 (Avogadro's number) molecules.

Problem 4.3 Calculate the molarity of an aqueous solution which contains 10.0 g of glucose $C_6H_{12}O_6$, in 542 ml of solution.

Problem 4.4 Find the number of moles of sucrose, $C_{12}H_{22}O_{11}$, present in 41.2 ml of 0.015 *M* solution.

2. We shall see that a whole class of substances such as K_2SO_4 is not molecular in form either in the crystalline solid state or in aqueous solution. For these substances it is a more precise use of language to speak of a solution containing one gram formula weight (174.2 g for K_2SO_4) of solute per liter as a *formal* solution rather than a molar solution. Thus we may refer to a 0.25 formal or 0.25 *F* solution of K_2SO_4, or we may say the formality is 0.25. It is convenient to refer to a unit containing 2 atoms of K, 1 atom of S, and 4 atoms of oxygen as a formula weight unit.

Problem 4.5 Find the formal concentration of 40.0 ml of a solution which contains 2.42 g of $BaCl_2$.

Problem 4.6 Calculate the number of gram formula weights present in

a) 104 g of $K_2Cr_2O_7$,   b) 240 ml of 0.20 *F* K_2SO_4,
c) 240 ml of 0.2 *F* $(NH_4)_2HPO_4$.

3. We shall have occasion to use the term *molal* solutions later in this chapter. A solution containing one mole of solute per 1000 g of solvent is said to be one molal or 1.0 *m* in concentration. For very dilute *aqueous* solutions, the numerical values of the molarity and molality become virtually identical. Why?

Problem 4.7 Find the molal concentration of a solution containing 6.4 g of sulfur, S_8, in 500 g of CCl_4.

4. Still another useful way of expressing concentration is that of the *mole fraction* or the ratio of the number of moles of one component to the total moles present. Thus, for a two-component solution

$$x_B = \frac{N_B}{N_A + N_B},\tag{4.1}$$

where x_B is the mole fraction of the solute, and N_A and N_B are the moles of solvent and of solute respectively. The mole fraction of the solvent x_A is expressed as $N_A/(N_A + N_B)$. For a two-component solution $x_A + x_B = 1$. The mole-fraction concept will be used in the discussion of Raoult's law (Section 4.6).

Problem 4.8   Find the mole fraction of solvent in a solution made by dissolving 2.0 moles of chloroform, $CHCl_3$, in 50.0 ml of CCl_4 at 20°C. The density of CCl_4 is 1.60 g/ml at this temperature.

Problem 4.9   Find the mole fraction of the solute in a solution containing 18.0 g of dextrose, $C_6H_{12}O_6$, in 1000 g of water.

4.4 COLLIGATIVE PROPERTIES OF SOLUTIONS

When small concentrations of molecular solutes are introduced into a solvent, a set of consequences results that at first seem unrelated but which careful analysis will reveal to be interdependent. The vapor pressure of the solvent falls below that for the pure solvent, the freezing point of the solution is less than that of the pure solvent, the boiling point of the solution is greater than that of the pure solvent, and the phenomenon of osmotic pressure develops (a description of osmotic pressure is given below). The fact that these four distinct properties of solutions have parallel effects is, however, only the most superficial of the regularities that may be detected. It would be quite reasonable to guess that the magnitude of vapor-pressure depressions, freezing-point depressions, boiling-point elevations, or osmotic pressures would be quite sensitive to the detailed chemical nature of the solute. (They might, for example, find explanation in the extent of chemical interaction between solute and solvent.) But this is not the case. Figure 4.5 collects representative data on freezing-point depression in aqueous solution for the three distinct solutes: dextrose, glycerol, and sucrose. Note that a single line fits all of the data when the concentrations of solutes are expressed as molal concentrations. The freezing-point depression appears to *depend on the molal concentration of the solute but not its*

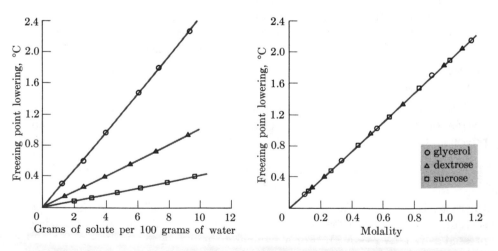

Fig. 4.5. Relation between freezing-point lowering and the concentrations of solutions.

chemical nature. The suggestion, which is realized generally at least for dilute solutions of nonelectrolytes, may be expressed by

$$\Delta T_f = K_f m, \tag{4.2}$$

where K_f is a constant characteristic of the solvent, m is the molal concentration, and ΔT_f is the freezing-point depression (defined as the freezing point of the pure solvent minus the freezing point of the solution). Thus a sufficiently extensive examination of dilute solutions of nonelectrolytes leads to the conclusions that (1) solution freezing points are always less than that of the pure solvent, (2) the extent of freezing-point depression for a given concentration of solute is determined by the nature of the solvent (e.g., greater in benzene than in water), but (3) is independent of the chemical nature of the solute.

Similar results follow from the examination of the effects of solutes on the boiling point of solutions with the only difference being that boiling points are elevated by solutes. We define the boiling-point elevation, ΔT_b, as the boiling point of the solution minus the boiling point of the pure solvent. We may then write an equation exactly analogous to Eq. (4.2) which describes boiling-point elevation

$$\Delta T_b = K_b m, \tag{4.3}$$

where K_b is again a constant characteristic for a given solvent and m is the molal concentration of a nonelectrolyte solute.

Table 4.1 collects the values of the constants for these equations for some common solvents.

Table 4.1

Molal freezing-point depression and boiling-point elevation constants for some common solvents

Solvent	*Freezing point,* °C	K_f	*Boiling point,* °C	K_b
Water	0.00	1.86	100.00	0.521
Benzene	5.51	5.09	80.2	2.61
Diethyl ether	−117	1.79	34.6	2.02
Acetone			56.5	1.71
Cyclohexanol	24.5	37.7		

Finally we may write an equation for the effect of a solute on the vapor pressure over a solution. This takes a particularly simple form if we limit attention to the case of nonvolatile solutes so that the measured vapor pressure is simply the vapor pressure of the solvent. Defining ΔP as the vapor pressure of the pure solvent minus the vapor pressure of the solution and K_p as a constant characteristic of the given solvent, we may write

$$\Delta P = K_p m. \tag{4.4}$$

Note that here again as with ΔT_f and ΔT_b, the colligative property, in this case ΔP, is independent of the chemical nature of the solute, but is affected by the number of moles m of the solute present per 1000 g of solvent.

The relationship expressed by Eq. (4.4) is inserted here because of its analogy with Eqs. (4.2) and (4.3). We shall see in a later section that Eq. (4.4) is derived by the more fundamentally based equation (4.6), which expresses Raoult's law.

We shall defer remarks on the parallel phenomenon of osmotic pressure to a later section, which will first describe what it is that is being observed in that case.

4.5 MOLECULAR WEIGHTS OF SOLUTES; INTERPRETATION OF DEPENDENCE ON *m*

Obviously, the equations given in the preceding section may be rearranged to give the value of an unknown molal concentration once the value of the constant for the solvent in question has been obtained. This represents an important advance in approaches to structural theory, since it permits the determination of *molecular weights* without requiring the *vaporization* of the sample. Clearly, if $m = \Delta T_f / K_f$ as in the case of freezing-point depression, it becomes possible to determine the molal concentration from the freezing point. But, the molal concentration is simply the number of moles per kilogram of solvent, and the number of moles is the weight of the solute sample divided by the molecular weight. This leads to the following results

$$m = \frac{\text{grams solute}}{MW} \times \frac{1000}{\text{grams solvent}},$$

$$MW = \frac{\text{grams solute} \times 1000}{m \times \text{grams solvent}}, \tag{4.5}$$

where MW designates molecular weight.

Problem 4.10   The freezing point of a solution containing 2.30 g of an organic compound in 50.0 g of benzene is found to be 4.48°C. Find the molecular weight of the solute.

Problem 4.11   Given that 0.491 g of a hydrocarbon which is 85.63% carbon is dissolved in 10.0 g of benzene. The freezing point of the solution is found to be 2.97°C. Find the molecular formula of the hydrocarbon.

Problem 4.12   Find the boiling point of a solution of 1.35 g of urea (CON_2H_4) in 100.0 g of water.

Problem 4.13

a) The vapor pressure of a solution of 10.8 g of glucose (a sugar) in 100.0 g of water is 23.434 mm of Hg compared to 23.690 mm for pure water at the same temperature. Given that K_p for water is 0.427 at this temperature, find the molecular weight of glucose.

b) The *general* simplest formula of sugars is CH_2O. Find the molecular formula of glucose.

Problem 4.14   The boiling point of a solution containing 11.4 g of sucrose in 100 g of water is 100.173°C. Find the molecular weight of sucrose.

In addition to the practical consequence of the development of a method for the determination of molecular weights in solution, it is significant to try to understand the implication of the dependence upon m in structural terms. If variation of vapor pressure, freezing point, boiling point, and osmotic pressure in dilute solutions do not reflect the chemical nature of the solute but only its molal concentration, then we may reasonably suggest that these phenomena depend on the *number of solute particles of molecular scale present* in the solution. Comparison of the values of m amounts to the comparison of relative *numbers* of solute molecules. There is a rough analogy here to the behavior of gases. The crucial factor is only the number of particles per unit volume of the solution (of course, not precisely the volume if the molal concentration scale is used). This is suggestive of Avogadro's principle.

4.6 RAOULT'S LAW

The presentation of the colligative property laws in the previous section relied on an argument based on empirical discovery of the regularities. This approach is not faithful to the historical origin of these laws. They were recognized in a theoretical context, the development of the laws of thermodynamics. The overall development is beyond the scope of this chapter, but it is interesting to try to capture the flavor of the *deductive* arguments from general principles advanced by F. M. Raoult and J. H. van't Hoff in the early 1880's. This can be done (in part). There is an added benefit of understanding something of the reasons for parallel behavior of boiling points, freezing points, and vapor pressures.

If the vapor-pressure equation is rewritten in terms of the *mole fraction*, it assumes a very simple form which is useful in solving many problems.

For many solutions, especially for those in which the solvent and solute are chemically similar, the relationship between vapor pressure and concentration is given by Raoult's law. It is expressed as follows

$$P_A = P_A^0 x_A = P_A^0 \frac{N_A}{N_A + N_B}, \tag{4.6}$$

where P_A^0 is the vapor pressure of pure component A, and P_A is the vapor pressure of component A from the solution. If component B is nonvolatile, then P_A is the total vapor pressure of the solution. The term x_A is the mole fraction of A and the terms N_A and N_B are the number of moles of A and B respectively.

It is often convenient to express Raoult's law in terms of ΔP, the lowering of the vapor pressure of the solution when the solute is nonvolatile. Then

$$\Delta P = P_A^0 - P_A = P_A^0 - P_A^0 \frac{N_A}{N_A + N_B},$$

$$\Delta P_A = P_A^0 \left(1 - \frac{N_A}{N_A + N_B}\right) = P_A^0 \frac{N_A + N_B - N_A}{N_A + N_B},$$

$$\Delta P_A = P_A^0 \frac{N_B}{N_A + N_B}. \tag{4.7}$$

Now for very dilute solutions, and especially for dilute *aqueous* solutions in which N_A is large because of the low molecular weight of water, N_B becomes small *in comparison with N_A*. Equation (4.7) then simplifies to

$$\Delta P_A = P_A^0 \frac{N_B}{N_A}. \tag{4.8}$$

Equation (4.8) can be related to Eq. (4.4) in the following way. Suppose that we consider a dilute solution of molality m. Then if N_A is the number of moles in 1000 g of the solvent A, it follows that $N_B = m$. Letting the molecular weight of A be $(MW)_A$, we find that

$$N_A = \frac{1000}{(MW)_A}$$

and

$$\Delta P_A = \frac{P_A^0 m (MW)_A}{1000},$$

$$\Delta P_A = \frac{P_A^0 (MW)_A}{1000} m.$$

Thus the expression $P_A^0 (MW)_A / 1000$ is identical with the constant K_p of Eq. (4.4).

Raoult's law (Eq. 4.6) is obeyed more or less accurately by *both solvents and solutes* in many cases. Precise agreement with the law over the full range of mole fractions is rare, but we may usefully introduce a reference concept— *ideal solutions*—those solutions obeying Raoult's law for both components.

Let us see what Eq. (4.6) implies about relationships among colligative properties. It asserts that the vapor pressure of the solvent depends *only* on the mole fraction of the *solvent*. Introducing N_B moles of *any* solute to N_A moles of the solvent will produce a *solvent* mole fraction $x_A = N_A/(N_A + N_B)$, *independent of the nature of the solute*.

Now observe that vapor pressure, melting, and boiling are studied as equilibrium phenomena. What we mean by the melting "point" is a temperature at which solid and liquid are in equilibrium. Heating leads to conversion of solid to liquid without change of temperature. Solids, as well as liquids, are demonstrably in equilibrium with the vapor. (Anyone who has opened a bottle of mothballs should be aware of vapor pressures of solids to his regret.) Suppose that at the melting-point temperature the vapor pressure over the solid was different from the vapor pressure over the liquid. In this circumstance, equilibrium could *not* be established between solid and liquid because there would be a mechanism for continued transfer from solid to liquid or the reverse in any closed system. The substance could vaporize from the phase having the higher vapor pressure and condense into the phase having the lower vapor pressure. It must be concluded that the vapor pressures of the liquid and solid *must be the same* at the melting point. Assuming that the solid which freezes out is always pure crystalline solvent, we find that the vapor-pressure curve for the solid will not change on forming a solution. For the liquid, the vapor pressure is reduced at each temperature. Thus the temperature at which the *solution* and the solid will have the same vapor pressure is lower than the temperature at which pure liquid solvent and the solid will have the same vapor pressure. The freezing point is "depressed."*

Boiling points are reached when vapor pressures reach one atmosphere. In the event that a nonvolatile solute lowers vapor pressures at all temperatures, it is clear that the boiling point will be elevated.

These relationships are exhibited in Fig. 4.6, which shows the vapor-pressure curve for a pure solvent, the solid form of that solvent, and the vapor-pressure curve over a solution of a nonvolatile solute in that solvent. The common origin of freezing-point depression and boiling-point elevation in vapor-pressure lowering emerges clearly from a reading of this graph.

Raoult's law allows us to see the reason for parallel behavior of freezing-point depression, boiling-point elevation, and vapor-pressure depression. But, you are perhaps now worried about the application of Eqs. (4.2), (4.3), and (4.4) to solutions which are *not* ideal. Fortunately, a more general proof of these equations (beyond the scope of our present discussion) may be given for solutions that are not ideal *so long as they are dilute*.

Problem 4.15 The vapor pressures of benzene C_6H_6 for 0.5° intervals between 79.5° and 83.0°C are given below. From these data construct the vapor-pressure curve over this temperature range. Making use of Raoult's law and any necessary molecular weights, calculate the vapor pressure for a 1.0 m solution of naphthalene $C_{10}H_8$ in

---

* Strictly speaking ΔT_f as indicated in Fig. 4.6 represents the lowering of the freezing point for the solute-solvent system in the absence of any external pressure such as that of the atmosphere. Fortunately, the effect of external pressure on ΔT_f is relatively small. See, for instance: W. J. Moore, *Physical Chemistry*, 2nd ed., Prentice-Hall, 1955, pp. 127–130.

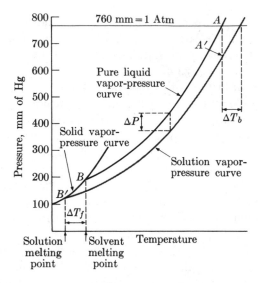

Fig. 4.6. Interdependence of ΔP, ΔT_b and ΔT_f in a solution.

benzene for each 0.5° interval and construct the vapor-pressure curve for the solution. From your graph, read off the boiling-point elevation ΔT_b for benzene for a 1.0-m solution of nonvolatile solute; hence give a calculated value of K_b. (Note that naphthalene has a very low vapor pressure at 83.0°, and we may therefore neglect it in this problem.)

Vapor pressure of benzene, 79.5 to 83.0°C

Temperature, °C	79.5	80.0	80.5	81.0	81.5	82.0	82.5	83.0
Vapor pressure, mm	746.1	757.7	769.3	781.2	793.4	805.6	817.9	830.5

4.7 OSMOTIC PRESSURE OF SOLUTIONS

Pauling has introduced the concept of osmotic pressure in an interesting way.

If red blood corpuscles are placed in pure water they swell, become round, and finally burst. This is the result of the fact that the cell wall is permeable to water but not to some of the solutes of the cell solution (hemoglobin, other proteins); in the effort to reach a condition of equilibrium (equality of water vapor pressure) between the two liquids water enters the cell. If the cell wall were sufficiently strong, equilibrium would be reached when the hydrostatic pressure in the cell had reached a certain value, at which the water vapor pressure of the solution equals the vapor pressure of the pure water outside the cell. This equilibrium hydrostatic pressure is called the *osmotic pressure* of the solution.*

---

* From Linus Pauling, *General Chemistry*, 2nd ed., W. H. Freeman and Co., San Francisco. Copyright 1953, p. 347.

A semipermeable membrane has a structure such that the solvent molecules can pass through it, but the passage of solute molecules is barred. Such membranes for use in aqueous solutions may consist of porous glass having very minute passages, of certain animal membranes, or of specially prepared membranes such as that of cupric ferrocyanide $Cu_2Fe(CN)_6$ deposited in the pores of unglazed porcelain for support. Accurate measurements have been made with this latter membrane at pressures up to more than 200 atm, corresponding to the support of a column of water over a mile high! In such a system the osmotic pressure is measured as that pressure which must be applied to the compartment containing the solution, which is just sufficient to prevent any passage of water from outside into this compartment.

The first direct quantitative measurements of osmotic pressure were made by Pfeffer in 1877. The significance of Pfeffer's measurements was first pointed out by van't Hoff ten years later. He showed that for dilute solutions of molecular solutes, the following equation holds to a good approximation

$$\pi V = N_B RT, \tag{4.9}$$

where π is the osmotic pressure, V the volume of the solution, N_B the moles of solute, and T the absolute temperature; R represents a proportionality constant, which can be evaluated from experimental data. If π, V, N, and T are measured and expressed in appropriate units, then in the limit of sufficiently dilute solution, $R = 0.082$ liter-atm/deg-mole. This limiting value is identical with the universal gas constant R discussed on page 34. It is a striking fact that Eq. (4.9) has the same form as the perfect-gas equation.

It can be shown by thermodynamics that Eq. (4.9) can be related quantitatively to the equations for ΔP, ΔT_f, and ΔT_b already discussed.

4.8 SATURATED SOLUTIONS AND SOLUBILITY

This chapter should perhaps conclude with some general remarks on dissolution. There are many pairs of substances, such as ethanol and water, which are miscible in all proportions to form a single homogeneous phase or solution. But it is a matter of common experience that the capacity of a solvent to dissolve a given solute is frequently limited, e.g., the systems water-sugar or water-salt. When excess sugar is stirred with water until a solution of constant composition is attained, then this solution is said to be saturated.

The concentration at saturation of a specific substance in a given solvent at a specified temperature is said to be its solubility. The solubility may range from infinity down to virtually zero, depending on the particular solute-solvent pair under consideration. It is sometimes convenient to express solubility in terms of solubility curves as with the aqueous solutions of a selected number of crystalline solids shown in Fig. 4.7.

There is no general law governing the solubility of liquid or solid solutes in solvents such as water or ethanol. On the other hand there is a definite law

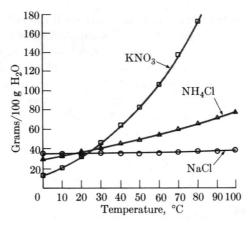

Fig. 4.7. Solubility curves for salts in aqueous solution.

in relation to the solubility of gases such as the components of air. Henry's law states that the solubility of these gases expressed in grams or moles is directly proportional to the partial pressure of the gas in equilibrium with the solution at a specified temperature.

Problem 4.16 Recognizing that the "solubility" of a gas is just the concentration in solution in equilibrium with a given vapor pressure, prove that Henry's law must hold for *any* ideal solution. Find the proportionality constant.

SUMMARY PROBLEMS

Problem 4.17 A solution of naphthalene $C_{10}H_8$ in cyclohexane C_6H_{12} contains 2.0% of solute by weight.

a) Find the molality and the mole fraction of solute present. What is the mole fraction of cyclohexane?
b) What additional information is needed to find the molarity of this solution?
c) Find the freezing point of a solution which contains 5.0 g of naphthalene in 50 g of benzene.

Problem 4.18 Some data on the freezing points of a series of solutions in benzene, C_6H_6, as a solvent are given in the table below. Complete the table by filling in the values of K_f. Assume that $\Delta T_f = mK_f$. This is the origin of the results in Table 4.1.

Solute	Formula	Grams of solute in 50.0 g C_6H_6	Freezing point of solution, °C	K_f for benzene
Biphenyl	$C_{12}H_{10}$	2.99	3.36°	
Urea	N_2H_4CO	1.56	2.72°	
Naphthalene	$C_{10}H_8$	1.98	3.82°	
Ethylenediamine	$C_2N_2H_8$	0.90	3.83°	

Problem 4.19

a) It was found that 2.93 g of benzoic acid when dissolved in 60.0 g of benzene C_6H_6 lowered the melting point 1.02°C. On the basis of Eq. (4.2), calculate the molecular weight of the solute benzoic acid in the benzene solution.

b) Compare your answer with that obtained from the formula C_6H_5COOH and the atomic weights of C, H, and O. Can you offer an explanation for the apparent anomaly?

READINGS AND EXTENSIONS

This chapter could quite logically follow directly after Chapter 1, since it shows how to extend methods for the determination of molecular weight to molecular substances that are not readily vaporizable. Actually, quite a few of the molecules mentioned in Chapters 2 and 3 are studied by the methods of this chapter. This chapter, then, has extended the scope of stoichiometry. An excellent treatment of the subjects of this chapter, which examines the ideas in more detail (especially in more practical detail), is to be found in L. K. Nash's paperback on stoichiometry cited at the end of Chapter 1.

CHAPTER 5

Solutions of Electrolytes

5.1 INTRODUCTION

In the preceding chapters, we have been concerned to a large extent with molecular substances both in the pure state and in solution. And when we have made generalizations, such as Avogadro's hypothesis or Raoult's law, which are related to the concentration of particles, we have with justification considered these particles to be molecular. The great majority of carbon-containing or organic substances would fall into this class of molecular substances.

A large class of compounds remains for our consideration. A large number of the members of this class dissolve in water, and the resulting solutions conduct electric current; these are electrolytic (or electrolyte) solutions. The solutes are called *electrolytes*. Examples of electrolytes are NaCl, K_2SO_4, HCl, NaOH, $Ca(OH)_2$, BaI_2, NH_4Cl, $PbCl_2$, and H_2SO_4. Note that the methods of Chapter 1 would allow us to establish these simple formulas as written. But we do not know the relations of simple formulas to molecular formulas. Taking NaCl as an example, we can agree from arguments based on percentage composition and atomic weights that the compound is composed of one Na atom per Cl atom. We can even argue that we should assign a *valence* number of *one* to Na and Cl. But is the molecule NaCl, Na_2Cl_2, something larger, or are the Na and Cl strictly linked in a molecule at all? What is the nature of the *valence of one?* Is it *one* bond in a molecule? Some of these substances such as HCl and H_2SO_4 *are* molecular in the pure state. It turns out that none of them exists as molecules in aqueous (water) solution.

The clues for development of a structural theory for electrolytes in solution come in large degree from the analysis of their electrical conductivity, although a full and convincing presentation of the theory depends on the confluence of several lines of evidence in addition to that derived from conductivity measurements. These additional lines of evidence result from a consideration of the colligative properties of aqueous solutions of electrolytes. The tools for this approach have been developed in Chapter 4 for molecular solutions. Our present task is to see what conclusions may be drawn from the considerably more complex behavior of the aqueous solutions of substances such as those listed by formula in the preceding paragraph.

The study of electrolyte solutions was one of the main enterprises of the first group of chemists who described themselves as "physical" chemists. Leading names in this late nineteenth century group include the now familiar van't Hoff, the German chemist Kohlrausch, who accumulated the key experimental data on the electrical behavior of electrolyte solutions, and a brilliant young Swedish chemist, Svante Arrhenius, whose insight and daring accomplished the synthesis of the various lines of investigation. Kohlrausch's contributions require attention first; they will be discussed in Section 5.2.

We shall see that the ideas of Arrhenius and the German school were entirely adequate for many solutions of electrolytes—in particular for the class known as weak electrolytes. However, it remained for succeeding generations of physical chemists, physical chemists across the Atlantic as well as in Europe, to refine these ideas, to make them applicable to all electrolytic solutions and, in particular, to the solutions of the so-called strong electrolytes.

5.2 THE CONDUCTIVITY OF ELECTROLYTES

The first experiments on the passage of electric currents through electrolyte solutions were carried out using direct current sources (e.g., batteries). One idea emerged promptly. Solutes which lead to electrical conductivity must function as sources of charged particles. It was suggested that the electric field might induce molecules to exchange positively and negatively charged parts along a chain. This current flow is accompanied by chemical reactions. This will be explored in Chapter 6, but for the present purpose such experiments are not fruitful. The solutions do not obey Ohm's law,

$$\mathcal{E} = RI. \tag{5.1}$$

The proportionality constant R is characteristic of the conductor and is called the resistance. (Note that a large value of R means that only a small current flows for a large applied voltage.) For wires, R increases with the length of the wire l and decreases as the cross-sectional area a is increased. Now, when electrolyte solutions are studied using an *alternating current* source, they do obey Ohm's law, and it is possible to define the resistance. This resistance also depends on the length and area through which current must flow, as shown by the equation,

$$R = \rho l/a, \tag{5.2}$$

where ρ is a constant called *specific resistance*, characteristic of the particular solute and solute concentration. It is the resistance of the solution measured between two inert metal electrodes of 1 cm^2 area separated by 1 cm. We may

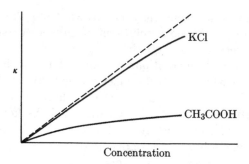

Fig. 5.1. Specific conductivity of aqueous potassium chloride and acetic acid as a function of concentration (conductivity of CH_3COOH is exaggerated relative to KCl).

equally well define the specific conductivity of the solution. Conductivity is the reciprocal of resistance. The *specific conductivity* $\kappa = 1/\rho$. This will be the quantity of interest for analysis of electrolyte structures. (The size of the "cell" in which conductance behavior is studied is an irrelevant variable but must be controlled. Thus the *specific* value is used.)

The first question to raise might easily be seen to be the dependence of specific conductivity on concentration. It seems reasonable to expect conductivity to increase with increasing electrolyte concentration, but is the relationship simple? Does doubling the concentration simply double κ, etc.? The results for typical electrolytes, potassium chloride and acetic acid, are shown in Fig. 5.1. The contemplated simple behavior would correspond to a straight line. Note that the experimental curves deviate from this kind of line, falling below it. In the case of the KCl, the deviation is slight. This behavior is characteristic of any *strong electrolyte*. The curve for acetic acid falls far below the expected straight line. This is typical behavior for a so-called weak electrolyte.

It is the difference in behavior leading to deviation from the straight line that is interesting. The implication of the shape of the KCl curve is that each mole of the KCl is a little less efficient as a carrier of current as its total concentration of the solution increases. In the case of acetic acid this falling off of efficiency at high concentration is much more profound.

To illuminate the interesting difference it is useful to define still another conductivity variable. The quantity is designated equivalent conductance, Λ. For a monovalent electrolyte (e.g., HCl, NaBr, KNO_3, etc.) the equivalent conductance is given by

$$\Lambda = \frac{\kappa}{M} \times 1000, \tag{5.3}$$

where M is the "molar" concentration.* Actually, Eq. (5.3) defines the molar

---

* It will emerge in the course of this chapter that it is difficult to say just what is a mole of an electrolyte. Nevertheless, it is customary to continue to calculate molar concentration as if the simplest formula represented a "mole" correctly.

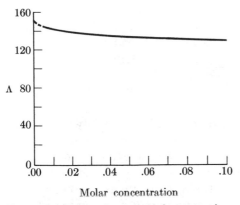

Molar concentration

Fig. 5.2. A plot of Λ vs. *M* for potassium chloride. The curve is extrapolated to *zero* concentration.

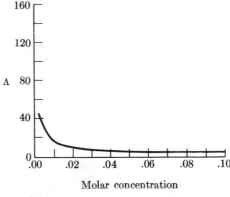

Molar concentration

Fig. 5.3. A plot of Λ vs. *M* for acetic acid. Note the difficulty in extrapolation to *zero* concentration.

conductance which is the same as equivalent conductance for a 1 to 1 electrolyte such as NaBr. If the valence type of the electrolyte is not 1 to 1, the molar concentration must be multiplied by the valence. Figures 5.2 and 5.3 show the equivalent conductance of the aqueous KCl and CH_3COOH* solutions. What these plots indicate more clearly than Fig. 5.1 is that the conductivity *per unit concentration* decreases as the concentration increases. This is the aspect of conductance behavior that has received most attention.

5.3 KOHLRAUSCH'S LAW

One convenient way to compare the conductance behavior of *different* electrolytes is through comparison of values of Λ at the same concentration. For fundamental purposes it is useful to make comparisons of the limiting values of Λ as the *concentration approaches zero*; Λ_0 is defined by Eq. (5.4) below. For strong electrolytes, it is measured by extrapolation of the experimental curves (see Fig. 5.2). Figure 5.3 shows that the estimation of Λ_0 for weak electrolytes is clearly uncertain. Attention should first be paid to strong electrolytes. In the case of strong electrolytes

$$\Lambda_0 = \lim_{M \to 0} \Lambda. \tag{5.4}$$

Kohlrausch was the first to make a systematic study of strong electrolytes and tabulate values of Λ_0. As a result of his collection of values of Λ_0, he was able to extract an important regularity. This is exhibited in Table 5.1. The difference in Λ_0 for *pairs* of electrolytes having a *common* group is *constant*.

* We shall refer to acetic acid so frequently as an example of a weak acid that, for convenience, we shall use the formula HOAc in place of CH_3COOH.

Table 5.1

NaCl	108.90	NaNO$_3$	105.33	NaBr	111.10
KCl	130.10	KNO$_3$	126.50	KBr	132.30
Difference	21.20		21.17		21.20

Thus, without regard to the combination partner, the difference between sodium and potassium salts is *constant*. The same observation can be made from various other pairs. This indirect comparison is the only way to compare sodium and potassium as charge carriers—the charge carriers are called *ions*—because sodium and potassium never occur without their counter ions like chloride or nitrate. However, Kohlrausch noted that the *constant difference* suggested that (at least at zero concentration—often called infinite dilution) sodium and potassium functioned to carry charge independently of the counter ion. The constant difference suggests that, for example, sodium functions as the same electrical conductor whether it is combined with chlorine, bromine, or nitrate. This idea is known as Kohlrausch's law of *independent ion mobility* (formulated in 1874).

5.4 ARRHENIUS' THEORY OF CONDUCTIVITY

In his doctoral thesis (1884) Svante Arrhenius (Swedish) presented a theory of the concentration dependence of Λ that was a natural extension of the ideas suggested by Kohlrausch's law. However, the thesis was accepted by Arrhenius' professors in Uppsala only with considerable misgiving, and his degree was awarded with reservation. This was because Arrhenius had the daring to suggest firmly that when electrolytes are dissolved in water, they dissociate *into independent charged particles, or ions*. All of his predecessors had believed that the presence of charged particles was *only transitory* in solutions which were *under the influence of an electric field*. They were unable to imagine a mechanism that would prevent electrically charged particles with opposite charges from combining. Arrhenius' theory of conductance involved the introduction of two postulates:

1. As any electrolyte solution is diluted, the degree of dissociation of the electrolyte into free separate ions increases, approaching 100% as the concentration approaches zero.

2. The fraction α of electrolyte dissociated at any concentration, may be calculated as the ratio of the equivalent conductance at that concentration to the equivalent conductance at zero concentration, $\alpha = \Lambda/\Lambda_0$.

This amounts to the suggestion that there is a simple reason for the decrease of Λ with increasing concentration. According to the theory of Arrhenius, all electrolytes are thought of as completely separated into charged particles in infinitely dilute solution. These solutions are maximally efficient as charge carriers. At higher concentrations, the fraction of uncharged molecules becomes higher, and, as a result, the electrolyte becomes a less efficient charge carrier.

5.5 ARRHENIUS' SYNTHESIS

The Arrhenius interpretation of the dependence of conductance on concentration may be legitimately viewed with skepticism if it stands alone. However, it need not. Once the conductivity data had suggested the interpretation of electrolyte structure in terms of the dissociation of molecules into free charged particles, or ions, on dissolution in solvents like water, it was possible to explore the utility of this model in the explanation of various types of data. In 1887, Arrhenius published a comprehensive paper on the structure of electrolytes in solution in the first volume of the new German physical chemistry journal, *Zeitschrift für physikalische Chemie.*

The key feature of the 1887 paper was a comparison of the behavior of electrolyte solutions with respect to conductance on the one hand and colligative properties on the other. To appreciate the argument, it is useful to look first at some nonelectrolyte solutions which cause trouble for the simple colligative-property laws and then turn to the colligative properties of electrolytes. For convenience, only freezing-point depression is discussed.

The application of freezing-point data to suit our purposes is illustrated by the set of experiments which supply the data for Problem 4.18. In that problem we test the validity of Eq. (4.2) by investigating the solutions in benzene of the following four solutes: (1) biphenyl, (2) urea, (3) naphthalene, and (4) ethylenediamine. It is possible to establish the molecular formulas for all four of these substances by the vapor-density method. Thus we can prepare benzene solutions of each of the four solutes, and from the known weights and molecular weights we can calculate the molality of each solution. We now proceed to measure ΔT_f for each, and from our experimental data, to calculate K_f from the relationship $K_f = \Delta T_f/m$. If the experiments are carefully done, we find essentially the same value for K_f with solutes 1, 2, 3, and 4 as listed above. This tells us that K_f for 1-m solutions in benzene solvent is 5.5°C, and that K_f is characteristic of the benzene solvent but independent of the chemical nature of the solute. For instance, if we were to repeat this experiment using *cyclohexane*, C_6H_{12}, as solvent, we should again get the same K_f value from each of the solutes, but the value would differ from that for the *benzene* solutions. That is, we do find that Eq. (4.2) expresses an experimental fact. From Eq. (4.2) it is seen that, for any given solvent, the value of ΔT_f is proportional to m, the molal concentration for different solutes. Of course, solutions having the same molal concentration must contain the *same number of solute molecules* per specified weight of solution. Hence we have argued that it is the number and not the kind of solute molecules that is the controlling factor in the freezing-point depression.

In the case of benzoic acid in benzene, it may be shown by experiment that the simple relationship expressed by Eq. (4.2) no longer holds. See Problem 4.19. The relationship we find here is

$$\Delta T_f = 0.5 m K_f. \tag{5.5}$$

How can we explain this deviation without discarding the simple law? First, we may observe that Eq. (5.5) does bear a formal relationship to Eq. (4.2). If we introduce a "fudge factor" i, we emerge with the equation:

$$\Delta T_f = i m K_f, \tag{5.6}$$

and in this particular case the value of i is 0.5. Now, how can this be interpreted? If molecules of benzoic acid were to dissociate (break up into fragments) in solution, then we would have more than the predicted number of particles, and the observed lowering might be expected to be greater than predicted by the simple equation (4.2). As a result, the factor i needed to bring the results into line would be greater than one. In fact, if the benzoic acid broke up completely into two particles per molecule, i would be 2. Similar considerations might lead to the concept of association. Thus the C_6H_5COOH might form double molecules, which we shall designate as $(C_6H_5COOH)_2$, the dimer. This would result in having only half as many double molecules or dimer molecules as we would have if the solute were present in the form of the monomer, C_6H_5COOH. And this would account for the value 0.5 for i.

Here again we have inferred that the observed ΔT_f value is dependent on the number of solute molecules per specific weight of solvent and not on the kind of solute molecules. It is significant that those aqueous solutions which are strongly electrolytic, i.e., conduct current well, are also the ones to which Eq. (4.2) fails to apply in its simple form. Rather we must introduce the "fudge factor" i, as we have done in Eq. (5.6), and moreover in these cases this i-factor is always *greater* than 1. This suggests that the formula-weight units* of solutes like KCl or K_2SO_4 must be split at least to some degree into two or more particles. No single value of a "fudge factor" i can put things straight in Eq. (5.6). This might seem a hopeless cause, but we do find we can make some sense out of it if we pursue the matter further. For all solutions of electrolytes, the value of i increases with dilution, approaching a small integer at "infinite"

---

* For molecular solutes such as ethyl alcohol, C_2H_6O, or glucose, $C_6H_{12}O_6$, we may properly refer to the formula-weight units as molecules. In the case of the alcohol, the formula-weight unit would consist of 2 carbon atoms and 6 hydrogen atoms and 1 oxygen atom to make up a molecule. For K_2SO_4, there is no good reason to believe that the formula-weight unit, comprised of two potassium atoms plus one sulfur atom plus four oxygen atoms would represent a molecule in any sense of the word. It is therefore good practice to refrain from talking about molecules of the solute. Despite this, it is conventional to talk about "molar concentrations." To be consistent, while we may refer to a 1.0-*molar* solution of alcohol, we should prefer to refer to 1.0-*formal* solution of K_2SO_4. This preference is not universally observed.

Table 5.2

Molality, M	0.10	0.05	0.01	0.005	0.001
i-value	2.32	2.45	2.69	2.77	2.84

dilution or "zero" concentration. We may formulate this statement as follows:

$$\lim_{m \to 0} i = n, \tag{5.7}$$

where m is the molality and n is a small integer 2, 3, 4, etc. For strong electrolytes such as NaOH, HNO_3, or HCl the value of n is 2.0. From this we may infer that a single formula unit of HCl or NaCl at very low concentration is dissociated into two particles, each of which contributes to the colligative properties of the solution. A limiting value of 3 as found for Na_2SO_4 or $Ca(OH)_2$ suggests three particles per formula unit, etc. The electrolytes may be labeled Valence Type 1-1, 1-2 or 2-1, 1-3 or 3-1, etc., according to the valence associated with the particle species involved. The relationship expressed by Eq. (5.7) is illustrated by the case of K_2SO_4, a 1-2 type, where the value of i at successively decreasing concentrations approaches the integer 3. Pertinent data in the case of K_2SO_4 is given in Table 5.2.

We might interpret the data of Table 5.2 as indicating an increasing degree of dissociation with decreasing concentration. Now, the treatment has reached the same conclusion as the analysis of conductance data. Is there quantitative agreement? It is necessary to compare i to the degree of dissociation ($\alpha = \Lambda/\Lambda_0$). If an electrolyte "dissociates" into *two* ions (e.g., NaCl $\to$ Na$^+$ + Cl$^-$), then the fraction dissociated has produced 2α particles, and the fraction remaining undissociated contributes $1 - \alpha$ particles per mole of solute dissolved. Hence

Table 5.3

Comparison of extent of dissociation of electrolytes calculated from conductance and freezing-point depression data. *

Substance	Formula	$\alpha(\Lambda/\Lambda_0)$	α (from i)
Ammonia	NH_3	0.01	0.03
Hydrochloric acid	HCl	0.90	0.98
Nitric acid	HNO_3	0.92	0.94
Potassium chloride	KCl	0.86	0.82
Barium chloride	$BaCl_2$	0.77	0.81
Acetic acid	CH_3COOH	0.01	0.03
Sugar	$C_{12}H_{22}O_{11}$	0.00	0.00

Note that the concentrations considered in each case are those for 10 g of solute dissolved in 1 liter of water.

* After S. Arrhenius, Z. *Physik. Chem.* **1**, 631 (1887).

$i = 2\alpha + (1 - \alpha) = 1 + \alpha$. If an electrolyte "dissociates" into three particles (e.g., $BaCl_2 \rightarrow Ba^{2+} + 2Cl^-$), then $i = 3\alpha + (1 - \alpha) = 1 + 2\alpha$. In general, if dissociation produces n particles per formula unit, $i = 1 + (n - 1)\alpha$. Arrhenius' 1887 paper contained a comparison of α for various electrolytes at particular concentrations calculated from *both* conductivity (by means of Λ/Λ_0) and freezing-point depression (by means of i). A small part of his table is reproduced in Table 5.3.

Arrhenius argued that the agreement was impressive. His interpretation of the concentration dependence of conductivity led to the suggestion of dissociation into ions. In considering the colligative properties, the calculated degree of dissociation is supported quite satisfactorily by a method of assessing dissociation which in *no way* depends upon charge and electrical behavior.

Problem 5.1 Calculate the limiting equivalent conductance at infinite dilution for acetic acid by application of Kohlrausch's law to the following data.

Compound	Λ_0
HCl	426.2
NaCl	126.4
NaOAc	91.0

Explain why this method for obtaining Λ_0 for acetic acid is superior to extrapolation of the experimental values of Λ for acetic acid.

Problem 5.2 The conductivity of a $0.01\,M$ solution of acetic acid is much less than the conductivity of $.01\,M$ trifluoroacetic acid. Predict the relative freezing points of the two solutions.

Problem 5.3 List, in order of increasing conductivity, $0.01\,M$ solutions of the following: $SrCl_2$, CH_3COOH, and NaCl.

Problem 5.4 A solution of 0.1499 g of KNO_3 in 30.00 g of water freezes at $-0.1739°C$.

a) Find the i-factor.
b) Find the degree of dissociation according to the Arrhenius theory.
c) Find the temperature at which this solution would boil.
d) If Λ_0 for KNO_3 is 144.9, predict the value of Λ for this solution on the basis of Arrhenius' theory.
e) Consider the accuracy of the freezing-point data. How nearly would a measured value of α be expected to agree with the answer to (d) if Arrhenius' theory is accurate in detail?
f) Arrhenius quotes α from conductance as 0.81 for KNO_3 and α from freezing-point depression as 0.67. Comment.

5.6 SOLVENT DIELECTRIC CONSTANT

The previous sections have pursued the development of the idea of dissociation of molecules of electrolytes into charged ions on dissolution of the electrolyte in water. Historically, it was a major hurdle for chemists to accept the idea that free charged particles might exist. It was difficult to understand why the attraction of opposite electric charges did not lead immediately to combination. This difficulty finds resolution in a property of those *solvents* in which electrolytes readily dissolve. This property of such solvents can be used to explain why they might accommodate charged particles.

Turning briefly to fundamental electrical theory, we quote the "law" which governs the force observed to act between two charged spheres (e.g., metal spheres mounted on insulating mounts and charged) in a *vacuum*. This is Coulomb's law. It is given by

$$F = \frac{q_1 q_2}{d^2}, \tag{5.8}$$

where F is the force, q_1 and q_2 are the magnitudes of the charges on the objects, and d is the distance between them. Now, it is found that immersion of the charged objects in *any* nonconducting material leads to a *decrease* in the force acting between them. This *decrease* is representable by a *number characteristic of the material medium* usually called the *dielectric constant, D*. We may modify Eq. (5.8) by introducing D:

$$F = \frac{q_1 q_2}{D d^2}. \tag{5.9}$$

By following the conventional procedure of putting D into the denominator of Eq. (5.9), it should be clear that we mean that any medium characterized by a large value of D will be one which accomplishes a substantial reduction in the force acting between charged particles. Now, it turns out to be *exactly those solvents of high dielectric constant that dissolve electrolytes*. Table 5.4 lists dielectric constants of some common liquids along with notes on the solubility of sodium chloride in each. Note that if the positive sodium ion and the negative chloride ion behave like large-scale charged spheres, the force between them in water will be only about 1.3% of the force in vacuum and only about 3.5% of the force between them in benzene.

One might well ask at this point what property of the solvent molecule results in a solvent exhibiting a large dielectric constant. Let us suppose that

Table 5.4

Dielectric constants

Liquid	Molecular formula	D	Solubility of NaCl
Benzene	C_6H_6	2.28	Insoluble
Diethyl ether	$C_2H_5OC_2H_5$	4.23	Insoluble
Methanol	CH_3OH	32.6	Slightly soluble
Water	H_2O	78.5	Very soluble

the solvent molecule, although uncharged, is polar. That is, imagine that the molecule has an *uneven* distribution of electric charge internally so that one end is positive and the other end is negative. This is indicated diagrammatically in Fig. 5.4(a). Let us look at the effect of such molecules on the force between two parallel charged plates. In Fig. 5.4(b) the plates are shown with nothing between them; in Fig. 5.4(c) they are shown with *polar molecules* intervening. The *polar molecules* will *tend* to line up (this lining up is exaggerated in the figure). The *net result* is a pairing of positive and negative ends of polar molecules throughout the solution *except* that a row of *positive ends* touches the negative plate and vice versa. This amounts to a *partial neutralization* of the

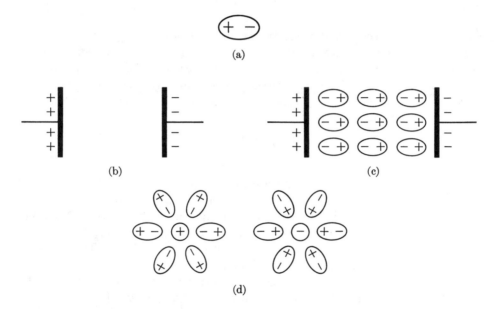

Fig. 5.4. Polar molecules: (a) a polar molecule; (b) a capacitor; (c) polar molecules in a capacitor; (d) polar molecules oriented around ions.

charges on the plates—the force is expected to be reduced. The arrangement of polar molecules about ions in solution ("solvation") is shown in Fig. 5.4(d). It is at least plausible to attribute high dielectric constant to *molecular polarity*.

Problem 5.5 The specific conductance of 0.01 M HCl is 41.2×10^{-4} in water. In a solvent mixture of 82% dioxane ($D = 2.21$) and 18% water, the specific conductance of 0.01 M HCl is 1.5×10^{-4}. Explain the decrease.

Problem 5.6 There is a time-honored rule that "like dissolves like." Can you suggest an explanation for the fact that polar (although uncharged) molecules tend to be more soluble in polar than in nonpolar solvents?

5.7 CRITICAL EVALUATION OF ARRHENIUS' THEORY

As the data in Table 5.3 may suggest, precise and critical evaluation of Arrhenius' views was not offered in 1887. The numbers are only in semiquantitative agreement and are reported for a single concentration. One of the first more careful tests proposed was an examination of the dissociation hypothesis from the point of view of an "equilibrium constant" to be discussed in some detail in Chapter 6. Prior to 1887, it had been found possible to make a generalization for all chemical reactions which reach equilibrium (no further net chemical change) with measurable concentrations of both reactants and products present. The generalization is frequently put in the form of the *equilibrium-condition* expression. Thus for a reaction $A + B \rightleftharpoons C + D$, it was possible to write:

$$K = \frac{[C][D]}{[A][B]}, \tag{5.10}$$

where K is a constant and the brackets denote molar concentration. Applying this to electrolyte dissociation yields

$$MX \rightleftharpoons M^+ + X^-, \qquad K = \frac{[M^+][X^-]}{[MX]}. \tag{5.11}$$

Now the concentration of M^+ ions, $[M^+]$, *must equal* the concentration of X^- ions, $[X^-]$, and is αC, where α is the fraction of MX dissociated and C is the total concentration of the electrolyte that was introduced. Similarly, the concentration of undissociated electrolyte, $[MX]$, is just $(1 - \alpha)C$, where $(1 - \alpha)$ is the fraction undissociated. We may rewrite (5.11) as

$$
\begin{aligned}
K &= \frac{(\alpha C)(\alpha C)}{(1 - \alpha)C} \\
&= \frac{\alpha^2 C^2}{(1 - \alpha)C} \\
&= \frac{\alpha^2 C}{1 - \alpha}.
\end{aligned}
\tag{5.12}
$$

If K is to be a *constant*, this equation *specifies* the form of the concentration dependence of the degree of dissociation, α. Equation (5.12) is known as the Ostwald dilution law after the prominent German physical chemist W. Ostwald, who first derived it shortly after Arrhenius' announcement of the theory of electrolytic dissociation. Ostwald was able to show that *weak* electrolytes obeyed the dilution law *very* precisely, *but strong electrolytes deviate markedly*. This presents a problem for the Arrhenius theory.

Problem 5.7 Find the value of the "equilibrium constant" K of the Ostwald dilution law for each concentration of the strong electrolyte KCl and the weak electrolyte acetic acid from the data given below. Comment on the fit to the law. (Recall $\alpha = \Lambda/\Lambda_0$.)

Concentration (C)	$\Lambda(KCl)$	$\Lambda(HOAc)$
0.0000	149.8	390.8
0.0001	148.9	134.6
0.0010	146.9	49.2
0.0100	141.27	16.18
0.0500	133.30	7.36
0.1000	128.9	5.20

Looking at the behavior of strong electrolytes a little more closely, we find that (1) freezing-point depression and conductivity data *are not* in precise agreement as to the degree of dissociation α and (2) the dependence of α of a strong electrolyte on concentration is simpler than the prediction of the Ostwald dilution law. To explore this second assertion rewrite the Ostwald law in terms of $\Lambda/\Lambda_0 = \alpha$.

$$K = \frac{(\Lambda/\Lambda_0)^2 C}{1 - \Lambda/\Lambda_0}. \tag{5.13}$$

It is clear that Λ is a complicated function of C. Experimentally, it turns out that *dilute* solutions of strong electrolytes conform quite satisfactorily to a simple equation involving the square root of the concentration, $\sqrt{C}$:

$$\Lambda = \Lambda_0 - B\sqrt{C}, \tag{5.14}$$

where B is a constant. This equation implies a straight line in the plot of Λ vs. $\sqrt{C}$. (It should be noted that departures from this straight-line behavior are always observed at higher concentration.)

Problem 5.8 Plot Λ vs. $\sqrt{C}$ for KCl. Use the data of Problem 5.7.

Problem 5.9 You have seen one method of finding Λ_0 (HOAc) in Problem 5.1. Now that we have developed the concept of the equilibrium condition and are making plots, we have an alternative approach to finding Λ_0 for a weak electrolyte independent of using Kohlrausch's law.

a) Show from Eq. (5.13) that $\Lambda^2 C = K\Lambda_0^2 - K\Lambda_0\Lambda$.
b) Plot the functions $\Lambda^2 C$ vs. Λ.
c) From the slope and intercept of the straight line obtained in (b), find Λ_0. Note that no extrapolation was necessary here.

5.8 THE STRUCTURE OF STRONG ELECTROLYTES

It has emerged that Arrhenius' hypothesis of an equilibrium between molecules and charged ions in solution is an eminently satisfactory theory of the behavior of solutions of *weak* electrolytes. However, close examination of strong electrolytes in aqueous solution reveals important discrepancies. We must ask if there is not an *alternative account* of their structure, which is more accurate in *detail*.

In the search for such an alternative, the data provided in Table 5.5 prove suggestive. This table lists values of i and α for *different* strong electrolytes at the *same* concentration. The notable feature is the *constancy* within a valence type. If an electrolyte forms M^+ and X^- ions, the degree of dissociation seems to be the same whatever the *chemical nature* of the ions M and X. An electrolyte giving M^+ and X^{2-} seems also to have a degree of dissociation determined only by the charge type. The American physical chemist A. A. Noyes was the first to point out (1904) that the degree of dissociation *seemed to be related only to the charges on the ions* and not to their *chemical nature*. He noted that this indeed represented strange behavior. It was unprecedented to suggest that the tendency of the bond in a molecule to break *was not dependent upon the atoms involved in the bond*.

Table 5.5

Values of i and α calculated from freezing-point depression in 0.005-M solutions

	Uni-uni-valent			
	HCl	KCl	KNO_3	$CsNO_3$
i	1.95	1.96	1.95	1.95
α	0.95	0.96	0.95	0.95

	Uni-bi-valent		
	Na_2SO_4	K_2SO_4	$Ba(NO_3)_2$
i	2.78	2.77	2.76
α	0.89	0.89	0.88

Noyes noted that if one assumed that *strong electrolytes in solution are always 100% dissociated into ions*, the results in Table 5.5 might be given a very reasonable interpretation. This implies that the α of Arrhenius is the apparent degree of dissociation, rather than the real degree. Noyes suggested that the observed deviation from $i = 2$ or 3 should be attributed *not* to the presence of molecules *but* to the effect on the behavior of any ion of the *electric field* produced by other ions. As the concentration of strong electrolytes increases, the number of ions increases and the electric field increases. This electric field is, of course, determined only by the charge on the ion, not by its chemical nature. Thus it seems useful to explore the alternative idea that *strong electrolytes exist only as ions and do not form molecules*.

5.9 CONDUCTIVITY—THE THEORY OF DEBYE-HÜCKEL AND ONSAGER

To explore the suggestion that strong electrolytes are *entirely dissociated*, it is necessary to find a way to describe the effects of the electric field of other ions on the behavior of an ion. This involves fairly sophisticated electrical theory beyond the scope of this discussion. The basic job was accomplished by Debye and Hückel in 1923 and applied explicitly to conductivity by Onsager. We shall attempt to describe the main feature of the Debye-Hückel and Onsager theory without giving the detailed justification.

Suppose that we consider an ion, say K^+ if we are discussing KCl, in an aqueous medium. In the neighborhood of that ion, we shall have other plus and minus ions and on the average, taking any one K^+ ion as a point of reference, we shall have one more negative ion than we shall have positive ions in the region immediately around it. Within the solution as a whole, electroneutrality must be preserved; that is, there will be no build-up of an excess of either plus or minus ions in the solution. If we are to consider conductance through the solution and if we now focus our attention on the migration of this K^+ ion toward the negative electrode under the duress of the applied electric field, then we may expect a certain amount of "drag effect," that is, a holding back of the ion because of the excess of negative ions around it. It is reasonable that this "drag effect" should assume greater importance the more concentrated the solution. The nearer together the plus and minus charges are, the greater will be the "drag effect" and conversely the farther apart they are, the less it will be. It is quite reasonable that the equivalent conductance would diminish with increasing concentration.

A second effect also operates. An ion moving through solution tends to carry the dipolar solvent molecules along with it. Thus positive ions are carrying solvent one way and negative ions carry solvent the other way. There is a "frictional drag" of solvent flowing past solvent.

Giving these two effects quantitative expression, Onsager was able to show that in dilute solutions the dependence of equivalent conductance on concentration should be of the form $\Lambda = \Lambda_0 - B\sqrt{C}$, where B is a constant. *This is identical to Eq. (5.14) which was suggested from the data.*

In comparing electrolytes as different chemically as HCl on the one hand and $CsNO_3$ on the other, it would seem that their only common trait is that each represents a system of unit plus and minus charges. We could not account for the data of Table 5.5 by saying that these are only partially ionized molecules, because if that were true, then surely we would not expect the degree of this postulated partial ionization to be the same for the widely different chemical species represented. We postulate 100% ionization, and we explain the deviation from the limiting values of either "i" as manifested in the freezing points or Λ as manifested in conductance, as coming solely from this electrostatic effect of the ion charges on each other. *The fundamental structural unit of a strong electrolyte is, then, not a molecule but an ion.*

5.10 ELECTROLYTES IN THE SOLID STATE

If we agree that *strong* electrolytes in solution are *completely* ionized, what is their structure in other environments? Consider HCl; it is a gas when pure. Vapor-density measurements establish the molecule HCl as the unit particle. What of the salts (e.g., NaCl, KNO_3, $BaSO_4$) which are normally solids and may be vaporized only with great difficulty? The simplest important observation is that *fused* salts conduct electric current in a manner similar to electrolyte solutions. It would appear that *melting* the *pure salt* makes available free ions to conduct current. Perhaps it is not a great inferential leap to suggest that the ions are also present in the *solid itself*. The most direct evidence on this point comes from study of the x-ray diffraction patterns of crystalline electrolytes (see Chapter 3). The arrangement of atomic centers in the crystals may be inferred from x-ray data. Sodium chloride is a typical case. Figure 5.5 shows the arrangement. Note that each chloride is surrounded by *six* sodiums equally spaced, and each sodium is surrounded by six chlorides equally spaced. There is no identifiable chloride which is the specific partner of a particular sodium. There is no sodium which is identifiable as the specific partner of a particular chloride.

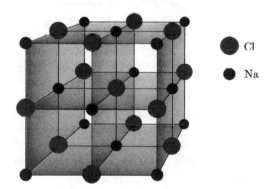

Fig. 5.5. Diagram of the atomic disposition in a sodium chloride crystal.

Thus there would seem to be *none* of the sodium and chloride *pairings* which would be recognized as *molecules*. Again, the independent sodium and chloride *ions* appear to be the fundamental structural units. Perhaps it is appropriate to conclude with a brief remark from W. G. Palmer's, *A History of the Concept of Valency to 1930.*

After nearly a century of atoms the (ionic) theory added ions to the chemists' repertory of the constituent units of chemical compounds . . . and . . . brought the electrical nature of chemical union finally and permanently into the forefront of chemical theory.

5.11 ACID-BASE NEUTRALIZATION: NET IONIC EQUATIONS

In Chapter 2 we discussed Liebig's theory of the reactions of acid plus base to yield salt plus water. He suggested that the key feature was the presence of a reactive hydrogen. Some typical reactions of acids with bases are:

$$HCl + NaOH \rightarrow NaCl + H_2O,$$
$$2HNO_3 + Ca(OH)_2 \rightarrow Ca(NO_3)_2 + 2H_2O,$$
$$HOAc^* + KOH \rightarrow KOAc^* + H_2O.$$

All of the acids, bases, and salts here are strong electrolytes except CH_3COOH. One of the important bits of supporting evidence that Arrhenius cited for the dissociation hypothesis was that the quantity of heat evolved per mole of H_2O formed when a strong electrolyte acid reacted with a strong electrolyte base was a *constant*. That is, HCl, HNO_3, etc., react with NaOH or KOH, or $Ba(OH)_2$ giving the *same heat evolution* per mole of H_2O produced. This is suggestive of the same reaction in *all cases*. That is:

$$H^+ + OH^- \rightleftharpoons H_2O.$$

This represents the *net* ionic reaction for any strong acid reacting with any strong base. The other ions of the strong electrolytes involved do not experience any chemical change. They may be called spectator ions. If, for example, the reaction is HCl + NaOH, it should be written:

$$H^+ + Cl^- + Na^+ + OH^- \rightleftharpoons Na^+ + Cl^- + H_2O.$$

It is clear that there is *no good reason* to include Na^+ and Cl^- ions in the equation *on both sides*. In fact, it is a convention which will be adopted *from this point on* to write only *net ionic equations*, which show only those species experiencing chemical change. Thus we shall omit the spectator ions in the equation for the net reaction. Note that for the *weak* electrolyte HOAc, the heat evolution on "neutralization" (reaction of the acid with a base) is a little different. Here there are acid molecules, and the net reaction is better written:

$$HOAc + OH^- \rightleftharpoons OAc^- + H_2O.$$

At this point, we may bring up-to-date the definitions of the important categories of chemical substances denoted by the terms "acid" and "base." An early definition might be: acids—"the sour principle"—react with bases to form salts—"which taste salty"—and water. A definition of this sort provides a way of recognizing acids and bases in the laboratory. If the categories can be recognized, a modern chemist will seek a structural interpretation. The

---

* The HOAc, KOAc, and OAc⁻, used here, represent acetic acid, potassium acetate, and the acetate ion respectively.

first structural interpretation that was on the right track was Liebig's, mentioned in Chapter 2 in connection with carboxylic acids.

Real success, however, was delayed until the development of the theory of electrolytes. Arrhenius suggested that any substance which produces H^+ ions is an acid, and any substance which produces OH^- is a base. (See the chemical equations at the beginning of this section.) H^+ reacts with OH^- to form water, and the remaining spectator ions (e.g., Na^+ and Cl^-) are the resulting salt. This is entirely satisfactory logically. In the main it is a matter of definition. But, when NH_3 is dissolved in water it behaves as a weak electrolyte base ($NH_3 + H_2O \rightleftharpoons NH_4^+ + OH^-$), and aqueous solutions of NH_3 react with aqueous solutions of HCl to give water and the salt $NH_4^+ + Cl^-$. This product also forms without the intervention of water in the reaction of gaseous NH_3 with gaseous HCl:

$$NH_3(g) + HCl(g) \rightleftharpoons NH_4^+ Cl^- \text{ (solid)}$$

as you are likely to witness as the white dust in any elementary chemistry laboratory. Is there a way to formulate the definition of acid and base to include this reaction as well as the one in water?

The answer lies in a generalization of the definition of acid and base that clearly recognizes that ions in solution are strongly associated with solvent molecules. We saw above that interaction of ions with polar solvent molecules was crucial to keeping them in solution. Suppose we write the dissolution of HCl (gas) in water as:

$$HCl + H_2O \rightleftharpoons H_3O^+ + Cl^-.$$

This emphasizes formation of solvated H^+ as a transfer of a *positively charged* hydrogen atom (H^+) from Cl^- to H_2O. The positively charged hydrogen atom is usually called a *proton*. The origin of this term is discussed in Chapter 7.

Now, one warning should be entered. H_3O^+ is not necessarily the correct formula for the solvated H^+ ion. It isn't clear that *only one* water molecule is strongly attached. The formula $H_9O_4^+$ in which four water molecules are attached has been persuasively advocated by M. Eigen of Gottingen, Germany, in part of his work on very fast chemical reactions for which he shared the Nobel Prize in chemistry in 1967. The formula H_3O^+ is only a convenient convention no more "incorrect" than our custom of writing Fe^{3+} and Ba^{2+} without indicating at all the solvent molecules with which they interact.

To return to the problem of the definition of acid, the definitions usually attributed to the Danish chemist Brønsted are widely used in the chemistry of electrolytes in solution. According to these definitions an acid is a proton donor and a base is a proton acceptor. Thus a reaction between an acid and a base is considered to involve the transfer of a positively charged hydrogen atom (or proton) from the acid to the base.

In Table 5.6 are given several reactions, which occur in aqueous solution.

Table 5.6

Acid-base reactions

1	2	3	4
$HOAc$ + H_2O $\rightleftharpoons$ H_3O^+ + OAc^-
H_2O + NH_3 $\rightleftharpoons$ NH_4^+ + OH^-
HCO_3^- + OH^- $\rightleftharpoons$ H_2O + CO_3^{2-}
NH_4^+ + H_2O $\rightleftharpoons$ H_3O^+ + NH_3
$acid_1$ + $base_2$ $\rightleftharpoons$ $acid_2$ + $base_1$

It is seen that each of the ionic or molecular species in column 1 donates a proton to the ion or molecule in column 2. Thus the species in column 1 are to be considered acids and those in column 2 are bases. We may also consider the reverse reactions in which the bases of column 4 receive protons from the acids of column 3. Using the first equation to illustrate, we refer to OAc^- as the conjugate base of the acid $HOAc$, and we refer to H_2O as the conjugate base of the acid H_3O^+. We call the system, as a whole, a conjugate acid-base system. Note that these equations are stripped of all spectator ions; in other words they express whatever change is occurring without introducing any nonparticipants.

It is fair to say that we might be interested in the spectator ions. For instance in the third reaction we would have introduced some positive ions to preserve the electroneutrality of the solutions (see the Appendix to Chapter 6). We might have used solutions of sodium bicarbonate $NaHCO_3$ and sodium hydroxide $NaOH$ to produce water and normal sodium carbonate Na_2CO_3. If we were to evaporate much of the water from the solution after mixing the reactants, then white crystals of Na_2CO_3 would appear when the concentration of the product solution reached a critical value which would represent the solubility of Na_2CO_3.

Problem 5.10 Identify the acid and base species on the left and the conjugate on the right of the following equations. Supply plausible *spectator ions* for any equations which are in the stripped-down net ionic equation form. What salts could you then obtain by evaporation of water?

a) $HCN + OH^- \rightleftharpoons CN^- + H_2O$

b) $CH_3NH_2 + H_3O^+ \rightleftharpoons CH_3NH_3^+ + H_2O$

c) $HSO_4^- + OH^- \rightleftharpoons SO_4^{2-} + H_2O$

d) $NH_3 + H_3O^+ \rightleftharpoons NH_4^+ + H_2O$

e) $H_2O + H_2O \rightleftharpoons H_3O^+ + OH^-$

(Conductivity measurements show that reaction (e) goes only to a very slight extent. In pure water at 25°C, the ion concentrations are $[H_3O^+] = [OH^-] = 10^{-7}\ M$. Note the absence of any "salt" in this case.)

READINGS AND EXTENSIONS

An excellent comprehensive review of the properties of electrolyte solutions is available in E. J. Margolis, *Formulation and Stoichiometry*, Appleton-Century-Crofts, New York, 1968. The subject is also developed in many textbooks of qualitative analysis, where it is treated along with the subject matter of Chapter 6 of this book to provide a theoretical framework for treatment of the chemistry of common inorganic species in aqueous solution. An example of such a book which is very careful in its treatment of the theory developed here is E. J. King's *Qualitative Analysis and Electrolytic Solutions*, Harcourt, Brace, and Company, New York, 1959.

Seminar: Coordination Compounds

We have examined the two basic kinds of chemical species, molecules and ions. It's worth noting that some of our ions were *molecular* in the sense of a polyatomic grouping which might be internally represented by a molecular bonding pattern similar to those in carbon compounds. Some examples are NO_3^-, SO_4^{2-}, and CO_3^{2-}:

$$\begin{bmatrix} O \\ \parallel \\ N \\ \diagup \quad \diagdown \\ O \qquad O \end{bmatrix}^{-} \quad \begin{bmatrix} O \\ \parallel \\ C \\ \diagup \quad \diagdown \\ O \qquad O \end{bmatrix}^{2-} \quad \begin{bmatrix} O \\ \parallel \\ O-S=O \\ \mid \\ O \end{bmatrix}^{2-}$$

There exists a family of even more complicated substances called "coordination," or sometimes "complex," compounds. The development of the structural theory for these requires the application of the techniques of reasoning displayed in Chapters 2 and 3 complemented by those in Chapter 5.

A typical example of these species is one that may be written $CoCl_3 \cdot 6NH_3$. It is complex in the sense that it is an adduct of two species which separately conform to normal valence rules, $CoCl_3$ and NH_3. Some reaction patterns of cobalt "complex" compounds are collected below. They are all compounds with the metal in the "plus-three oxidation state" which, with respect to its *charge*, is to be considered similar to a Co^{3+} ion. Some results on solution conductivity are also provided. Propose a structural theory for these compounds with a careful account of the way your structural theory would explain all the data given. It would be a good idea to discuss other possible theories and to suggest additional experiments which would help to support the interpretation you give these observations.

1. The following "complexes" may be formed by the Co^{3+} ion and NH_3 or ethylenediamine (en, $= H_2NCH_2CH_2NH_2$, which appears to behave like the equivalent of two NH_3 molecules).

Formula	Color	Formula
$CoCl_3 \cdot 6NH_3$	yellow	$CoCl_3 \cdot 3en$*
$CoCl_3 \cdot 5NH_3$	purple	—
	red	$CoCl_3 \cdot 2en \cdot NH_3$⎱ Two
	slightly deeper red	$CoCl_3 \cdot 2en \cdot NH_3$⎰ isomers
Two ⎧$CoCl_3 \cdot 4NH_3$	green	$CoCl_3 \cdot 2en$ ⎱ Two
isomers ⎩$CoCl_3 \cdot 4NH_3$	violet	$CoCl_3 \cdot 2en$*⎰ isomers
$Co(NO_2)_3 \cdot 3NH_3$	yellow	

* These compounds can be resolved into optically active enantiomers by resolution using the D-tartrate ion.

2. Some reaction patterns of the compounds in (1) include:

A. $CoCl_3 \cdot 6NH_3$ $\xrightarrow{\text{Concentrated HCl}}$ No reaction*

$\xrightarrow{\text{Moist Ag}_2O}$ $3AgCl(ppt) + Co(OH)_3 \cdot 6NH_3$
a strong base
similar to NaOH

B. $CoCl_3 \cdot 5NH_3$ $\xrightarrow{\text{Moist Ag}_2O}$ $CoCl(OH)_2 \cdot 5NH_3$ (if separation completed rapidly)

$\xrightarrow{H_2O}$ $CoCl_3 \cdot 5NH_3 \cdot H_2O$ ΔT_f study suggests $i = 3.7$

Slowly heat solid sample

C. $CoCl_3 \cdot 4NH_3$ $\xrightarrow[\text{then freeze liquid}]{\text{Dissolve in H}_2O}$ $CoCl_3 \cdot 4NH_3 \cdot H_2O$ $\xrightarrow[\text{then evaporate to dryness}]{\text{Dissolve in H}_2O}$ $CoCl_3 \cdot 4NH_3$
Green   Pink   Violet

3. Some values of molar conductivities:

Compound	Approximate $\Lambda_0(ohm^{-1}\ cm^{-2})$
$CoCl_3 \cdot 6NH_3$	400
$CoCl_3 \cdot 5NH_3$	230
$CoCl_3 \cdot 4NH_3$	100
$Co(NO_2)_3 \cdot 3NH_3$	1.6
NaCl	120
$BaCl_2$	260
$CeCl_3$	408

* Recall $NH_3 + HCl \rightarrow NH_4^+ + Cl^-$.

Useful Reading

F. Basolo and R. Johnson, *Coordination Compounds*, W. A. Benjamin, New York, 1964. Chapter 1 and Section 3.4.

D. F. Martin and B. B. Martin, *Coordination Compounds*, McGraw-Hill, New York, 1964. Chapters 2 and 3.

H. J. Emeleus and J. S. Anderson, *Modern Aspects of Inorganic Chemistry, 3rd. Ed.*, Routledge and Kegan Paul, London, 1960. Chapter 5.

R. K. Murmann, *Inorganic Complex Compounds*, Reinhold, New York, 1964. Chapter 1.

CHAPTER 6

Electrochemistry and Energetics of Chemical Reactions

In the previous chapter our efforts to understand the structural difference between substances like sodium chloride and substances like ethanol led us into an extended discussion of the *charge carriers* in electrolyte solutions and the behavior of the electrical resistance of those solutions. In doing this, we bypassed some of the earlier discoveries of electrochemistry. We shall return to a consideration of them now because they can help to develop an insight into the *physical nature* of chemical bonds and because they can lead to the development of a connection between chemical reactions and the concept of *energy*.

6.1 FARADAY'S LAWS OF ELECTROLYSIS

In everyday experience we encounter flashlights, portable radios, terrifyingly mobile toys, and even portable razors. All of these are operated by batteries. Batteries are the most obvious examples of the connection between chemical reactions and electricity. Batteries use chemical reaction to *produce* electrical effects. A discussion of batteries and related devices forms the main subject of this chapter. To set the stage for the discussion, it is useful to recall a key characteristic of battery-operated electric circuits. A battery causes electric charge to move through a wire or electrical device. As is fairly obvious in battery-operated devices, there is a *direction* to this flow. If the battery is connected via the wrong terminals, the device does not work. Thus we are dealing with *direct current* (which was avoided in the last chapter). Direct currents are conventionally considered to be a flow of so-called *positive* charge from the positive terminal of the battery to the negative. All problems concerning the application of battery circuits may be solved using such an approach, but we shall see that it is equally satisfactory to regard current flow in the wire as a flow of *negative* charge from the negative to the positive battery terminal. These two viewpoints are equivalent and indistinguishable in the context of the experiments to be described. The latter is chosen for reasons to be explained in Chapter 7.

The current must flow inside the battery as well as outside for a complete circuit to exist. The flow inside is through an electrolyte medium. We have argued in the last chapter for description of the charge carriers as ions. This idea is to be carried over to the present discussion. It was noted in Chapter 5, however, that direct current passing through an electrolyte solution leads to a situation where Ohm's law is not simply applicable. Direct-current experiments were therefore avoided. Now that we know something about the structure of electrolyte solutions, we can try to understand direct-current experiments. The focus now will lie on the *chemical reactions* occurring at electrodes and their relation to two electrical quantities: (1) the amount of *charge* passed (current $\times$ time) and (2) the *voltage* (or amount of work required to pass a unit charge through the circuit).

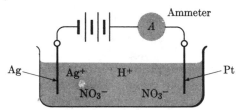

Fig. 6.1. An electrolysis cell.

The effects of passing an electric current through an electrolyte solution were first investigated by the young Michael Faraday as his earliest experimental investigations in electricity (1830's). He explored the behavior of a cell similar to the one shown in Fig. 6.1, which is composed of Ag and Pt plates (electrodes) dipping into a solution containing the Ag^+ ion and H^+ ions and connected to an external source of power. With an applied voltage (requirements varying according to the nature of the metal) a current flows. Bubbles are observed to form at the Pt electrode, and after some time it becomes obvious that the Ag electrode is dissolving. Attaching a battery has caused a chemical reaction to occur. We should explore this reaction quantitatively. If the gas is collected and the Ag electrode dried and weighed, it emerges that the evolution of 0.100 g of H_2(gas) is accompanied by the dissolution of 10.8 g of Ag. That is, the electrolytic process which produces 0.0500 mole of H_2(gas) dissolves just 0.100 mole (gram-atom) of Ag. We recognize in this that the electrolysis of *two monovalent* hydrogen ions to produce one hydrogen molecule was accompanied by the equivalent electrolysis of *two Ag atoms* to produce two monovalent Ag^+ ions. If the experiment is repeated replacing Ag with Cu in the electrode, evolution of 0.100 g of H_2 is accompanied by dissolution of 3.18 g of Cu. The 0.05 mole of H_2 evolved corresponds to 0.05 gram-atom (mole) of Cu dissolved. Electrolytic evolution of two *monovalent* hydrogen ions is accompanied by dissolution of the equivalent one *bivalent* (Cu^{2+}) ion. The relationship indicated here is general. *Chemically equivalent quantities react at the two electrodes.*

Problem 6.1   Compute (a) the number of moles and (b) the weights of the following metal ions produced by electrolysis at the other electrode of an electrolysis which produces 0.0173 g of H_2(gas) at one electrode: Au^{3+}, Hg^{2+}, Pt^{4+}.

Problem 6.2   Electrolytic cells containing zinc sulfate, silver nitrate, and copper sulfate were connected in series and electrolyzed until 1.45 g of Ag had been deposited. (a) How many grams of zinc and copper were deposited? (b) Describe how the hook-up of these cells would differ from that shown in Fig. 6.1.

Problem 6.3   Water may be electrolyzed at a pair of Pt electrodes with H_2 being evolved at the negative electrode and O_2 being evolved at the positive electrode. Predict the ratio of the volume of oxygen to the volume of hydrogen produced.

A second important question arises in these experiments. What is the connection between electrical quantities and the amount of a substance electrolyzed? With the cell illustrated in Fig. 6.1, the simplest experiment to interpret would involve setting up the source so that a constant current (measured with an ammeter) passed through the cell. A given current is found to electrolyze a given weight of silver (a given number of moles) in a given time. It is also found that cutting the current in half leads to doubling of the time required for electrolysis of the given weight of silver. In general, *the product of the constant current times the time of passage is proportional to the weight electrolyzed.* Now, the product of a constant current times the time of passage is a simple quantity. Current is defined as the charge passing a point per unit of time so the product is simply the *total electrical charge* passed (the most common unit of charge is the coulomb). A common way to measure the amount of charge passing through an electrical circuit in which the current is not constant is to measure the amount of a metal (often silver) deposited or dissolved in an electrolytic cell in the circuit referred to as a coulometer. This was actually the way Faraday measured charge in the classic experiments.

Problem 6.4*   It is important to evaluate the quantity of charge required to electrolyze one mole of a monovalent substance. This might be called, with some justice, the "mole of electricity." Dissolution of Ag is a good specific case and was used in recent (1960) precise determinations at the National Bureau of Standards. In a typical experiment a current of 0.5090 amp (coul/sec) was passed for 7200 sec, and the silver electrode lost 4.095 g in weight. Calculate the number of coulombs required to dissolve one mole of silver (or any other monovalent metal). This quantity is called the *faraday* (symbol F).

Problem 6.5   On the basis of the above result, calculate the number of moles of metal and Cl_2(gas) that would be formed if a current of 2.50 amp were passed for 45.0 sec through each of the following molten salts: (a) KCl, (b) $SnCl_2$, (c) $AlCl_3$.

Faraday summarized the regularities that are observed in electrolysis experiments in two laws.

---

* The solution of this problem is important as a preparatory step for the reading of the rest of this chapter.

1. *The quantity of a chemical substance electrolyzed is proportional to the amount of charge* (number of coulombs) *passed.*

2. *The weights of different elements electrolyzed when a given amount of charge is passed are in the ratio of the equivalent weights of those elements.*

These laws are of great significance in the Daltonian context. They have something to say about fundamental *units*, just as the laws of definite and multiple proportions do.

6.2 THE ELECTRON—A FIRST APPROACH

Michael Faraday's work on electrolysis was carried out in the 1830's. This was well prior to Cannizzaro's clarification of the problem of the determination of atomic weights, and Faraday was one of those who was skeptical of the atomic theory. Yet he was clearly aware that his result that a fixed quantity of electricity liberated one equivalent weight (or what we would call a mole of a monovalent substance) of any species implied a strong connection between electricity and the atomic theory. In his *Experimental Researches in Electricity* (1839) he closed the discussion of the laws of electrolysis with the remark:

The equivalent weights of bodies are simply those quantities of them which contain equal quantities of electricity. Or, if we adopt the atomic theory . . . the atoms of bodies which are equivalent to each other in ordinary chemical actions, have equal quantities of electricity associated with them.

Faraday here observes that the statement of his second law in the form, electrochemical deposition of one equivalent weight of any element requires a fixed amount of charge, may be interpreted by assuming that there is associated with one valence unit a definite charge. Translating this into the terms of Daltonian theory, we may say that the *amount of charge associated with one valence unit on any atom* (of any kind) *is the same. There is a microscopic unit of charge associated with the valence of atoms.*

Faraday's skepticism seems to have caused him to withdraw from following this line of reasoning to its conclusion. In 1874, however, G. Johnstone Stoney read a paper before the British Association for the Advancement of Science in which he took the final step. Stoney suggested that the reason that a fundamental unit of charge is associated with the valence unit of an atom in electrolysis is that charge itself comes in "atoms." (From a rough approximation for Avogadro's number obtained from the kinetic theory of gases he obtained 10^{-20} coul as the value of the charge unit—about $\frac{1}{16}$ of the currently accepted value.) The argument for an atom of electricity derived from Faraday's laws was repeated in 1881 by the great German physicist Helmholtz in a lecture before the Chemical Society of London. This latter presentation drew much wider attention in the contemporary scientific community than Stoney's paper, but it was Stoney who gave us the name used for this electrical atom—the *electron.*

Problem 6.6   Using the results of Problem 6.4 and Avogadro's number, calculate the magnitude of the charge of the electron.

In subsequent sections of this chapter, we shall employ the idea of an electron to discuss a scheme of classification of valence which is very useful for electrolytes and some other substances. We shall even go so far as to assume that the *atom of electricity has a negative charge*. It should be clearly understood that this involves jumping well ahead of what we have established. The discussion of the electron so far rests on a single line of evidence. The demise (in the nineteenth century) of a particle theory of light, which had explained many experiments, illustrates reasons to be skeptical of a particle theory of electricity based on a single line of evidence. We shall return to the problem in Chapter 8 and consider further justification of an atomic picture of electricity.

6.3 GALVANIC CELLS

Consider a chemical reaction that occurs spontaneously. If a strip of zinc metal is dipped into a beaker containing copper sulfate solution, a chemical reaction will take place. The zinc will "dissolve," and copper metal will plate out according to the equation

$$Zn(\text{metal}) + Cu^{2+}(\text{aq}) \rightarrow Zn^{2+}(\text{aq}) + Cu(\text{metal}). \tag{6.1}$$

Such reactions are actually very closely related to the electrolytic processes we considered in Section 6.1. This is seen if advantage is taken of a remarkable property of these reactions. The process of plating out the copper may be *separated in space* from the process of "dissolution" of the zinc. An appropriate apparatus is shown in Fig. 6.2. When a wire connects the zinc and copper strips with an ammeter in the circuit (to detect a current flow), the zinc electrode "dissolves," and copper is plated out on the copper electrode. Analysis reveals that the concentration in solution of copper sulfate decreases, and the concentration of zinc sulfate increases. The chemical reaction is accompanied by a flow of current in the external circuit as evidenced by the reading of the ammeter.

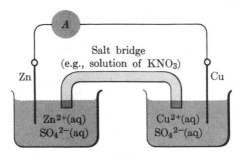

Fig. 6.2. A galvanic cell.

The overall chemical reaction (6.1) has been divided into two "half"-reactions occurring in separate beakers. In one, Zn(metal) loses two electrons to form Zn^{2+} ions [Eq. (6.2a)]. In the other, Cu^{2+} ions acquire two electrons to form Cu(metal) [Eq. (6.2b)]. The electrons flow from their source to the site at which they are required through the wire as the ammeter reading indicates, giving the overall (sum) reaction [Eq. (6.2c)] equivalent to (6.1):

$$Zn \rightleftharpoons Zn^{2+} + 2e^-, \tag{6.2a}$$

$$Cu^{2+} + 2e^- \rightleftharpoons Cu, \tag{6.2b}$$

$$\overline{Cu^{2+} + Zn \rightleftharpoons Cu + Zn^{2+}.} \tag{6.2c}$$

The overall reaction is one in which two species have changed charge type, one gaining electrons, the other losing electrons. Such charge transfer reactions (called redox processes for reasons to be explained) may be broken down into two half-reactions. These may (at least in principle and usually in practice) form the two parts of a cell producing a current flow in an external circuit. The tube connecting the two compartments and labeled a *salt bridge* is an important part of this apparatus. As shown in Chapter 5, the circuit is completed through the solution by *migration of ions*. For a *complete* electrical circuit, the excess of *positive charge* (e.g., Zn^{2+} ions) generated near the Zn electrode must flow toward the Cu side and the *excess* of negative ions (e.g., SO_4^{2-}) generated near the Cu electrode, as Cu^{2+} ions are *used up*, must flow back. A complete circuit through the solution is essential for the cell to function. Removal of the salt bridge *stops* the flow of current in the wire. The cell is *incapable* of building up *measurable* excesses of positive and negative ions in *one compartment*. (This is called the *principle of electroneutrality*.) The ions used to complete the circuit are chosen to be chemically *inert* to prevent the mixing of the reactants from the two sides.

This set-up is an example of a galvanic cell and is the opposite case to the electrolytic cell of Section 6.1. In the cell described in Section 6.1, an electric current *impressed* from an external source produced a chemical reaction. In this case the *spontaneous* occurrence of a chemical reaction produces an electric current in an external circuit.

Problem 6.7

a) Describe what would be expected to happen to the current flow if the Cu^{2+} solution were brought into direct contact with the zinc electrode by means of a stirring device which circulated solution through the connecting tube.

b) In the completed circuit of Fig. 6.2 what do the results of Chapter 5 suggest are the species in the solution which carry the current?

c) Explain in your own words the reason for separating the two "reactant" compartments by the tube containing an unreactive electrolyte (in this case KNO_3), known as a "salt bridge."

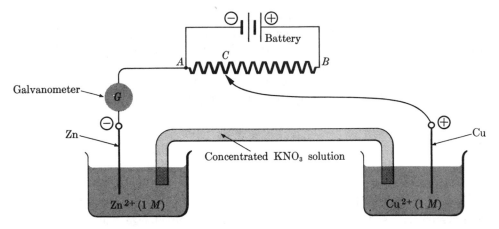

Fig. 6.3. Modified Zn-Cu cell equipped with potentiometer.

Now, an electric current may be employed to do useful work, for example, by using it to drive an electric motor. We have here a connection between a naturally occurring (spontaneous) chemical reaction and mechanical energy. This is a situation which deserves close analysis. After all, even the language of contemporary popular culture reflects the conviction that energy is a very fundamental physical quantity.

The difference between the galvanic cell in Fig. 6.2 and the electrolytic cells of Section 6.1 may be regarded as a difference of degree, not a difference of kind. Suppose we reconstruct the apparatus for the galvanic cell so that it is possible to attach a source of variable voltage in opposition to the spontaneous direction of current flow (i.e., try to make the cell an electrolytic cell with the reaction running opposite to its direction in a galvanic cell). Such a device is shown in Fig. 6.3. For the quantitative considerations to follow, solution concentrations are important. We choose values near $1\ M$ for all species.* The variable voltage applied to the cell is obtained by applying a voltage from a battery across a resistance AB and attaching leads from the cell to point A and to a sliding contact, C. The voltage applied to the cell is the voltage across AC, which may be calculated if we know the resistance of segments of AB. The current flowing through AB is determined by the voltage of the battery, and the total resistance of AB through use of Ohm's law [E(volts) $= I$(amps) $\times R$(ohms)]. A second application of Ohm's law gives the voltage applied to the cell if the current through AB (and therefore AC) is known from the first application of the law and the known resistance of AC. Note that the half-reaction Zn + $Zn^{2+} + 2e^-$ in the galvanic cell suggests that negative charge is built up at the Zn electrode and that it is the negative terminal of the cell. The Zn terminal

---

* We say near $1\ M$ because the numbers to be cited refer to an idealized $1\ M$ solution. The difference may be ignored in this discussion and $1\ M$ regarded as a standard concentration.

is connected to A, which is connected to the negative terminal of the battery. Thus the variable applied voltage is attached to the galvanic cell negative to negative and positive to positive or in *opposition*. The galvanometer in the circuit is a sensitive detector of a current flow. It can deflect in either direction showing reversals of the direction of current flow.

Starting observations on the apparatus with the variable contact pushed up against A (zero applied voltage), we observe a current flow indicating electrons flowing from zinc to copper. Chemical changes could be detected by analysis of electrodes and solution to indicate dissolution of zinc to form zinc ions and plating of copper ions to form copper. The quantities reacted could of course be calculated using Faraday's laws from observation of current and time. As the applied voltage is increased by moving C to the right, the galvanometer deflection decreases indicating a smaller current flow. When the applied voltage reaches a value of 1.1 volts, the galvanometer reads *zero*. The current flow has ceased. But, this means that *the chemical reaction must also have stopped*. Its tendency to proceed spontaneously is limited. If, in order to occur, it must drive current through the external circuit *against a voltage of* 1.10 V or more, it cannot proceed. Now this observation may be rephrased in terms of an energy criterion. The concept of voltage is very closely related to electrical work (a form of energy of course). The volt may be defined in terms of the amount of work required to transport charge between two points in a circuit. If one joule [the standard metric unit of energy—force (in newtons) × distance (in meters)] is required to transport one coulomb of charge between two points, the voltage (or potential difference) between those points is said to be one volt. Thus, work done = volts × coulombs. The maximum voltage against which our reaction will transport charge is 1.10 V. If we wish to convert one mole of Zn(metal) into Zn^{2+} ions in a 1 M solution at the expense of converting Cu^{2+} ions in 1 M solution to a mole of copper metal, Faraday's laws show that we must transport two faradays of charge or $2 \times 96,500$ coul. The *maximum work*, then, that we can make the reaction do is:

$$1.10 \text{ V} \times 2 \times 96,500 \text{ coul} = 212,000 \text{ joules,}$$

or 212,000 joules *per mole*. This is the energetic limit on the spontaneously occurring reaction. No more energy can be derived from the spontaneous tendency of the reaction to go.

If the applied voltage on the cell in Fig. 6.3 is increased beyond 1.10 V, the galvonometer now shows a deflection in the opposite direction. The original current flow has been reversed. The cell is now operating *electrolytically*. An external applied voltage is producing an opposite current flow, *and the chemical reaction must now be reversed*. If an external source of energy with a capacity in excess of the maximum ability of the spontaneous process to do work is applied, the reaction may be forced backward, and this is the essence of the electrolytic cell.

The applied potential which a galvanic cell will exactly balance, the potential reflecting the limit on the amount of work that a spontaneous chemical reaction can be harnessed to do, is called the potential for the reaction, and if all species in solution are at 1 M concentration, it is called the *standard potential* (usually designated ε^0). We write that the reaction $Zn + Cu^{2+}$ (1 M) → $Cu + Zn^{2+}$ (1 M) has the standard potential (ε^0) of +1.10 V. The *positive sign* indicates that the reaction is spontaneous *in the direction written*. If we wrote the reaction Zn^{2+} (1 M) $+ Cu → Cu^{2+}$ (1 M) $+ Zn$, this would have a *standard potential* of -1.10 V indicating that the reaction runs *backward* spontaneously. A table of reactions accompanied by a list of standard potentials, then, tells in which direction the reactions tend to proceed when all reactants and products in solution are brought together at 1 M concentration in contact with appropriate solid reactants and products.

One important question remains in this first discussion of galvanic cells and work obtainable from chemical reactions. We have talked about the maximum amount of work a reaction could be made to yield when it is opposed and just balanced by an applied potential. Does the reaction yield this much work when it is not significantly opposed? (We must again recall that work is volts times coulombs.) Clearly, the amount of charge required to dissolve one mole of Zn and plate one mole of Cu does not change. It must be twice Faraday's number of coulombs. But the voltage may change. Suppose that we just put a voltmeter across the cell in Fig. 6.2 which is driving a large current in the external circuit as indicated by the ammeter. Would the voltmeter read 1.10 V? An experiment shows that, in fact, the reading is a little less. The reaction running along smoothly with little opposition does less work than the maximum of which it is capable. It is for this reason that the balancing procedure is important for determining the limit of the tendency of the reaction to proceed. *The unbalanced running reaction is always a little inefficient.*

6.4 OXIDATION-REDUCTION REACTIONS

In the first three sections of this chapter we have introduced the idea of an atom of electricity, the electron, a new aspect of chemical equivalence (Faraday's laws), and a measure of the energy output of a chemical reaction. All of these ideas have emerged from an examination of the class of chemical reactions in which species change charge, the reactions which may be run in electric cells. Before we pursue the *extensive* consequences of the new ideas, we must pause to develop a systematic terminology for cells and their associated reactions.

Each of the half-reactions in an electrochemical cell involves either an increase or decrease of positive charge on some species. Any reaction in which a species *loses electrons* and *increases in positive charge* is termed an *oxidation*. The paired reaction in which the lost electrons are consumed by some other species with a corresponding *decrease in positive charge* is called a reduction. Since phenomena in solution suggesting the presence of free electrons are not observed, it follows that every oxidation must be accompanied by a reduction. The electrons lost in oxidation must be used for a reduction. The class of reactions involving oxidation and reduction, whether run by simply mixing reagents or in an electrochemical cell, are often called *redox* reactions. Equations (6.3) give several examples of redox reactions (The product which has undergone oxidation is marked O. The product which has suffered reduction is marked R.):

$$\overset{\text{O}}{} \overset{\text{R}}{}$$
$$2Na + Cl_2 \ \rightarrow \ 2Na^+ + 2Cl^-,$$

$$2Cu^{2+}(aq) + 2Br^- \ \rightarrow \ 2Cu^+(aq) + Br_2,$$

$$Sn^{2+} + 2Fe^{3+} \ \rightarrow \ Sn^{4+} + 2Fe^{2+}, \tag{6.3}$$

$$Fe + 2Ag^+ \ \rightarrow \ Fe^{2+} + 2Ag.$$

Problem 6.8 Consider the group of equations (6.3). Break each one down into half-reactions, like those in Eq. (6.2), showing electron loss and gain. Check that each pair of half-reactions adds up to the original reactions. Which half-reaction is an oxidation in each case?

In a consistent, and only apparently almost reverse, usage, the species which undergoes oxidation is thought of as causing the reduction of the other and is called a *reducing agent*. The species which undergoes reduction is held responsible for the oxidation of its partner and therefore is called an *oxidizing agent*.

Problem 6.9 Identify the oxidizing agent and the reducing agent in each of the following reactions:

a) $2K + Cl_2 \ \rightarrow \ 2KCl$,
b) $Cd + Pb(NO_3)_2 \ \rightarrow \ Cd(NO_3)_2 + Pb$,
c) $Mg + 2H^+ \ \rightarrow \ Mg^{2+} + H_2$,
d) $Br_2 + 2I^- \ \rightarrow \ 2Br^- + I_2$.

A nomenclature common to both electrolytic and galvanic cells may now be introduced for cell processes. The junction between the external electric circuit and the electrolytic solution, the metal strip in the examples, is always called an *electrode*. In both types of cells, the *electrode* at which electrons *"enter" the external circuit*, and hence the one at which oxidation takes place in the cell, is called the *anode*. Since oxidation is increasing positive charge, in the solution, negative ions or *anions* must be migrating toward this electrode for the solution to remain electrically neutral with equal numbers of particles of each charge in all regions. The electrode at which electrons *"enter" the solution*, and hence the one at which reduction is occurring, is called the *cathode*. To maintain electro-neutrality in the face of reduction, in the vicinity of the *cathode*, positive ions

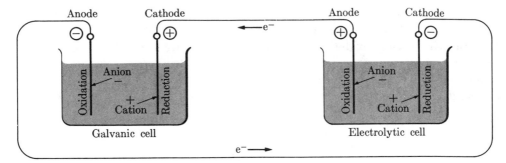

Fig. 6.4. Nomenclature for electrochemical cells.

or *cations* must be migrating toward the cathode. These usages are shown on the diagram in Fig. 6.4, which shows a galvanic cell used as a power source to drive an electrolytic cell. Note that the sign of the charge on an electrode reverses on going from a galvanic to an electrolytic cell. The galvanic cell has the apparent anomaly of positive ions migrating toward positive electrodes and negative ions migrating toward negative electrodes. But, it must be remembered that the electrical fields are *not* driving the galvanic cell. The spontaneous chemical reaction is producing the electrical effects in the external circuit.

Problem 6.10 Suppose that each of the reactions in Problem 6.8 is to be utilized to construct a galvanic cell. Which half-reaction will be the cathode reaction, which the anode reaction? All but the first might occur in aqueous solution. Suggest a candidate for the migrating cation and anion. (Note that in some cases the migrating ion is not a participant in the electrode reaction and hence does not appear in the equations. It may be one of the spectator ions which are usually omitted.)

6.5 THE OXIDATION STATE CONCEPT OF VALENCE

Faraday's laws of electrochemical equivalence can serve to define the concept of the valence of an element in much the same way as the law of combining proportions was shown to do in Chapter 2. Once again, the fact is that electro-chemical reactions of various substances occur in discrete units; the number of moles of Cu plated by a fixed current is related by a small integer to the number of moles of Ag plated by the same current. The interpretation is the same here as for combining proportions. We regard the quantity of electricity as fixed by the characteristic number or combining power of the element. But the fundamental structural unit in a strong electrolyte solution is *not a molecule but an independent ion. The valence number corresponds to the number of units of electric charge on the ion* and not the number of *chemical bonds* in a molecule. Since a well-established set of atomic weights has been at hand from the begin-ning of this discussion, there is no new information in the valence numbers obtained. The formulas of the compounds were known from the beginning, and the fact that electric charge was associated followed from the conductance results. But confirmatory evidence does come from electrolysis results, which illustrate clearly a situation parallel to the origin of the idea of valence.

It is no more evident *a priori* why Cu atoms normally lose two electrons to become Cu^{2+} ions or Cl atoms normally acquire one to become Cl^- ions than it is evident why carbon forms four bonds directed to the corners of a tetrahedron, but the concept of ionic valence *does* seem physically different from the idea of chemical bonds, despite the common occurrence of the same valence numbers. The number of elementary units of charge on an ion means something different from the number of "bonds" in a molecule. Perhaps this justifies the existence of *two distinct valence scales.* In opposition to valence as the number of bonds formed by an atom in a molecule, we shall now define the *oxidation-number* concept of valence which is a *generalization* of the idea of the number of units of charge on an ion, or ionic valence. The concept was introduced in 1919 by one of the leading figures in early twentieth century inorganic chemistry, Alfred Stock.

When an element, for example, sodium, reacts at an electrode to lose an electron, it has been oxidized. Any electrode oxidation process involves increasing the positive charge on the species. Conversely, any electrode reduction process involves reducing the positive charge (increasing the negative charge) on a species. For simple monatomic ions we *define* the *oxidation state* of the element as the signed number of units of charge on the ion. (That is, the *oxidation state* is the ionic valence number with the appropriate sign included.) Now, what about an ion like NO_3^-? In the ionic solid BaO, oxygen bears the charge -2; therefore we might say that in the NO_3^- ion the three oxygens contribute an overall charge of -6, meaning that nitrogen must bear a charge of $+5$ for the NO_3^- ion to bear only one unit of negative charge. Now, in NO_3^- oxygen is not any longer a free O^{2-} ion, but we may *agree* by convention to continue to regard its oxidation state as -2. If we adopt some conventions for expressing the oxidation states of elements which occur commonly in monatomic ions, *we may then assign oxidation states to all other elements which occur in polyatomic ions or neutral molecules* by simply insisting that the sum of oxidation numbers add up to the overall charge on the species (zero for neutral molecules).

The usual oxidation-state conventions in diminishing order of priority (they cannot always *all* be applied consistently) are:

1. The alkali metals (Li, Na, Rb, Cs) occur in compounds in the $+1$ oxidation state.

2. The alkaline earth metals (Be, Mg, Ca, Sr, Ba) occur in compounds in the $+2$ oxidation state.

3. Hydrogen in its compounds is $+1$.

4. Oxygen in its compounds is -2.

5. The halogens (F, Cl, Br, I) in their compounds are *frequently* -1.

Problem 6.11   Classify the following sulfur compounds according to the oxidation state of sulfur: SF_6, SO_2, SO_3^{2-}, H_2S, H_2SO_4, SO_4^{2-}, ZnS, and SCl_4.

Problem 6.12   Assign oxidation states to each atom in each of the following compounds: $KMnO_4$, Fe_2O_3, V_2O_5, HNO_2, Na_2O_2, LiH.

Problem 6.13   What does the answer to the previous problem suggest about the electrode at which H_2 is evolved in electrolysis of fused LiH?

The oxidation-state concept makes all elements into "as if" ions in their compounds.   Compound formation, in general, is viewed "as if" ionic.   Such a pattern of thought has proved sufficiently felicitous in inorganic chemistry that the uniform conventions for naming of inorganic compounds make basic use of the notion of oxidation state.   In a binary compound, the name of the metal is given with its oxidation state.   The nonmetallic element is given second with the suffix *ide*.   The following examples are illustrative:

VCl_2,   vanadium(II) chloride,

VCl_4,   vanadium(IV) chloride,

Cu_2O,   copper(I) oxide,

CuO,   copper(II) oxide,

HgS,   mercury(II) sulfide.

Beyond the applications in which the oxidation state has even a remote connection to an actual free monatomic ion, the system remains useful as a device for the bookkeeping of charges and balancing oxidation-reduction reactions because one way of expressing the idea that the number of units of charge is *conserved* in a redox process is to say that the sum of oxidation numbers for reactants must equal the sum of oxidation numbers for products.

As an example, let us balance the equation for the reaction

$$MnO_4^- + Fe^{2+} \rightarrow Mn^{2+} + Fe^{3+}.$$

First, let us identify those species undergoing oxidation and reduction and the oxidation states before and after reaction.

$$\overset{\displaystyle \text{Mn(VII) to Mn(II)}}{\overbrace{MnO_4^- + Fe^{2+} \rightarrow Mn^{2+}}} + Fe^{3+}$$
$$\underset{\displaystyle \text{Fe(II) to Fe(III)}}{\underbrace{\phantom{MnO_4^- + Fe^{2+} \rightarrow Mn^{2+} + Fe^{3+}}}}$$

We see that Mn has undergone a decrease in oxidation state of 5, while Fe has been oxidized one unit.   Accordingly, to make the number of electrons equal on both sides, five times as much Fe as Mn must be involved.   We write:

$$MnO_4^- + 5Fe^{2+} \rightarrow Mn^{2+} + 5Fe^{3+}.$$

Now the balancing may be completed by properly accounting for the role of the *solvent* (water) by introducing appropriate H^+, OH^- and/or H_2O to obtain hydrogen and oxygen balance:

$$MnO_4^- + 5Fe^{2+} + 8H^+ \rightarrow Mn^{2+} + 5Fe^{3+} + 4H_2O.$$

Problem 6.14   Balance the following redox reactions occurring in *acid* solution:

a) $MnO_4^- + H_2S \rightarrow Mn^{2+} + S$,
b) $S_2O_3^{2-} + I_2 \rightarrow I^- + S_4O_6^{2-}$,
c) $Zn + NO_3^- \rightarrow Zn^{2+} + NH_4^+$.

(Don't forget presence of H^+ and H_2O in these solutions.)

Problem 6.15   Balance the following redox reactions occurring in *basic* solutions:

a) $Al + NO_3^- \rightarrow Al(OH)_4^- + NH_3$,
b) $N_2H_4 + Cu(OH)_2 \rightarrow N_2 + Cu$,
c) $Fe + Fe(OH)_3 \rightarrow Fe(OH)_2$.

(Don't forget presence of OH^- and H_2O in these solutions.)

6.6 STANDARD HALF-CELL POTENTIALS

Let's now return to cells. Often, a useful kind of electrode is one in which the metal of the electrode itself is inert and not involved in the electrochemical reaction. The cell in Fig. 6.5 includes such an electrode, the hydrogen electrode. Hydrogen gas is bubbled over a platinum (Pt) strip. The Pt strip serves as a connection to the external circuit but is *not involved in the reaction*. The overall reaction which is responsible for this cell producing electric current is the familiar dissolution of zinc by acid:

$$Zn(metal) + 2H^+ \rightarrow H_2(g) + Zn^{2+}.$$

The half-reaction occurring at the inert platinum electrode is the reduction of hydrogen ions:

$$2H^+ + 2e^- \rightarrow H_2(g).$$

The hydrogen electrode is one example of a fairly large class of electrodes which may be constructed in which the actual metal of the strip merely mediates a redox reaction between two species, a gas and a solute or two solutes. The

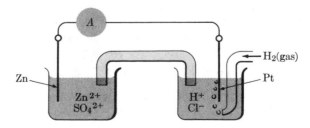

Fig. 6.5. The zinc-hydrogen cell.

potential for a cell involving a gas depends on the pressure of the gas. Conventions dictate the choice of a hydrogen pressure of 1.00 atm if the standard potential is to be measured.

The standard potential (ε^0) of a cell oxidizing zinc at the expense of reducing hydrogen ions is 0.76 V. If we were to construct a corresponding cell in which the zinc strip was replaced by silver, and $ZnSO_4$ in solution by $AgNO_3$ in solution, the second cell would have a standard potential (ε^0) of -0.80 V. That is, the reaction oxidizing silver at the expense of reducing H^+ is not spontaneous. The reaction proceeds in the direction opposite to:

$$Ag + H^+ \rightarrow Ag^+ + \tfrac{1}{2}H_2, \qquad \varepsilon^0 = -0.80 \text{ V}. \qquad (6.4)$$

The spontaneous process is reduction of Ag^+ ions at the expense of oxidation of H_2 molecules, the reverse of Eq. (6.4).

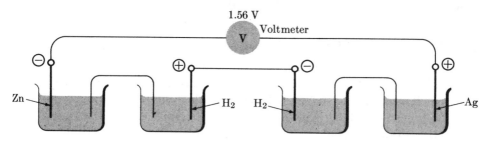

Fig. 6.6. Zn-H_2 and H_2-Ag cells connected in series.

Consider now the potential expected if these two cells are hooked up in series as in Fig. 6.6. As is not at all surprising, the standard potential of the combination is 1.56 V. The potentials are additive. Perhaps we have a right to be a little more surprised that the potential of a cell made up with a zinc electrode and a silver electrode, omitting hydrogen altogether, is still 1.56 V. This second result indicates a more profound kind of additivity. The overall reaction in the final cell is oxidation of zinc and reduction of Ag^+ ions. This reaction is occurring *directly*. In the situation where two cells were coupled in series, zinc was oxidized and H^+ reduced to H_2. Then, H_2 was oxidized back to H^+ and Ag^+ reduced to Ag. Overall H^+ finally wound up as H^+, and the reaction was the same as in the simpler cell. *When the same overall redox reaction has occurred, the detailed pathway is irrelevant to the determination of the standard potential.* To rephrase these verbal results algebraically, reaction (6.5a) and reaction (6.5b) may be added to give reaction (6.5c). A corresponding addition of standard potentials is possible:

$$
\begin{array}{llll}
Zn + 2H^+ & \rightarrow & Zn^{2+} + H_2, & \varepsilon^0 = 0.76 \text{ V}, & (6.5a) \\
H_2 + 2Ag^+ & \rightarrow & 2Ag + 2H^+, & \varepsilon^0 = 0.80 \text{ V}, & (6.5b) \\
\hline
Zn + 2Ag^+ & \rightarrow & Zn^{2+} + 2Ag, & \varepsilon^0 = 1.56 \text{ V}. & (6.5c)
\end{array}
$$

A consequence of additivity is that if a table of potentials of cells with all electrodes coupled to the hydrogen electrode (or any other for that matter) is available, then the potential for *any* arbitrary combination can be calculated. Often it has been interesting to consider a cell as two half-reactions, an oxidation and a reduction. Can any significance be given to separate potentials for the half-reactions? If potentials *are additive*, then the potential of any cell is just the sum of the potentials for its two electrode half-reactions. If, *by convention*, we assigned some particular electrode reaction a standard potential of *zero*, then potentials of cells with that electrode and any other would give standard *half*-cell potentials. These could be added to give the potential of any cell, a procedure which is equivalent to adding two cells in series to eliminate the common electrode which has been assigned the zero potential. In practice, the hydrogen electrode is assigned a standard potential of *zero*, and consequently the fact that

$$Zn^{2+} + H_2 \rightarrow Zn + 2H^+, \qquad \mathcal{E}^0 = -0.76 \text{ V},$$

is interpreted to signify

$$Zn^{2+} + 2e^- \rightarrow Zn, \qquad \mathcal{E}^0_{1/2} = -0.76 \text{ V}.$$

A positive half-cell potential corresponds to an electrode reduction which will occur *spontaneously* at the expense of *oxidation of hydrogen*. A negative half-cell potential corresponds to an electrode oxidation which will occur spontaneously at the expense of the *reduction of* H^+ *to* H_2. Thus the most extensively oxidizable species have the most negative standard potentials, and the most extensively reducible species have the most positive standard potentials. Half-cell potentials based on the hydrogen zero *convention* are collected in Table 6.1. Note that it is conventional to list half-cell potentials as reduction potentials (e.g., $M^{s+} + pe^- \rightarrow M^{(s-p)+}$).

Problem 6.16 Using the table of half-cell potentials, calculate the potential of a cell consisting of 1 M zinc sulfate in contact with a zinc electrode coupled by a salt bridge to a solution of 1 M KI saturated with I_2* and in contact with a Pt electrode. What reaction is occurring spontaneously?

Problem 6.17 Consider the alternate formulations

$$I_2 + 2e^- \rightarrow 2I^-,$$
$$\tfrac{1}{2}I_2 + e^- \rightarrow I^-.$$

Show that the choice doesn't make any difference for the standard potential. [*Hint:* potential is work *per unit charge*.] Hence, show that any half-cell equation may be multiplied by a constant without affecting the potential.

---

* This corresponds to the conditions for which the standard potential is defined.

Table 6.1

Standard half-cell potentials

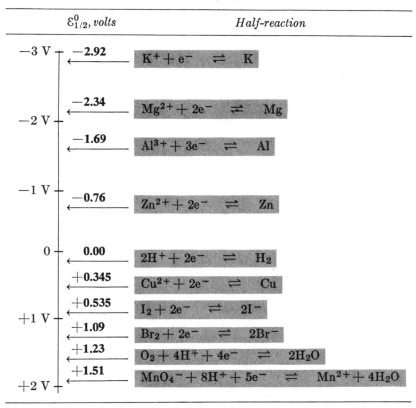

$\mathcal{E}^0_{1/2}$, volts		Half-reaction
-3 V	-2.92	$K^+ + e^- \rightleftharpoons K$
-2 V	-2.34	$Mg^{2+} + 2e^- \rightleftharpoons Mg$
	-1.69	$Al^{3+} + 3e^- \rightleftharpoons Al$
-1 V	-0.76	$Zn^{2+} + 2e^- \rightleftharpoons Zn$
0	0.00	$2H^+ + 2e^- \rightleftharpoons H_2$
	$+0.345$	$Cu^{2+} + 2e^- \rightleftharpoons Cu$
$+1$ V	$+0.535$	$I_2 + 2e^- \rightleftharpoons 2I^-$
	$+1.09$	$Br_2 + 2e^- \rightleftharpoons 2Br^-$
	$+1.23$	$O_2 + 4H^+ + 4e^- \rightleftharpoons 2H_2O$
$+2$ V	$+1.51$	$MnO_4^- + 8H^+ + 5e^- \rightleftharpoons Mn^{2+} + 4H_2O$

Problem 6.18   Calculate the standard potentials for oxidation of Br^-, H_2, Zn, and water with acidic MnO_4^-.

Problem 6.19   All the following reactions might take place in an acidic aqueous solution of potassium bromide:

$$2K^+ + 2Br^- \rightleftharpoons 2K + Br_2,$$
$$2H^+ + 2Br^- \rightleftharpoons H_2 + Br_2,$$
$$2H_2O \rightleftharpoons O_2 + 2H_2.$$

a) Show that all have negative standard potentials.
b) Arrange the reactions in order of the increasing value of an externally applied potential required to cause them to proceed. Determine which products would be produced most easily *electrolytically*.
c) If the standard potential of $Cl_2 + 2e^- \rightarrow 2Cl^-$ is $+1.36$ V, what would be the products of electrolyzing the analogous KCl solution?

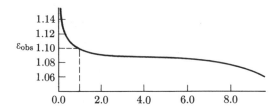

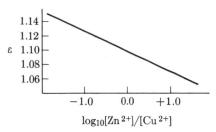

Fig. 6.7. Observed potential of the zinc-copper cell as a function of the concentration ratio $[Zn^{2+}]/[Cu^{2+}]$ at 25°C.

Fig. 6.8. Reconstruction of Fig. 6.7 using log $[Zn^{2+}]/[Cu^{2+}]$ at 25°C.

6.7 THE EFFECT OF CONCENTRATION ON POTENTIALS

So far our discussion has been limited to *standard potentials*. This implies a limitation to cases in which all reactants and products are present either *in solution* at approximately *1 M concentration* or as *pure solid* or *liquid phases* or as gases at *1-atm pressure*. Anyone who has used a battery will have observed one thing about electrochemical cells. They run down. As the reactants are exhausted, the voltage changes. When the chemical reaction no longer proceeds spontaneously (i.e., has reached equilibrium), the potential is zero and the battery is dead. As a result of these considerations, we can argue that *potentials depend on concentration*, but we must turn to *quantitative* experiments to define in detail the nature of this dependence.

Consider, again, the zinc-copper cell which runs spontaneously, with the oxidation of zinc and the reduction of copper. How does this cell vary when Zn^{2+} and Cu^{2+} concentrations are varied? Some qualitative predictions concerning the effects of modifying the concentrations are easy to make. The reaction proceeds toward an equilibrium position with high Zn^{2+} ion concentration and low Cu^{2+} ion concentration. Thus, increasing the $[Zn^{2+}]/[Cu^{2+}]$ ratio moves the system toward equilibrium and should *decrease* the potential observed. Conversely, increasing the $[Cu^{2+}]/[Zn^{2+}]$ ratio should increase the observed potential. The result of a quantitative study of the role of the $[Zn^{2+}]/[Cu^{2+}]$ ratio is shown in Fig. 6.7. Note that the relationship is qualitatively as expected but does not appear to be of a simple form. A generally useful approach to experimental data is to search for a presentation which leads to a straight-line plot. In this case, a straight line is obtained if we choose to plot the logarithm of the $[Zn^{2+}]/[Cu^{2+}]$ ratio. This is shown in Fig. 6.8. The slope of the line is negative (increasing $[Zn^{2+}]/[Cu^{2+}]$ corresponds to decreasing ε). Its magnitude is -0.03 if the temperature is 25°C. Of course, ε is equal to ε^0 (the standard potential) when $[Zn^{2+}]/[Cu^{2+}] = 1.00$ (i.e., log $[Zn^{2+}]/[Cu^{2+}] = 0.00$). These results are easily translated into an equation:

$$\varepsilon = \varepsilon^0 - 0.03 \log [Zn^{2+}]/[Cu^{2+}]. \tag{6.6}$$

Note that zinc and copper metal do not play an explicit role in this equation. So long as they are present in pure solid phases there is nothing corresponding to concentration to vary. Experimentally, the size (i.e., mass) of a pure solid phase is irrelevant to the potential. However, it should be kept in mind that metallic zinc and copper are involved in the reaction and have a role in determining ε^0.

For a generalized oxidation-reduction reaction run at 25°C in which a moles of A react with b moles of B to produce c moles of C and d moles of D:

$$aA + bB \rightarrow cC + dD.$$

The generalization of Eq. (6.6) is

$$\varepsilon = \varepsilon^0 - \frac{0.06}{n} \log \frac{[C]^c[D]^d}{[A]^a[B]^b}, \tag{6.7}$$

where n is the number of electrons used in balancing the half-reactions or, experimentally, the number of faradays required to run one unit of the equation (i.e., to react a moles of A with b moles of B). Note the similarity of the concentration quotient to the one which was suggested in Chapter 5 as the governing equation for the relationship among concentrations at equilibrium.

An equation of the form of Eq. (6.8) was introduced on the basis of theoretical considerations by Walter Nernst in 1889. He was able to interpret the constant (0.06 at 25°C) in terms of more fundamental quantities. Since the logarithm of a concentration ratio is dimensionless, the constant must have the dimensions of potential (volts). Nernst's term gives it as energy/charge. The Nernst equation is:

$$\varepsilon = \varepsilon^0 - \frac{2.3\,RT}{n\mathfrak{F}} \log \frac{[C]^c[D]^d}{[A]^a[B]^b}, \tag{6.8}$$

where RT is the gas constant times the absolute temperature (dimensions of energy) and $\mathfrak{F}$ is the faraday of charge. The factor 2.3 is required to convert the logarithmic term to base 10 or common logarithms, since the theoretical argument leads to an equation in natural logarithms (base $e = 2.718 \ldots$).

Problem 6.20   Consider the reaction $Fe^{3+} + I^- \rightarrow Fe^{2+} + \frac{1}{2}I_2$. Write the half-reactions. Show that the Nernst equation can be written as a sum of terms for the two half-reactions utilizing the standard half-cell potentials. Write the Nernst equation for the Fe^{2+}/Fe^{3+} half-cell.

Problem 6.21   Which of the following oxidizing agents becomes stronger as the H^+ concentration of the solution increases? Which are unchanged? Which become weaker?

a) Br_2,
b) Fe^{3+},
c) MnO_4^-,
d) H^+,
e) $Cr_2O_7^{2-}$ (reacts to Cr^{3+}).

[*Hint:* Look for H^+ in the balanced half-equation for the redox reaction.]

6.8 CELL POTENTIALS AND REACTION EQUILIBRIUM

In Section 6.7 we observed that as a chemical reaction nears its equilibrium position, it is losing its tendency to proceed spontaneously and therefore losing its ability to produce an electrical potential. When the chemical composition of a system reaches the point where there is no further tendency for spontaneous chemical change (reaches chemical equilibrium), the potential of a cell harnessing the reaction must be *zero*. In terms of the Nernst equation for the generalized reaction of a moles of A with b moles of B to give c moles of C and d moles of D, at 25°,

$$\mathcal{E} = 0 = \mathcal{E}^0 - \frac{0.06}{n} \log \frac{[C]^c[D]^d}{[A]^a[B]^b}, \tag{6.9}$$

which may be rewritten:

$$\mathcal{E}^0 = \frac{0.06}{n} \log \frac{[C]^c[D]^d}{[A]^a[B]^b}. \tag{6.10}$$

But, $\mathcal{E}^0$ is a *constant* characteristic of the reaction so the concentration ratio on the right *must also be constant at equilibrium*. There exists an *equilibrium constant* governing the *ratios of concentration at equilibrium*, which has the form:

$$K_{eq} = \frac{[C]^c[D]^d}{[A]^a[B]^b}. \tag{6.11}$$

This result is identical to the form suggested in Section 5.7 as an empirical result for reactions of all types. It is not, in fact, limited to oxidation-reduction processes.

The best way to illustrate the significance of an equilibrium constant is a sample calculation. Consider the reaction in a silver-mercury cell:

$$2Ag^+ + 2Hg \rightleftharpoons 2Ag + Hg_2^{2+}$$

The standard potential for this process is 0.010. Thus we may calculate K_{eq}:

$$\mathcal{E}^0 = (0.06/2) \log K_{eq} = 0.010,$$

$$K_{eq} = 2.15 = [Hg_2^{2+}]/[Ag^+]^2.$$

Note that the pure solid Ag and pure liquid Hg do not appear. They behave as constant concentrations and are assigned a value of unity. Suppose, now, that liquid mercury is brought to equilibrium with a solution initially 1 M Ag^+. Let x represent the concentration of Hg_2^{2+}; the concentration of Ag^+ must be the original value minus $2x$ since 2 moles of Ag^+ are required to produce one mole of Hg_2^{2+}. Thus

$$K_{eq} = \frac{x}{(1.00 - 2x)^2} = 2.15.$$

Solving the resulting quadratic equation we get

$$x = [\text{Hg}_2{}^{2+}] = 0.31\ M.$$

If the initial concentration of Ag^+ ion were made 2.00 M instead, the equilibrium concentration would *differ*. In particular, $x = .72\ M$. But, despite the changes, *the ratio* $[\text{Hg}_2{}^{2+}]/[\text{Ag}^+]^2$ *at equilibrium remains a constant.*

This much was implied in Chapter 5. Note that there is an additional significant fact here. The value of the equilibrium constant is closely related to $\mathcal{E}^0$. The potential which represents the maximum electrical work obtainable from the reaction when all concentrations are 1 M (and quantitatively represents the tendency of the reaction to proceed spontaneously at unit concentration) determines the equilibrium concentration ratios. A large positive $\mathcal{E}^0$ corresponds to a large equilibrium constant. Thus a large positive $\mathcal{E}^0$ indicates a reaction which goes well to the right in the direction written, a reaction which goes nearly to completion.

Problem 6.22   A cell is constructed in which a Pt wire dips into a solution of 1 M Fe^{3+} and 1 M Fe^{2+}. The other electrode consists of thallium metal dipped into a Tl^+ solution of 1-M concentration. Given the half-cell potentials:

$$\text{Tl}^+ + \text{e} \rightarrow \text{Tl}, \qquad \mathcal{E}^0 = -0.34\ \text{V},$$
$$\text{Fe}^{3+} + \text{e} \rightarrow \text{Fe}^{2+}, \qquad \mathcal{E}^0 = \ \ \ 0.77\ \text{V},$$

a) determine which electrode is negative,
b) determine which electrode is the cathode,
c) describe how decreasing the Tl^+ ion concentration would affect the potential.

Problem 6.23   From the standard potentials:

$$\text{Cu}^{2+} + \text{e} \rightarrow \text{Cu}^+, \qquad \mathcal{E}^0 = 0.15\ \text{V},$$
$$\text{Cu}^+ + \text{e} \rightarrow \text{Cu}, \qquad \mathcal{E}^0 = 0.52\ \text{V},$$

calculate the equilibrium constant of the reaction

$$\text{Cu} + \text{Cu}^{2+} \rightleftharpoons 2\text{Cu}^+.$$

Would you expect to be able to form much Cu^+ from the reaction? CuCl is quite insoluble. How would addition of chloride affect the possibility of reducing Cu^{2+} with Cu? (*Hint:* What would you expect to be the effect on $\mathcal{E}^0$ of coupling the reaction to one that has a spontaneous tendency to go far as written.)

6.9 HEAT, WORK, FREE ENERGY, AND SPONTANEOUS REACTION: PART I

We now know that $\mathcal{E}^0$, or the maximum *work* obtainable, determines the tendency for a reaction to go. It is interesting to look into some of the early attempts to formulate an energetic criterion of spontaneous chemical reaction. There are several energetic aspects of chemical reaction, all of which are inter-

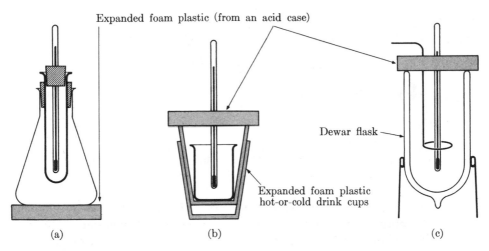

Fig. 6.9. Cross-sectional views of three simple designs for calorimeters. The components are a reaction chamber, a thermometer, and adequate insulation to prevent heat transfer.

esting. Contrasting electrical work obtainable to other measures of energy will clarify our criterion for spontaneous chemical change.

One of the earliest aspects studied was the transfer of heat accompanying chemical reaction. If a reaction is run in an insulated vessel and the temperature change noted, it is only necessary to know the heat capacity of the vessel and the reaction mixture to calculate the amount of heat liberated or absorbed in the course of the chemical reaction. Simple forms of the device used in such a study, called an adiabatic calorimeter, are illustrated in Fig. 6.9. Experimentally, it emerges that there are some differences between heat of reaction in systems open to the atmosphere and hence at *constant pressure* and heat of reaction in closed systems of *constant volume*. The former, constant-pressure heats of reaction, are called *enthalpy* changes of reaction (denoted ΔH). If heat is evolved, ΔH is assigned a negative sign, and the reaction is called *exothermic*. If heat is absorbed, ΔH is assigned a positive sign and the reaction is called *endothermic*. In 1840, Hess enunciated a law of constant heat summation for enthalpies of reaction. This law states that a chemical reaction carried out by means of several steps will result in the same total ΔH as will result from carrying out the reaction in a single step. As an example, consider burning carbon (graphite) in excess oxygen to produce CO_2. The enthalpy change is usually measured in calories (one calorie is the heat required to raise one gram of water from 14.5°C to 15.5°C). It is found to be $-94,050$ cal/mole of graphite at 25°C and 1-atm pressure. When graphite is burned in limited O_2 to give CO, conversion of one mole of graphite yields $-26,420$ cal. Burning CO in O_2 to give CO_2 gives $\Delta H = -67,630$ cal/mole. The two stages add up to the overall reaction:

$$\begin{array}{llll}
C + \tfrac{1}{2}O_2 & \rightarrow & CO, & \Delta H = -26,420, \\
CO + \tfrac{1}{2}O_2 & \rightarrow & CO_2, & \Delta H = -67,630, \\
\hline
C + O_2 & \rightarrow & CO_2, & \Delta H = -94,050.
\end{array} \qquad (6.12)$$

Hess' law permits calculation of heats of reactions for which direct experimental evaluation would be difficult. Consider the reaction at 25°C:

$$C(\text{solid}) + 2H_2(\text{gas}) \rightarrow CH_4(\text{gas}).$$

We can calculate ΔH as follows:

$$
\begin{array}{llr}
C(\text{solid}) + O_2(g) & \rightarrow CO_2(g), & -94{,}100 \text{ cal,} \\
2 \times [H_2(g) + \tfrac{1}{2}O_2(g)] & \rightarrow H_2O \text{ (liquid)}, & 2 \times (-68{,}300) \text{ cal,} \\
CH_4(g) + 2O_2(g) & \rightarrow CO_2(g) + 2H_2O \text{ (liquid)}, & -(-212{,}800) \text{ cal,} \\
\Delta H & = -17{,}900 \text{ cal.} &
\end{array}
\quad (6.13)
$$

The measurement of reaction heats attracted the attention of two prominent nineteenth century theoretical chemists. In 1854 Julius Thomsen summed up his conclusion from studies of reaction heats in the statement: "Every reaction, simple or complex, of a purely chemical nature is accompanied by the evolution of heat." (That is, ΔH is negative.) In 1873, Berthelot, without acknowledging Thomsen, wrote, "Every chemical change accomplished without the intervention of external energy tends to the production of that substance or systems of substances which evolves the maximum quantity of heat." These are attractive propositions, suggesting simply that reactions run "downhill," liberating energy in the form of heat. They were supported by an extensive set of observations but unfortunately not by a complete set. There are spontaneous chemical reactions which present exceptions. A fairly large class of spontaneous *endothermic* reactions is the dissolution of a number of common salts in water. Exothermicity is a rough but not entirely reliable guide to the direction of reaction.

Returning to the electrical potential criterion, let us reformulate it in terms *parallel* to the heat criterion. A positive cell potential indicates a reaction which proceeds spontaneously. The maximum useful work w obtainable from the reaction is:

$$n\mathfrak{F} \times \mathcal{E} = w. \tag{6.14}$$

Let us identify the work done with a decrease in what we shall call the *free energy* G of the system. A reaction which proceeds spontaneously is capable of doing work (w positive); hence its *free-energy change* ΔG is negative. The spontaneous reaction is *exothermic* in the sense of *free energy*. A reaction which is endothermic in free energy does not proceed spontaneously. Above we calculated work in joules. If preferred, we may convert this to calories. One calorie is equal to 4.18 joules. Thus:

$$\Delta G(\text{cal}) = -\frac{1}{4.18}\, n\mathfrak{F}\mathcal{E}. \tag{6.15}$$

Corresponding to the standard potential $\mathcal{E}°$ for the reaction between all components at 1-M concentration, there is a standard free-energy change,

ΔG^0, which has the opposite sign, negative for the spontaneous direction of reaction. It follows that equilibrium constants may be calculated from standard free-energy changes (ΔG^0):

$$\Delta G^0 = -n\mathcal{F}\mathcal{E}^0 \quad \text{and} \quad \mathcal{E}^0 = \frac{2.3\,RT}{n\mathcal{F}} \log K_{eq}.$$

Thus

$$\Delta G^0 = -2.3RT \log K_{eq}. \tag{6.16}$$

This last equation not only gives a method of obtaining K_{eq} from known ΔG^0 but also suggests that *free-energy changes* for reactions not susceptible to incorporation into electrochemical cells might be calculated from equilibrium constants. Finally, we should note that since cell potentials are additive, we should expect the *free energy changes* to be additive and obey the analog of Hess' law.

A sample calculation best shows the utility of this result. Problem 6.23 drew attention to the effect of chloride on the reduction of Cu^{2+} to Cu^+ by Cu metal. Suppose we have data at 25°C as follows:

$$Cu + Cu^{2+} \rightarrow 2Cu^+, \qquad \mathcal{E}^0 - -0.37 \text{ V},$$

and the equilibrium constant for the reaction

$$CuCl(\text{solid}) \rightleftharpoons Cu^{\,|} + Cl^-$$

is $K = [Cu^+][Cl^-] = 1 \times 10^{-6}$. From (6.15), ΔG^0 for the first reaction is

$$\Delta G^0 = -\frac{1}{4.18}\, n\mathcal{F}\mathcal{E}^0 = -\frac{1}{4.18}\, (2)\,(96{,}500)(-0.37) \text{ cal}$$

$$= +17{,}100 \text{ cal}.$$

For the second reaction, Eq. (6.16) yields ΔG^0:

$$\Delta G^0 = -2.3RT \log K_{eq} = (-2.3)(1.98)(3.0 \times 10^2)(-6.0)$$

$$= 8200 \text{ cals},$$

where the value of 1.98 cal/deg-mole has been used for R. Now we can add the two reactions:

$Cu + Cu^{2+} \rightarrow 2Cu^+,$	$\Delta G^0 = +17{,}100,$
$-2 \times (CuCl \rightarrow Cu^+ + Cl^-),$	$-2 \times \Delta G^0 = -16{,}400,$
$Cu + Cu^{2+} + 2Cl^- \rightarrow 2CuCl,$	$\Delta G^0 = +700 \text{ cal}.$

For the final reaction:

$$\mathcal{E}^0 = -4.18\, \Delta G^0/n\mathcal{F} \text{ V} = -4.18 \times 700/2 \times 96{,}500 = -0.015 \text{ V},$$

and:

$$\log K_{eq} = -\Delta G^0/2.3RT = -700/1360 = -0.52$$
$$K_{eq} = 0.3$$

Comparing the result of Problem 6.23 we see that addition of chloride favors reduction of Cu^{2+} by Cu but that the reaction still is not strongly favored. When all participants are present at 1-M concentration it will still go the other way. However, if no CuCl(solid) is initially present, some will be formed at equilibrium.

The most *important point* of this calculation, however, is that it is possible to *combine* a free-energy change calculated from an electrochemical reaction with a free-energy change calculated from an equilibrium constant calculated for a *nonelectrochemical* reaction to *correctly* predict another electrochemical reaction. An experimental value for the $\mathcal{E}^0$ we calculated is 0.016 V! This justifies general application of Eq. (6.16). A ΔG^0 from a K_{eq} *predicted* a new $\mathcal{E}^0$!

Problem 6.24   Consider the following reactions for which values of K_{eq} are given.
a) Write the expression in terms of concentrations which corresponds to K_{eq}.
b) Does the reaction proceed *as written* under *standard* conditions?
c) Determine the sign of ΔG^0 (and of $\mathcal{E}^0$ if applicable).
d) If you mixed the reactants in the *absence* of product, would the system reach equilibrium after forming (i) little product, (ii) substantial amounts of product, (iii) almost the stoichiometrically limiting amount of product?

$$
\begin{aligned}
CH_3COOH(aq) &\rightarrow CH_3COO^- + H^+, &\quad K_{eq} &\simeq 10^{-5}, \\
AgCl(s) &\rightarrow Ag^+ + Cl^-, &\quad K_{eq} &\simeq 10^{-10}, \\
H_2(g) + Cl_2(g) &\rightarrow 2HCl(g), &\quad K_{eq} &\simeq 10^7 \ (425°C), \\
N_2(g) + 3H_2(g) &\rightarrow 2NH_3(g), &\quad K_{eq} &\simeq 10^{-5} \ (500°C), \\
Zn + Cu^{2+} &\rightarrow Zn^{2+} + Cu, &\quad K_{eq} &\simeq 10^{37}.
\end{aligned}
$$

6.10 HEAT, WORK, FREE ENERGY, AND SPONTANEOUS REACTION: PART II

The quantities ΔG and ΔH introduced in the last section deserve a little further exploration. Especially, let us look at the additive property. The fact that we can add quantities for steps and find the value of ΔH or ΔG for the *net reaction* is a significant circumstance, since it means that numbers designating G and H are *conserved* or constant. Suppose we consider carrying out a reaction from A to C through the *intermediate* compound B.

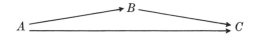

We now know that ΔH or ΔG for the reaction $A \rightarrow B$ plus ΔH or ΔG for reaction

$B \to C$ gives the value for the *direct* reaction $A \to C$. Suppose that we think of carrying out the direct reaction *backwards* $C \to A$. This closes the cycle.

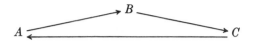

Where are we when we get back to A? Clearly ΔG and ΔH for $C \to A$ are just the *negative* of ΔG and ΔH for $A \to C$, which we could calculate by going through B. Thus, on closing the cycle we should be right *back where we started*. If we go "from A to A" (by way of C) any G or H expended on the way out is recovered on the return (or vice versa). The *total ΔG or ΔH around the cycle* is exactly *zero*. This is what we mean when we say some quantity is *conserved*. We also say that G and H are *functions of state*. By this we mean that if you say a system is in state A (a certain substance at a certain temperature and pressure), it will have a certain value of G or H and it does not matter what processes it has undergone on its way to reaching that state. The state has no notion of its "history."

There is a very practical application of the *conservative* (additivity) property of state functions. We can tabulate energy data on chemical reactions in a very abbreviated form. Just as the collection of "half-cell" potentials made possible calculation of potentials for all cells, a list of standard reactions, one for each compound of interest, will make possible calculation of ΔG and ΔH quantities for all reactions. We shall consider the following limited circumstances: (1) All reactions are here considered only at the temperature of 298°K (25°C). (2) We consider only standard conditions (i.e., approximately 1-M solutions, 1-atm pressure gases, a pure solid or liquid "insoluble" reactants). That is, we are considering ΔG^0 and ΔH^0. (Recall that it is ΔG^0 that is related to K_{eq} at the temperature under consideration.) The *standard reaction* will be *formation* of a compound from its elements under the stated conditions. As an example we could consider the formation of CO_2

$$C(\text{solid}) + O_2(\text{gas}) \to CO_2(\text{gas})$$

at 25°C (298°K). This reaction (all gases at 1-atm pressure) involves evolution of 94,050 cal of heat per mole of CO_2. ΔH^0 for this reaction is called ΔH_f^0, the standard *heat of formation*. Thus $\Delta H_f^0 = -94.05$ kcal/mole. The work extractable is $-94,266$ cal/mole or $\Delta G_f^0 = -94.26$ kcal/mole. (The convenient notation "kcal" is introduced for one thousand calories, since the energy changes are large.) Similarly, formation reactions for CO and H_2O would be written:

$$C(s) + \tfrac{1}{2}O_2(g) \to CO(g), \qquad \Delta H_f^0 = -26.41 \text{ kcal/mole},$$
$$\Delta G_f^0 = -32.81 \text{ kcal/mole},$$

$$H_2(g) + \tfrac{1}{2}O_2(g) \to H_2O(g), \qquad \Delta H_f^0 = -57.80 \text{ kcal/mole},$$
$$\Delta G_f^0 = -54.64 \text{ kcal/mole}.$$

The use of *formation* reactions for calculation of energetics of another arbitrary reaction may be illustrated with these. Consider a reaction of some technological importance called the water-gas reaction:

$$C(\text{solid}) + H_2O(g) \rightarrow H_2(g) + CO(g).$$

This may be written as the formation of $CO(g)$ *minus* the formation $H_2O(g)$

$$
\begin{aligned}
C(s) + \tfrac{1}{2}O_2(g) &\rightarrow CO(g), \\
H_2O(g) &\rightarrow H_2(g) + \tfrac{1}{2}O_2(g), \\
\hline
C(s) + H_2O(g) &\rightarrow H_2(g) + CO(g).
\end{aligned}
$$

For the "water-gas" reaction, ΔH^0 is $\Delta H_f^0(CO) - \Delta H_f^0(H_2O)$ or $+31.39$ kcal/mole. Similarly, ΔG^0 is $\Delta G_f^0(CO) - \Delta G_f^0(H_2O)$ or $+21.83$ kcal/mole. As this example suggests, ΔG^0 or ΔH^0 for any reaction will be the sum of ΔG_f^0 or ΔH_f^0 for the *products* minus the sum of ΔG_f^0 or ΔH_f^0 for the *reactants*.

Problem 6.25   Find ΔH^0 and ΔG^0 for the following reactions from values of ΔH_f^0 and ΔG_f^0 in Table 6.2 and the examples in the text:

$$
\begin{aligned}
NH_3(g) + HCl(g) &\rightarrow NH_4Cl(g), \\
CH_4(g) + 2O_2(g) &\rightarrow CO_2(g) + 2H_2O(g), \\
C_2H_4(g) + H_2(g) &\rightarrow C_2H_6(g).
\end{aligned}
$$

Table 6.2

Heats and free energies of formation of some typical compounds

Compound	ΔG_f^0, kcal/mole	ΔH_f^0, kcal/mole
$H_2(g) + \tfrac{1}{2}O_2(g) \rightarrow H_2O(g)$	-54.64	-57.80
$\tfrac{1}{2}N_2(g) + \tfrac{3}{2}H_2(g) \rightarrow NH_3(g)$	-3.98	-11.04
$\tfrac{1}{2}N_2(g) + 2H_2(g) + \tfrac{1}{2}Cl_2(g) \rightarrow NH_4Cl(g)$	-50.35	-71.76
$\tfrac{1}{2}H_2(g) + \tfrac{1}{2}Cl_2(g) \rightarrow HCl(g)$	-22.77	-22.06
$C + 2H_2(g) \rightarrow CH_4(g)$	-12.14	-17.89
$2C + 2H_2(g) \rightarrow C_2H_4(g)$	16.28	12.50
$2C + 3H_2(g) \rightarrow C_2H_6(g)$	-7.86	-20.24

6.11 THE ENTROPY FUNCTION

What about the quantity which is the difference between ΔG and ΔH? The enthalpy change which is measured by heat evolved is perhaps the most obvious energy change accompanying chemical reaction. But, ΔG, or the maximum work extractable from a reaction, is the measure of the direction and extent of chemical reaction. Is it fruitful to ask if there is a factor which can be combined with enthalpy change to give the free-energy change? Or, to look at the matter another way, what is it that is the difference between the maximum amount of useful work we can extract from a reaction in a cell under conditions of maximum efficiency and the amount of heat evolved when the reaction just "burns itself out" in a calorimeter?

We would like to find the difference in terms of a *state function* which has conservative behavior like ΔH and ΔG. The desideratum is another quantity for which cyclic calculations are possible. At the risk of "pulling the rabbit out of a hat" we may write an equation:

$$\Delta G = \Delta H - T \, \Delta S. \tag{6.17}$$

In this equation we introduce the required function S, known as *entropy*. The role of absolute temperature T in this equation is dictated by the desire to make S a state function, i.e., ΔS cyclically conservative, as described. Focusing on the difference between ΔH and ΔG we write:

$$\Delta S = \frac{\Delta H - \Delta G}{T}. \tag{6.18}$$

What is ΔS, the entropy change accompanying a reaction? First the trivial part of the answer, its units: It is clearly the ratio of an energy to a temperature so that it may be given as cal/deg-mole (or, if you prefer, joules/deg-mole). To tackle the serious question of what the "entropy factor" represents, let us try to build a molecular model of a process in which the entropy change is the dominant factor and try to understand why it is that the process *goes* spontaneously.

A good example of such a process is the dissolution of solid naphthalene in liquid benzene. This process occurs spontaneously and therefore *must* have a *negative* value of ΔG. However, calorimetric measurements show that ΔH is *essentially zero*. If $\Delta H = 0$, we rewrite Eq. (6.17):

$$\Delta G = -T \, \Delta S. \tag{6.19}$$

The *negative* ΔG must correspond to a *positive* ΔS and no more. What is happening? Figure 6.10 purports to show the two states, separated phases and solution, from a molecular point of view. The separated phases appear in

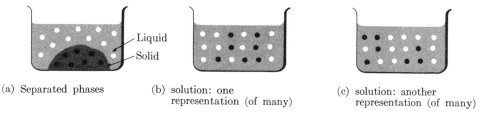

(a) Separated phases   (b) solution: one   (c) solution: another
                                  representation (of many)   representation (of many)

Fig. 6.10. Molecular representation of a dissolution process.

Fig. 6.10(a). The solution is given two of *many* possible representations in Figs. 6.10(b) and 6.10(c).

Accepting the plausible notion that $\Delta H = 0$ represents the fact that the very similar naphthalene and benzene molecules do not experience any important new forces in the solution from those experienced in the separated phases, we indicate the crucial new factor in the solution by having two drawings for the solution as compared to one drawing for the separated phases. Clearly, to represent the separated phases, solid naphthalene and liquid benzene, we must restrict the way we distribute black circles and white circles. The black circles representing solid must be all together at the bottom and the white circles representing liquid must be segregated above to represent separated phases. There are a great many more ways to dispose ("mix-up") black and white circles to represent the solution. The arrangements of molecules indicated by Fig. 6.10(b) and 6.10(c) are only a small start toward indicating the number of *different* ways of disposing the molecules that we would recognize as the *solution*. These *many* ways are to be contrasted to the *very small* number of different ways of disposing the same set of molecules to produce the separate phases. If molecules were to be disposed *at random*, the solution would be the *much more probable result.*

This result is comparable to the situation after a jar containing white balls and black balls is shaken. If they were initially separated, we would expect shaking to mix them. However, if they were initially *mixed*, we would hardly expect shaking to *sort them out again*. This is not because there is anything especially wrong with the sorted arrangement. It is only because there are so many more different mixed arrangements than sorted arrangements that we expect any random process (shaking) which leads from one arrangement to another to lead *to* a mixed arrangement.

Chemically, again, the interesting point is that all these *different* "mixed-up" arrangements correspond to only *one observed* arrangement; what we call solution. A very much smaller number of "ordered" arrangements correspond to the one observable arrangement we call "separated phases." If we can imagine— as well we must if we know anything of diffusion—that there are ways that molecules can get shaken together, we should regard it as likely that they would mix.

What should now be said about a positive ΔS? It appears to correlate with the system moving toward what might be loosely called its *"most probable"* or *"most disordered"* state if these terms are carefully understood in terms of the number of different molecular arrangements corresponding to a given *observed* situation. Is it surprising that this factor is involved in defining spontaneous direction in natural processes?

Positive ΔS is, then, associated with change to more "probable" or "disordered" states. *Positive ΔS is a contribution to negative ΔG and hence to* spontaneity of reaction in the indicated direction. Positive ΔS is expected in various disordering situations which can be predicted. For example, positive

Table 6.3

An entropy "bestiary"

Substance	ΔS^0 (melting), eu	ΔS^0 (vaporization), eu
O_2	$+1.95$	$+18.1$
N_2	$+2.73$	$+17.2$
NH_3	$+6.92$	$+23.3$
H_2O		$+26.1$
CCl_4		$+20.0$

Some reactions	ΔS^0, eu
$2H_2(g) + O_2(g) \rightarrow 2H_2O(\text{liquid})$	-65.3
$AgI(s) \rightarrow Ag^+(aq) + I^-(aq)$	$+16.2$
$CaCO_3(s) \rightarrow CaO(s) + CO_2(g)$	$+38.4$

ΔS is anticipated for:

1. the transition: solid $\rightarrow$ liquid,

2. the transition: liquid $\rightarrow$ gas,

3. reactions in which the number of molecules increases.

To illustrate these, an entropy "bestiary" is presented in Table 6.3. This is a small "zoo" of examples to give you some feeling for ΔS. Recall that entropy units (eu) are cal/deg-mole.

Problem 6.26 If excess solid NaCl is added to a 1 M solution of NaCl, more NaCl *dissolves* but the solution is observed to become *cooler*. Determine the sign of ΔG^0 for the reaction

$$NaCl(s) \rightarrow Na^+(\text{liquid}) + Cl^-(\text{liquid}).$$

Find the sign of ΔH^0. Can you give a molecular picture of the reason the reaction goes contrary to its ΔH^0 value?

6.12 THE EFFECT OF TEMPERATURE ON CHEMICAL REACTIONS

So far we have been discussing the energetics of chemical reactions at only *one* temperature. We must not forget that K_{eq}, $\mathcal{E}^0$, ΔG^0, ΔH^0, and ΔS^0 *do* change with temperature. It is very important to realize that the analysis in the foregoing pages may be applied only if all the reactions are considered at a single temperature, for example, room temperature (298°K). But, it is interesting to examine Eq. (6.17) with a view to making some approximate remarks about the effect of changing temperature on the tendency for a reaction to go. To recapitulate,

$$\Delta G = \Delta H - T \, \Delta S \tag{6.17}$$

relates ΔG, the quantity determining the direction of reaction to ΔH and ΔS *multiplied* by T. Clearly, unless ΔH changes much faster as T changes than ΔS, the fact that ΔS is multiplied by T means that ΔS becomes *more important* as temperature increases. As a useful rough rule, we may say reactions at low temperature are mostly ΔH controlled, whereas reactions at high temperature are mostly ΔS controlled. Consider again the "water-gas" reaction mentioned on p. 172:

$$C(s) + H_2O(g) \rightarrow H_2(g) + CO(g).$$

We found (at room temperature) $\Delta G^0 = +21.8$ kcal/mole. This means the reaction does not go well in the direction written. The data provided show that this is largely the result of the fact that $\Delta H^0 = +28.2$ kcal/mole. Since ΔH^0 is *more unfavorable* than ΔG^0, we can conclude that ΔS^0 must be *favorable*. This is reasonable. The reaction converts a mole of solid and a mole of gas into *two moles of gas*. The ΔS^0 should be at least approximately comparable to an entropy of vaporization (see Table 6.3). All that is necessary for ΔG^0 to become favorable is that $T \Delta S^0$ become larger than ΔH. Clearly, this could happen at some sufficiently high temperature. The "water-gas" reaction is one of industrial significance, *but not at room temperature*. Table 6.4 generalizes the results suggested by this case. It shows that whenever ΔH and ΔS are opposed, there will be appropriate temperatures to make the reaction go.

Table 6.4

Temperature effects

ΔH	ΔS	*Reaction proceeds*
< 0 (negative)	> 0 (positive)	At *all* temperatures
> 0 (positive)	> 0 (positive)	At sufficiently *high* temperatures
< 0 (negative)	< 0 (negative)	At sufficiently *low* temperatures
> 0 (positive)	< 0 (negative)	At *no* temperature

Problem 6.27 Consider the following reactions in terms of the rules on p. 175 and the examples in Table 6.3. Try to decide whether ΔS^0 is positive or negative. Next consider them in the light of the rules in Table 6.4. Determine whether they would be more favorable at low or high temperature.

a) $2NaHCO_3(s) \rightarrow Na_2CO_3(s) + CO_2(g) + H_2O(g)$
b) $CO_2(g) + H_2O(l) + Na^+(aq) + NH_3(aq) \rightarrow NaHCO_3(s) + NH_4^+(aq)$

Note: (a) and (b) are the reactions used, in reverse order, in the Solvay process for industrial preparation of carbonates. You might like to think about conditions for the successive steps in the Solvay process.

c) $NH_3(g) + HCl(g) \rightarrow NH_4Cl(s)$
d) $N_2O_4(g) \rightarrow 2NO_2(g)$

Industrially, Carborundum® is prepared by the reaction:

e) $SiO_2(s) + 3C(s) \rightarrow SiC(s) + 2CO(g)$.

SiC is the desired product, Carborundum.
  The metallurgy of nickel is not atypical. We start with the sulfide:

f) $2NiS(s) + 3O_2(g) \rightarrow NiO(s) + 2SO_2(g)$.

The NiO is reduced with "water-gas" through two reactions:

g) $NiO(s) + H_2(g) \rightarrow Ni(s) + H_2O(g)$,
h) $NiO(s) + CO(g) \rightarrow Ni(s) + CO_2(g)$.

Seminar: Bond Energies

We have noted that the changes of energy quantities associated with reactions, ΔH (the heat absorbed at constant pressure) and ΔG (the maximum available work), are quantities that are *additive* like $\mathcal{E}$'s (potentials) so that values for net processes may be computed by adding values for a series of steps which, taken together, accomplish the overall reaction. Interestingly, the idea may be extended to make the "component steps" simply *bond* formation rather than whole molecule formation. We *define* average bond energies as follows.
  Consider the reaction:

$$CH_4 \text{ (gas)} \rightarrow C \text{ (gas, atom)} + 4H \text{ (gas, atom)}.$$

The $\Delta H°$ for this reaction is called the enthalpy of *atomization* of methane. The average C—H bond energy, ϵ_{CH}, is defined by

$$\epsilon_{CH} = \frac{\Delta H° \text{ (atomization)}}{4} = 99.3 \text{ kcal mole}^{-1}.$$

Similarly, other molecules may be atomized to find bond energies:

$$\epsilon_{CC} = 78, \quad \epsilon_{C=C} = 148, \quad \epsilon_{CO} = 82,$$

$$\epsilon_{C=O} \cong 169, \quad \epsilon_{OH} = 111,$$

where all values are in kcal mole^{-1}.

® Carborundum is a registered trademark of Carborundum Corporation, Niagara Falls, Ontario.

Remember that these values are not meant to be specified for a particular molecule in which the bond occurs but are suggested to be generally applicable. Consider the enthalpies of formation, ΔH_f°, of the following molecules:

Molecules	ΔH_f°, kcal/mole^{-1}
$2H_2$ (gas) $+$ C (solid) $\rightarrow$ CH_4 (gas)	-17.9
$3H_2$ (gas) $+$ 2C (solid) $+ \frac{1}{2}O_2$ (gas) $\rightarrow$ C_2H_5OH (liquid)	-66.4
$5H_2$ (gas) $+$ 4C (solid) $\rightarrow$ $CH_3CH_2CH_2CH_3$ (gas)	-29.8
$5H_2$ (gas) $+$ 4C (solid) $\rightarrow$ $\genfrac{}{}{0pt}{}{CH_3}{CH_3}\rangle CH{-}CH_3$ (gas)	-31.5
$2H_2$ (gas) $+$ 2C (solid) $+ O_2$ (gas) $\rightarrow$ CH_3COOH (liquid)	-116.4

Try to *calculate* these enthalpies of formation from the bond energies given *and* the ΔH° values for the following reactions.

Reactions	ΔH°, kcal/mole^{-1}	
H_2 (gas) $\rightarrow$ 2H (gas, atom)	104.2	Heats of
C (solid) $\rightarrow$ C (gas)	171.7	atomization
O_2 (gas) $\rightarrow$ 2O (gas, atom)	118.2	
CH_3COOH (gas) $\rightarrow$ CH_3COOH (liquid)	-10.5	Heats of
C_2H_5OH (gas) $\rightarrow$ C_2H_5OH (liquid)	-10.1	liquefaction

Now, consider three questions.

1. How accurately does the concept of a "bond energy" independent of the molecule in which the bond occurs seem to work?

2. What does the notion of bond energy add to our concept of bond as derived from organic structural chemistry?

3. What is the implication of the fact that bond energies only approximately represent molecules?

Useful Reading   (Look up bond energy in the following.)

B. H. Mahan, *University Chemistry*, Addison-Wesley, Reading, Mass., 1965.
B. H. Mahan, *Chemical Thermodynamics*, W. A. Benjamin, New York, 1964.
J. Waser, *Basic Chemical Thermodynamics*, W. A. Benjamin, New York, 1966.
W. J. Moore, *Physical Chemistry*, Prentice-Hall, Englewood Cliffs, N. J., several editions.

Calculations from Some Equilibrium Constants

6A.1 SOME FURTHER EXAMPLES OF THE EQUILIBRIUM CONDITION

Equation (6.11) expresses the equilibrium condition for the reversible reaction.

$$aA + bB + \cdots \rightleftharpoons cC + dD + \cdots \tag{6A.1}$$

Equation (6.11) was derived from the Nernst equation which is applied to oxidation reactions involving electron transfer between ions in aqueous solution. It turns out that Eq. (6.11) is universally true whether or not the substances A, B, C, and D are involved in electron transfer. Indeed it is not a necessary condition that they be ions just so long as equilibrium is achieved. One or even all the reacting species may consist of molecules rather than ions.

To illustrate, we may cite the equilibrium, in aqueous solution, between molecular acetic acid (written HOAc here) and its ions, as well as that involving the acidic dissociation of the ammonium ion:

$$\begin{aligned} HOAc &\rightleftharpoons H^+ + OAc^- & \text{(aqueous solution),} \\ NH_4{}^+ &\rightleftharpoons NH_3 + H^+ & \text{(aqueous solution).} \end{aligned} \tag{6A.2}$$

Reactions involving uniquely molecular substances are illustrated by the hydrolysis of the ester ethyl acetate in the liquid phase and by the thermal dissociation of hydrogen iodide in the gaseous phase:

$$\begin{aligned} CH_3COOC_2H_5 + HOH &\rightleftharpoons CH_3COOH + C_2H_5OH, \\ 2HI &\rightleftharpoons H_2 + I_2. \end{aligned} \tag{6A.3}$$

It can be shown for all these cases that the equilibrium condition of Eq. (6.11) is an experimental fact at any specific temperature, whether involving ions in aqueous solution or molecules in the liquid or vapor phase. Thus we write

$$\frac{[H^+][OAc^-]}{[HOAc]} = K_{eq} \quad \text{or} \quad \frac{[H_2][I_2]}{[HI]^2} = K_{eq}, \tag{6A.4}$$

where K_{eq} is the equilibrium constant and brackets denote concentration at equilibrium. This constant has a specific value for each reversible reaction provided the temperature does not vary.

If we wish to be entirely specific in our notation, we may use subscripts such as $HOAc_{(aq)}$, $NH_4^+{}_{(aq)}$, $C_2H_5OH_{(liq)}$, $CH_3COOH_{(liq)}$, $HI_{(gas)}$ and $I_{2(gas)}$ to indicate whether a particular species is found in aqueous solution or in the liquid or gaseous state. The corresponding concentrations would then be written $[NH_{4(aq)}]$, $[C_2H_5OH_{(liq)}]$, $[I_{2(gas)}]$, etc. This complete notation is rather unwieldy to write or print. And as we become familiar with the various systems under consideration we tend to use the simpler notation of Eqs. (6A.2) and (6A.4).

At this point it is important to reiterate the statement on page 170 to the effect that Eqs. such as (6A.4) might apply to reactions not susceptible to incorporation into electrochemical cells. In other words, any equilibrium-condition expression (whether it involves ion concentrations, molecular concentrations, or both) is derivable from thermodynamic theory, as well as being experimentally verifiable.

6A.2 APPLICATIONS OF THE EQUILIBRIUM CONDITION TO ACID-BASE REACTIONS

How can we put to good use the type of expression illustrated by Eqs. (6A.4)?

We shall discuss in some detail the equilibrium condition as applied to ionic aqueous solutions and, in particular, to solutions of *weak* acids and *weak* bases. It turns out that such ionic systems are of special importance in the study of, among other things, physiology.

We shall find that the concept of an equilibrium condition is not a useful one in the case of solutions of strong acids and bases. Strong electrolytes presumably are completely ionized, and hence there is no equilibrium set up between a molecular solute species and its ions.

Molecular equilibria similar to those illustrated by Eqs. (6A.3) are of great importance in the field of industrial chemistry. For instance, the synthesis of ammonia is effected at high temperatures by the reaction

$$3H_{2(gas)} + N_{2(gas)} \rightleftharpoons 2NH_{3(gas)}.$$

Interesting as they are, such systems will not be considered further in this book.

Returning now to a consideration of aqueous solutions of acids and bases, it is first important to recognize that there are always at least two reactions involved. Consider the example of acetic acid in water. Not only does acetic acid react with water to give H^+ and its conjugate base, acetate ion, but also water ionizes:

$$HOAc \rightarrow H^+ + OAc^-, \qquad H_2O \rightarrow H^+ + OH^-.$$

These two reactions are not independent. They share H^+ as a common product. When HOAc is added to water and produces H^1, this H^+ must shift the equilibrium position of the second reaction to the left. Thus, it has the effect of reducing the OH^- concentration in the aqueous solution. Now, both reactions must be at equilibrium if net chemical change has ceased to occur in the solution, and one cannot be at equilibrium unless the other is, since both will change the concentration of hydrogen ion.

The equilibrium constant for ionization of water is usually written

$$K_w = [H^+][OH^-]; \tag{6A.5}$$

K_w has a value of 1.00×10^{-14} at 25°C. This equation expresses a condition which must be satisfied by all aqueous solutions which are at equilibrium.

Consider for a moment two reasonably dilute solutions of strong electrolytes that are acids and bases respectively, for example 0.1-M HNO_3 and NaOH. It is clear that the $[H^+]$ is about 10^{-1} in the first and that $[OH^-]$ is about 10^{-1} in the second. From Eq. (A6.5) it follows that

$$[OH^-] = K_w/[H^+] \qquad \text{and} \qquad [H^+] = K_w/[OH^-].$$

Applying these results to the two dilute solutions, we see that $[OH^-] = 10^{-13}$ in the first and that $[H^+] = 10^{-13}$ in the second. These values are very small because K_w is small. Between these two solutions that represent extreme conditions, $[OH^-]$ varies from 10^{-13} to 10^{-1} and $[H^+]$ varies from 10^{-1} to 10^{-13}. Among other things, it would not be at all convenient to plot a graph with a linear scale encompassing such a variation. The graph would be much more manageable, however, if a *logarithmic* scale for $[H^+]$ were employed. On such a scale, the variation from 10^{-13} to 10^{-1} would be -13 to -1. Let us define the symbol pH as

$$pH = -\log_{10} [H^+]. \tag{6A.6}$$

The negative log is chosen to make pH values for common solutions come out positive. The solution of $[H^+] = 10^{-13}$ has pH = 13, and the solution of $[H^+] = 10^{-1}$ has pH = 1. As H^+ decreases, pH increases. Correspondingly, pH increases as $[OH^-]$ increases.

Problem 6A.1 Find the pH of the following solutions: $[H^+]$ = 1.00, 0.0010, 0.0137, 3×10^{-8}; also, $[OH^-]$ = 0.00010, 2.5×10^{-7}, 0.300.

Problem 6A.2 From the equation $K_w = [H^+][OH^-]$ and the definition $pOH = -\log_{10} [OH]$, show that $pH + pOH = 14$ and that $pH = 14 - pOH$.

There is another notable advantage to pH. The potential of a cell varies with the *log* of concentration according to the Nernst equation. An electrode whose potential depends upon $[H^+]$ will form a cell whose potential depends directly on pH. This is the basis of the so-called pH meter.

Table 6A.1

Acidity constants

Acid	Conjugate base	K_a
CH_3COOH	CH_3COO^-	1.8×10^{-5}
HNO_2 (Nitrous)	NO_2^- (Nitrite)	4×10^{-4}
HCN (Cyanic)	CN^- (Cyanide)	7.2×10^{-10}
NH_4^+	NH_3	5.5×10^{-10}
$CH_3NH_3^+$	CH_3NH_2	2×10^{-11}
$C_6H_5NH_3^+$ (Anilinium)	$C_6H_5NH_2$ (Aniline)	2.1×10^{-5}

We can best display the usefulness of the concept of the equilibrium condition by illustrating with a series of problems in which we calculate concentrations in solution at equilibrium. The first problems will deal with dilute solutions of the order of 0.1 M. For the solution of these problems it is necessary to know, in each case, the value of the acidity constant K_a. These values have been determined and are to be found in various compilations of scientific data. An abbreviated list is given in Table 6A.1. The exact details of the procedures used to determine K_a values will be discussed in a later section after we have developed a better feeling for the relationships involved.

Problems in Aqueous Solutions

Example A. *The case of the weak acid HA.* Let us consider a dilute solution of acetic acid. Here we shall say that $C_a = 0.100$ when the solution was prepared from 0.100 mole of the pure acid dissolved in one liter of solution. The notation for K_a and K_w has already been introduced. We then have

$$C_a = 0.100, \qquad K_a = 1.8 \times 10^{-5} \qquad \text{and} \qquad K_w = 1.0 \times 10^{14}.$$

All concentrations are expressed in moles per liter. The variable species present in the solution are H^+, OH^-, OAc^-, and $HOAc$. Water is of course a species also present, but with increasing dilution its molecular concentration approaches the value $1000/18 = 55.5\ M$, which would obtain for pure water. This value is so high that it is virtually unchanged by any small quantity of water that may be used up or produced in any reactions occurring in the 0.1 M solution. Therefore $[H_2O]$ does not occur in any of the equilibrium constant expressions used here.

Having four variables, $[H^+]$, $[OH^-]$, $[OAc^-]$, and $[HOAc]$, we need four equations for the solution of the problem. These four are listed below.

$$
\begin{aligned}
1) \qquad & [H^+] = [OAc^-] + [OH^-] \quad \text{(e.n.)}, \\
2) \qquad & C_a = [HOAc] + [OAc^-] \quad \text{(m.b.)}, \\
3) \qquad & [H^+][OH^-] = K_w = 10^{-14}, \\
4) \qquad & [H^+][OAc^-] = K_a[HOAc] = 1.8 \times 10^{-5}\,[HOAc].
\end{aligned}
\qquad (6A.7)
$$

The equation relating $[H^+]$, $[OAc^-]$, and $[OH^-]$ is derived from the principle of electroneutrality, and is labeled e.n. This principle states that the sum of all the unit positive charges in a solution must equal the sum of all the negative charges, the unit negative charge being the charge on the electron. For univalent ions this means that the total cation and anion concentrations must be equal.

The second equation expresses the mass balance for acetic acid. It is obvious that the acid dissolved in a solution must be present either in the molecular or the ionized form, and hence $[HOAc] + [OAc^-]$ represents the number of moles of the pure acid from which a liter of aqueous solution was prepared.

With the values of the constants C_a, K_a, and K_w known and four equations available, the problem of solving for the unknown concentrations $[H^+]$, $[OH^-]$, $[OAc^-]$, and $[HOAc]$ could become a matter for algebra without any use of our chemical judgment. We could combine the equations to get a polynomial equation in terms of the $[H^+]$. This may be shown to be

$$
[H^+]^3 + K_a[H^+]^2 - (K_w + K_aC_a)[H^+] - K_wK_a = 0. \qquad (6A.8)
$$

Unless we resort to computer techniques, the solution of this equation, while possible, is rather formidable and, to say the least, cumbersome.

With a certain amount of chemical judgment we can make some simplifying assumptions which permit us to bypass the rather complex algebra. Under favorable conditions, we can, in this way, get close approximations to the true values of the four unknowns. Having completed the solution of the problem by the approximation method we shall be in a position to see how good these approximations are, by substituting the values obtained for the four unknowns into the first equation in (6A.7) and examining the result. Now let's try the simplifications.

We may begin with Eq. (6A.4) with an assist from Eqs. (6A.1) and (6A.2). Acetic acid is the essential acid in vinegar. Vinegar tastes sour. So we may conclude that there is at least some $[H^+]$ here as compared to a neutral solution. Suppose $[H^+] \geqq 10^{-4}$; then $[OH^-] \leqq 10^{-10}$, which is indeed far less than either the $[H^+]$ or the $[OAc^-]$ present. So we simplify Eq. (6A.1) by omitting $[OH^-]$, which is negligibly small in comparison to the other ionic species. We now have $[H^+] \cong [OAc^-]$ to a close approximation. If we can establish a simple relationship between $[HOAc]$ and $[H^+]$, we shall have an equation in one unknown, namely the $[H^+]$. From Eq. (6A.2), $[HOAc] = C_a - [OAc^-]$ and hence $[HOAc] = C_a - [H^+]$. Our expression for K_a reads

$$
\frac{[H^+][OAc^-]}{[HOAc]} \cong \frac{[H^+][H^+]}{C_a - [H^+]} = K_a. \qquad (6A.9)
$$

Solving this by algebra gives

$$[H^+]^2 + K_a[H^+] - K_aC_a = 0. \tag{6A.10}$$

This equation is certainly simpler than Eq. (6A.8), and it can be solved with a certain amount of number work by the quadratic formula

$$[H^+] = \frac{-b \pm \sqrt{b^2 - 4ac}}{2a},$$

where a, b, and c are 1, K_a, and $-K_aC_a$, respectively, or the coefficients of $[H^+]^2$, $[H^+]$, and 1 in Eq. (6A.10).

But we can make life still easier for ourselves if we are willing to compromise still further. Suppose we guess that the 0.10-M acetic acid solution is no more than one or two percent ionized. In this case, we can make the approximation that $[H^+]$ is very small in comparison with [HOAc], and then Eq. (6A.9) assumes the simple quadratic form

$$[H^+]^2/C_a = K_a, \tag{6A.11}$$

and

$$[H^+]^2 = K_aC_a \quad \text{or} \quad [H^+] = \sqrt{K_aC_a},$$

and

$$[H^+] = \sqrt{1.8 \times 10^{-5} \times 10^{-1}} = 1.34 \times 10^{-3} \ M.$$

It then follows from Eq. (6A.5) that

$$[OH^-] = \frac{10^{-14}}{1.34 \times 10^{-3}} = 7.46 \times 10^{-12} \ M,$$

$$[OAc^-] = 1.34 \times 10^{-3} \ M,$$

$$[HOAc] = 0.10000 - 0.00134 = 0.0987 \ M.$$

Checking back we see that we are quite justified in neglecting $[OH^-]$ as compared to $[H^+]$, since these two ions are estimated to be in the ratio of 0.00000000000747 M to 0.00134 M. The simplification of dropping out $[H^+]$ in the expression $C_a + [H^+]$ is less justifiable, but still does not result in an intolerable error, since the ratio $[H^+]/C_a$ is only 0.0134 or 1.34 percent.

Having spelled out the method of attack for the case of the dilute weak acid HOAc, we may now proceed to other similar problems with more confidence and less detailed discussion. Our objective will be to calculate the concentrations of all the ionic or molecular species present in a number of aqueous solutions of weak acids or bases. We shall see that certain cations such as the NH_4^+ ion can act as proton donors and are therefore acids on the basis of the Brønsted concepts discussed in Chapter 6. Molecules of NH_3 and OAc^- ions, on the other hand, are proton acceptors and therefore bases. Their conjugate acids are NH_4^+ and HOAc, respectively. We shall see that we can make use of the K_a values for these acids even in solving problems concerned with solutions of their conjugate bases.

The general method of attack is similar in all these cases to that already used for HOAc. It can be separated into the four following steps.

Step 1. List all the unknown ionic or molecular species present in the aqueous solution.

Step 2. Write down a number of simultaneous equations involving these species; the number should equal the number of species present.

Step 3. Solve these equations by chemical intuition if you can or by more complex algebra if you must.

Step 4. Check to see whether your answers fit the equations, especially the equation for the acidity constant K_a.

We may illustrate by several examples.

Example B. Find the concentrations of all the solute species in a dilute aqueous solution of NaOAc. Say $C_b = 0.100\ M$. This means that 0.100 moles of NaOAc crystals are dissolved in water to make one liter of solution.

NaOAc is a strong electrolyte and presumably completely ionized into Na^+ and OAc^- ions. On the other hand, OAc^- is a weak base reacting with H_2O as follows:

$$OAc^- + HOH \rightleftharpoons HOAc + OH^-. \tag{6A.12}$$

Step 1. Species present: Na^+, H^+, OAc^-, OH^-, HOAc.

Step 2.

$$\text{e.n.} \qquad [Na^+] + [H^+] = [OAc^-] + [OH^-], \tag{1}$$

$$\text{m.b.} \qquad [Na^+] = C_b = 10^{-1}, \tag{2}$$

$$\text{m.b.} \qquad [OAc^-] + [HOAc] = C_b = 10^{-1}, \tag{3}$$

$$K_w \qquad [H^+][OH^-] = K_w = 10^{-14}, \tag{4}$$

$$K_a \qquad \frac{[H^+][OAc^-]}{[HOAc]} = K_a = 1.8 \times 10^{-5}. \tag{5}$$

Step 3. Once we have written the five simultaneous equations we could maintain with some justification that the chemical problem is solved. However, we want *the numbers.* Let's try our chemical intuition. We note that some OH^- ions are produced by the reaction of OAc^- with water. This will make the solution at least slightly basic. Suppose the $[OH^-]$ is about $10^{-5}\ M$; then it would follow that $[H^+] = K_w/[OH^-] = 10^{-14}/10^{-5} = 10^{-9}\ M$. This means that $[H^+] \ll [OH^-]$, so we drop $[H^+]$ out in Eq. (1). With this one assumption we can use Eqs. (1), (2), and (3) to show that

$$[OAc^-] \cong C_b - [OH^-] \qquad \text{and} \qquad [HOAc] \cong [OH^-].$$

It will be convenient to substitute these values in the equilibrium condition expression for Eq. (6A.12):

$$\frac{[HOAc][OH^-]}{[OAc^-]} = K'. \tag{6A.13}$$

The numerical value of K' can be evaluated from Eqs. (4) and (5). It is 5.55×10^{-10}. Throwing Eq. (6A.13) into functions of $[OH^-]$ only, we have

$$\frac{[OH^-]^2}{0.100 - [OH^-]} = 5.56 \times 10^{-10}.$$

The 0.1 M NaOAc solution is by no means strongly basic. We have already supposed that the $[OH^-]$ is about 10^{-5}. This makes admissible the further simplifying assumption that we may with impunity neglect $[OH^-]$ in comparison to 0.1000 M. We then have a simple quadratic equation which is easily solved to give $[OH^-] \cong 7.45 \times 10^{-6}$ M. We may now use Eqs. (1), (2), (3), and (4) to show that $[H^+] \cong 1.34 \times 10^{-9}$ M, $[HOAc] = 7.42 \times 10^{-6}$ M and $[OAc^-] = 0.100000 - 0.0000074 \cong 0.100$ M.

Step 4. Substituting these values into Eq. (5), we get $K_a = 1.82 \times 10^{-5}$. Thus we infer that very little error has resulted as a consequence of the algebraic liberties we have taken. We can see more specifically from Eq. (1) that

$$[OAc^-] = C_b + [H^+] - [OH^-]$$

or

$$[OAc^-] = 0.100 + 0.00000000135 - 0.00000742.$$

Thus the neglect of $[H^+]$ and $[OH^-]$ compared to C_b really has a negligible effect on the answer.

Example C. Find the concentrations of all the solute species in a solution prepared by dissolving 0.10 mole of HOAc and 0.10 mole of NaOAc to make one liter of solution. The two solutes here comprise an acid-base system with $C_a = 0.10$ and $C_b = 0.10$.

The solute species are: Na^+, H^+, OH^-, OAc^-, and HOAc.
We need five equations, and we have

$$\text{e.n.} \qquad [Na^+] + [H^+] = [OH^-] + [OAc^-], \qquad (1)$$
$$\text{m.b.} \qquad [Na^+] = C_b \qquad (2)$$
$$\text{m.b.} \qquad [HOAc] + [OAc^-] = C_a + C_b, \qquad (3)$$
$$[H^+][OH^-] = K_w \qquad (4)$$
$$\frac{[H^+][OAc^-]}{[HOAc]} = K_a. \qquad (5)$$

It is apparent that we would like to evaluate $[OAc^-]$ and $[HOAc]$ leading to an equation in $[H^+]$ only. We have

$$[OAc^-] = C_b + [H^+] - [OH^-]$$

and

$$\begin{aligned} [HOAc] &= C_a + C_b - [OAc^-] \\ &= C_a + C_b - C_b - [H^+] + [OH^-] \\ &= C_a - [H^+] + [OH^-]; \end{aligned}$$

then

$$\frac{[H^+][OAc^-]}{[HOAc]} = \frac{[H^+](C_b + [H^+] - [OH^-])}{C_a - [H^+] + [OH^-]} = K_a. \qquad (6A.14)$$

Here we have a polynomial equation convertible to one unknown which can be solved mathematically with some difficulty. What about simplifying assumptions in this case? Suppose we assume that both $[H^+]$ and $[OH^-]$ are small compared to C_a and C_b. What is the intuitive basis for making these assumptions? Perhaps it need not be altogether intuitive but rather based on a semiquantitative foundation. Suppose you test the pH of the solution under consideration with the appropriate indicators. You could demonstrate that the pH is somewhere within the limits of 4 to 6, meaning that the $[H^+]$ lies between 10^{-4} and $10^{-6} M$ with corresponding $[OH^-]$ values of 10^{-10} and 10^{-8}. Comparing these values with C_s or $C_a = 10^{-1} M$, we see that the $[H^+]$ and $[OH^-]$ may be dropped out. We then have the simple expression

$$\frac{[H^+]C_b}{C_a} = K_a \qquad \text{or} \qquad [H^+] = K_a \frac{[C_a]}{[C_b]}, \qquad (6A.15)$$

and in the present problem

$$[H^+] = 1.8 \times 10^{-5} \times \frac{0.100}{0.100} = 1.8 \times 10^{-5} M.$$

If our assumptions were even approximately correct, then $[H^+]$ is indeed small compared to 0.100. The $[OH^-]$ would be approximately

$$\frac{10^{-14}}{1.8 \times 10^{-5}} \cong 10^{-9},$$

which is surely negligible in Eq. (6A.14).

Problem 6A.3   Find the concentrations of all the species present in a 0.18-M aqueous solution of NH_3. [*Hints:* The equilibrium involved here may be considered to be

$$NH_3 + H_2O \rightleftharpoons NH_4^+ + OH^-,$$

and the equilibrium condition expression is

$$\frac{[NH_4^+][OH^-]}{[NH_3]} = K';$$

K' can be evaluated from K_w and K_a for NH_4^+ found in Table 6.1. Use the general method of attack already illustrated in Examples A, B, and C, evaluating $[NH_4^+]$ and $[NH_3]$ in terms of $[OH^-]$. Make the appropriate simplifying assumptions to avoid cumbersome calculations. Finally test your result.

This problem is closely similar to Example B, where the weak base was OAc^- instead of the weak base NH_3 involved here.

Problem 6A.4 Find the $[H_3O^+]$ and NH_3 concentrations in a solution prepared by weighing out 2.14 g of NH_4Cl crystals and dissolving them in sufficient water to produce a final volume of 200.0 ml. NH_4Cl is a strong electrolyte completely ionized to NH_4^+ and Cl^-. NH_4^+ is a weak acid. See K_a for NH_4^+ in Table 6.1.

The Experimental Determination of the Acidity Constant

The relationship expressed in Eq. (6A.14) or (in approximate form) in Eq. (6A.15) suggests a good experimental procedure for evaluating the acidity constant. Suppose, for instance, that we make up a solution by dissolving 0.06 mole (4.44 g) of propionic acid $CH_3CH_2CO_2H$ and 0.06 mole (5.76 g) of sodium propionate $CH_3CH_2CO_2Na$ in water to make one liter of solution. Then by Eq. (6A.15),

$$K_a = [H^+].$$

If we now measure the $[H^+]$ by means of a hydrogen electrode system, the numerical value for K_a is approximately determined. If we are unhappy about the approximations involved in Eq. (6A.15), we can use Eq. (6A.14), which was derived without any simplifying assumptions. The value for K_a is then given by

$$K_a = 1.3 \times 10^{-5} \frac{(0.060 + 1.3 \times 10^{-5} - 7.7 \times 10^{-10})}{(0.060 - 1.3 \times 10^{-5} + 7.7 \times 10^{-10})} = 1.3006 \times 10^{-5}.$$

All this nonsense with figures boils down to the fact that $[H^+]$ and $[OH^-]$ after we found them from the *measured* pH could reasonably be disregarded when *added* to or *subtracted* from 0.060 as in Eq. (6A.15).

Problem 6A.5 A convenient alternative to weighing the acid and its conjugate base (that is, HA and NaA) makes use of a volumetric technique. Suppose we have a sample of the acid resulting from the ozonolysis of the alkene $C_6H_5CH{=}CH_2$ and the subsequent oxidation of the 7-carbon fragment to an acid. We also have at our disposal a 0.1000-M NaOH solution, a burette, and an apparatus for measuring pH, but no analytical balance for accurate weighing.

Devise an experimental procedure for determining K_a for the acid.

Buffer Solutions

The solution resulting from dissolving a weak acid HA and its conjugate base NaA has the useful property that it can absorb moderate quantities of strong acid or base without appreciable change in pH. Such a solution is known as a buffer solution and is said to have buffering action against strong acids or bases. The buffering reactions are

$$HA + OH^- \rightarrow H_2O + A^-,$$
$$A^- + H^+ \rightarrow HA.$$

We see from Example C that a simple relationship exists between the $[H^+]$ and K_a for the acid:

$$[H^+] = K_a \frac{C_a}{C_b}.$$

If equimolar quantities of HA and NaA are used, then

$$[H^+] \cong K_a \qquad \text{and} \qquad pH \cong pK_a,$$

where $pK_a = -\log K_a$.

Thus if we want a buffer to operate at pH 7, we choose a weak acid of $pK_a \cong 7$ or $K_a = 10^{-7}$. The $[H^+]$ of a buffer of any given acid is proportional to C_a/C_b, and hence the $[H^+]$ and pH of a buffer are not changed upon dilution, since C_a and C_b are equally affected.

Human blood is buffered in a rather complex manner because it contains the conjugate acid-base mixture $H_2PO_4^-$, HPO_4^{2-}, as well as H_2CO_3 and HCO_3^-. Dissolved proteins, which also act as weak acids or bases, also contribute to the buffering action. Your pH is 7.2 and you had better keep it that way, otherwise you would be very sick indeed. Fortunately you can eat a pickle, even a very large one, whenever you like without lowering your pH beyond the danger point. Your built-in buffer system will take care of that. But don't try swallowing the equivalent of the acetic acid in the pickle in the form of a concentrated HCl solution or of a pellet of NaOH.

Ion Concentrations in Solutions of Strong Acids or Bases

In Chapter 5 we considered the evidence to support the notion that strong electrolytes are completely ionized. Typical would be HCl, NaOH, and NaCl, representing the classes *acids*, *bases*, and *salts*, respectively. HCl and NaOH differ from NaCl in that each has an ion in common with water. In dilute solutions of HCl, even as dilute as 10^{-5} molar, the ionization of the solvent water contributes to only an insignificant degree to the $[H^+]$. However, for even more dilute solutions, the $[H^+]$ from the water might become relatively significant. It will be interesting to see the extent to which this would affect the total $[H^+]$.

Let us consider a 10^{-6} M solution of HCl. If it were not for the ionization of water, the $[H^+]$ would be 10^{-6} M. However, water is a weak acid and it must contribute to the total $[H^+]$. How can we find this total $[H^+]$ of the 10^{-6} M HCl solution? We have only three solute species H^+, OH^-, and Cl^-. We may write three simultaneous equations

$$\text{e.n.} \qquad [H^+] = [Cl^-] + [OH^-], \qquad (1)$$

$$\text{m.b.} \qquad [Cl^-] = 10^{-6} \, M, \qquad (2)$$

$$K_w \qquad [H^+][OH^-] = K_w. \qquad (3)$$

Then
$$[H^+] = 10^{-6} + \frac{10^{-14}}{[H^+]},$$

or
$$[H^+]^2 - 10^{-6}[H^+] - 10^{-14} = 0.$$

By the quadratic formula,

$$[H^+] = \frac{+10^{-6} \pm \sqrt{10^{-12} + 4 \times 10^{-14}}}{2},$$

$$[H^+] = 1.01 \times 10^{-6} M.$$

If no HCl were present, the $[H^+]$ would be $10^{-7} M$. If this quantity is added to the $10^{-6} M$ $[H^+]$ due to the HCl, we would have a total $[H^+]$ of $1.10 M$. However, the ionization of the water is repressed here by the $[H^+]$ from the HCl, with the result that we have only $0.01 M$ $[H^+]$ contributed from the water.

Problem 6A.6   Calculate the $[H^+]$ in a 10^{-7}-M HCl solution.

Similar calculations could be made to find the $[OH^-]$, and hence the $[H^+]$ and pH, of very dilute solutions of strong bases such as NaOH or KOH.

Problem 6A.7   You might devise a problem for a very dilute KOH solution, then solve it.

6A.3 SOLUBILITY EQUILIBRIA OF SLIGHTLY SOLUBLE SALTS

Another, similar type of ionic equilibrium problem which is subject to simple treatment is the dissolution of slightly soluble salts like silver chloride. In water the reaction is

$$AgCl(solid) \rightarrow Ag^+(aq) + Cl^-(aq). \tag{6A.16}$$

Since the concentration of the *solid* AgCl in contact with the solution doesn't vary, the equilibrium constant for the reaction is written simply

$$K_{sp} = [Ag^+][Cl^-], \tag{6A.17}$$

where "sp" denotes "solubility product." These equilibrium constants are called solubility products because the equations involve only products of concentration of the aqueous ions raised to appropriate powers since the concentration of the solid salt is not a variable.

A good first problem is the question of the solubility of AgCl in water. If there is no source of $Ag^+(aq)$ other than the salt, calculation of the solubility reduces to determination of the concentration of silver ion. Given K_{sp} ($= 10^{-10}$ for AgCl) we write (6A.17) and some second equation because there are two unknowns. Since the dissolved Ag^+ ions come only from the dissolution of AgCl crystals, a second equation which is a *mass balance* but might also be regarded as an electroneutrality equation for this solution is $[Ag^+] = [Cl^-]$. From this it follows by substituting Ag^+ for Cl^- in (6A.17) that

$$K_{sp} = [Ag^+][Ag^+] = [Ag^+]^2$$

or

$$Ag^+ = \sqrt{K_{sp}} = \sqrt{10^{-10}} = 10^{-5}\, M,$$

which is the solubility of the salt in water.

Problem 6A.8 Find the solubility of $CaCO_3$ in water. K_{sp} for this salt is 8.7×10^{-9}.

Problem 6A.9 Find the solubility of AgCl in a solution 0.1 M in NaCl. Note that finding the solubility is still equivalent to finding Ag^+ but that there would now be an electroneutrality equation: $Na^+ + Ag^+ = Cl^-$. Also, it will be a good approximation to assume that Na^+ is much larger than Ag^+. (Why?)

Problem 6A.10 Find the solubility of AgCl in a 0.1 M solution of the soluble silver salt $AgNO_3$. Note that in the presence of the excess silver salt the solubility is no longer equal to the concentration of silver ion but is equal to the concentration of chloride ion. (Why?)

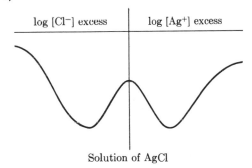

Fig. 6A.1. Solubility of AgCl in presence of excess common ions.

Problems 6A.9 and 6A.10 illustrate an obvious and important prediction of the solubility product concept. It is called the common ion effect. An excess of either of the ions produced by the dissolution of the salt (common ions) should reduce the solubility of the salt. This is reasonable. It amounts to suggesting that addition of sources of the products of Eq. (6A.16) shifts the equilibrium position to the reactant side. However, Fig. 6A.1 shows that there are serious limitations on these predictions, at least in the case of AgCl. Small amounts of added Ag^+ and Cl^- are found *experimentally* to reduce the solubility of AgCl but larger amounts seem to do just the reverse! What's wrong?

First, any calculation of the equilibrium concentrations must take into consideration all relevant reactions. In the analysis of AgCl solubility this has not been done. For example, the reaction

$$AgCl(\text{solid}) + Cl^-(\text{aq}) \rightarrow AgCl_2^-(\text{aq})$$

does occur to a significant extent in the presence of excess chloride ion and modifies the solubility of AgCl(solid). In fact one important way which has been used to show that this reaction must occur is the study of the solubility of AgCl in the presence of excess chloride with the assumption that appropriate equilibrium constants must apply. But there is a very important and very general additional difficulty of an entirely different sort.

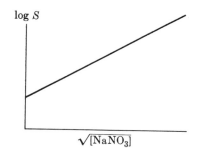

Fig. 6A.2. Solubility of AgCl in presence of inert salts like NaNO₃.

Figure 6A.2 shows the solubility (on a log scale) of AgCl as a function of the concentration of added $NaNO_3$. Neither the Na^+ nor the NO_3^- ion is involved in any reaction with Ag^+, Cl^-, or AgCl(solid). Yet these simple *spectator* ions are seen to increase the solubility of AgCl. Is the concept of equilibrium constant, and consequently the Nernst equation from which it was derived, faulty? The answer must be a qualified yes. The equations as we have treated them are accurate only over narrow concentration ranges for reactions in which the number of ions is changing. There is an important effect which has so far been relegated to an obscure footnote on page 152 that renders the preceding discussion only a *first approximation*.

6A.4 IONIC ATMOSPHERE EFFECTS

Why should ions which are "not involved" influence the position of equilibrium? Recall that in the theory of strong electrolytes an effect of any ion on any other, depending only on charge and concentration, was recognized to account for details of conductance and colligative properties. This effect follows from the electrostatic forces between ions and the fact that, on the average, it is more probable that ions near any given ion will have opposite charge, not the same charge. A portion of an electrolyte solution might look something like the model in Fig. 6A.3. Note the "atmosphere" of ions about the central ion. On the average, more ions of negative charge are close to the central positive ion. The actual distances will depend on concentration. The larger the concentrations, the shorter, on the average, the distances between ions. The question which we should ask if we wish to assess the effect of the ionic atmosphere on the equilibrium position of reactions which produce or consume ions in solution is, "Is it easier to form an ion in the presence of an ion atmosphere than in its absence?" This question reduces to that of the free energy of an ion in the presence of an ionic atmosphere. Does the presence of the atmosphere lower the free energy of an ion and hence make its formation more favorable? If so, higher ionic concentrations which produce closer ion atmospheres will favor ion production.

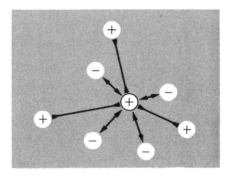

Fig. 6A.3. Possible distribution of ions about a specified positive ion. Note shorter average distance to negative ions.

Since ions of opposite charge are closer to each other than ions of like charge, it should be clear that the ionic atmosphere *lowers* the potential energy of an ion. It is not obvious that this corresponds to a *lower* free energy, but detailed analysis beyond the scope of this discussion shows that it does indeed. The result described is quite general. Any reaction (dissolution, acid-base) in which ions are *formed* will be *favored* by increasing the concentrations of any kind of ions in the solution and thereby bringing in the ionic atmosphere to shorter average distances of separation.

Perhaps you would guess that these effects can be given quantitative explanation in the case of dilute solutions by the Debye–Hückel theory of electrolyte solutions. The linear plot in the square root of the concentration of the "inert electrolyte" in Fig. 6A.2 is suggestive of the conductance equation described in Section 5.9. This is the case, once one adopts an appropriate way of computing the concentration of charge, called the *ionic strength*, in the solution.

Now, what can be said for the computations made in the examples of this Appendix? They are certainly "rough approximations" until the effects of ionic strength have been considered. Fortunately, such rough approximations are often very useful. Also, the techniques of calculation illustrated are the first step in more sophisticated analysis.

Seminar: A Diprotic Acid Calculation

We have dealt, so far, with such monoprotic acids as HOAc and NH_4^+, and with bases such as NH_3 and OAc^- which can donate or accept only one proton per molecule or ion. It is interesting to extend these ideas to the more complex case of the diprotic acid and its conjugate ion bases.

The weak acid H_2CO_3 ionizes in two steps with the consequence that there are two equilibrium constants, K_{a_1} and K_{a_2}.

Step 1:

$$H_2CO_3 \rightleftharpoons H^+ + HCO_3^-, \qquad \frac{[H^+][HCO_3^-]}{[H_2CO_3]} = K_{a_1} = 4.30 \times 10^{-7};^*$$

Step 2:

$$HCO_3^- \rightleftharpoons H^+ + CO_3^{2-}, \qquad \frac{[H^+][CO_3^{2-}]}{[HCO_3^-]} = K_{a_2} = 5.61 \times 10^{-11}.^*$$

The conjugate bases CO_3^{2-} and HCO_3^- can each accept a proton according to steps 1 and 2 below. This process is frequently referred to as the two stages of hydrolysis of Na_2CO_3.

Step 1:

$$CO_3^{2-} + H_2O \rightleftharpoons HCO_3^- + OH^-, \qquad \frac{[HCO_3^-][OH^-]}{[CO_3^{2-}]} = K_{b_1};$$

Step 2:

$$HCO_3^- + H_2O \rightleftharpoons H_2CO_3 + OH^-, \qquad \frac{[H_2CO_3][OH^-]}{[HCO_3^-]} = K_{b_2}.$$

Show that

$$K_{b_1} = K_w/K_{a_2} = 1.78 \times 10^{-4} \qquad \text{and} \qquad K_{b_2} = K_w/K_{a_1} = 2.33 \times 10^{-8}.$$

CALCULATIONS FOR SODIUM CARBONATE (Na_2CO_3)

Now suppose we want to find the $[H^+]$ in a solution of Na_2CO_3; say the concentration of Na_2CO_3 (C_b) is 0.100 M. You may attack this problem along the general lines outlined in Example B, page 185, but with the difference that you are dealing here with the base CO_3^{2-} which can accept *two* protons as in steps 1 and 2 shown above.

There are six variable species in the 0.100 M Na_2CO_3 solution. Write six equations involving these species. [*Hint:* You have one equation for e.n., two for m.b., and the expressions for K_w, K_{a_1}, and K_{a_2}. (You may substitute the

---

* Values for 25°C. Recall also that $K_w = 10^{-14}$ at 25°.

expressions for K_{b_2} and K_{b_1} if expedient; however, these are derived from K_w, K_a, and K_{a_2} and do not represent independent relationships.)] You may now proceed by one of the two alternatives suggested in step 3, page 185.

1. Method of Simplifying Assumptions. This is based on chemical judgment and experience. Look at the values for K_{b_1} and K_{b_2}. What is the ratio of K_{b_1}/K_{b_2}? Argue that step 2 of the hydrolysis takes place to a very small extent compared to step 1. We might then neglect step 2 altogether and base our solution of the problem on step 1. *Find the pH of the solution using this approximation.* [*Hint:* Look at the previously derived equation for OAc⁻.]

The above method has minimized the mathematics by making maximum use of chemical knowhow. Since you are just in the process of acquiring this knowhow, you may find that you would have been at a loss without the help which has been set forth here in leading you through this problem. There is *another way.* This minimizes the need for chemical insight and makes maximum use of algebra.

2. Method Based on the Exact Solution of the Simultaneous Equations. You have written six simultaneous equations involving the six variable species in a sodium carbonate solution of any specified concentration C_b. (Perhaps at this point you should find out from a reference or your instructor whether your equations are correct.) With these equations you can forget about chemistry and proceed to the right answer by means of algebra. By a rather complicated application of the methods for the solution of simultaneous equations which you learned in high school, you can boil your six equations down to a single equation in one variable. *You are invited to do this.* It is convenient to find the equation in terms of [H⁺]. For sodium carbonate of concentration 0.1 M, this is the fourth-degree equation

$$[H^+]^4 + (0.2 + K_{a_1})[H^+]^3 + (0.1\,K_{a_1} + K_{a_1}K_{a_2} - K_w)[H^+]^2 -$$
$$K_wK_{a_1}[H^+] + K_{a_1}K_{a_2}K_w = 0. \qquad (6A.18)*$$

You may find it convenient to substitute simpler symbols such as $K_{a_1} = C_1$, $K_{a_2} = C_2$ and $K_w = C_3$ while carrying out the algebra. For any specified concentration C_b of the Na_2CO_3 solution, substitute 0.2 C_b and 0.1 C_b for 0.2 and 0.1 in the second and third terms of the equation.

If you have access to a computer it is easy to program this equation for computer solutions. Using the values $K_{a_1} = 4.30 \times 10^{-7}$ and $K_{a_2} = 5.61 \times 10^{-11}$, the computer gave us the results shown in the table on p. 196.

---

* We are indebted to Mr. Roger Ingraham of Voorhees College, Denmark, South Carolina, for the derivation of Eqs. (6A.18) and (6A.19) from the simultaneous equations for Na_2CO_3 and $NaHCO_3$. Mr. Ingraham also wrote the computer program for solving Eqs. (6A.18) and (6A.19), which was used to obtain the results given on page 196.

C_b	$[OH^-]$, moles/liter	$[H^+]$, moles/liter
0.10	4.13×10^{-3}	2.42×10^{-12}
0.08	3.69×10^{-3}	2.71×10^{-12}
0.06	3.18×10^{-3}	3.14×10^{-12}
0.04	2.58×10^{-3}	3.87×10^{-12}
0.02	1.80×10^{-3}	5.55×10^{-12}
0.01	1.25×10^{-3}	8.00×10^{-12}
0.001	3.24×10^{-4}	2.92×10^{-11}

How do your values calculated by Method 1 compare with the data of this table?

The concepts and equations which have been developed here would apply to other diprotic acids. Method 2 would be rigidly applicable. Some of the simplifying assumptions of Method 1 may require adjustment to fit the specific case. For instance, could we neglect $[OH^-]$ in comparison to C_b for Na_2S?

CALCULATIONS FOR THE HYDROLYSIS OF OTHER ION BASES: S^{2-}, $AsO_4{}^{3-}$, $PO_4{}^{3-}$

H_2S ionizes in two stages to give HS^- and S^{2-} and conversely Na_2S, or better, the S^{2-} ion hydrolyzes to give HS^- and H_2S. K_{a_1} and K_{a_2} for H_2S are given respectively as 9.1×10^{-8} and 1.1×10^{-12}. By computer solution of Eq. (6A.18), we find 0.1 M Na_2S that $[OH^-] = 2.60 \times 10^{-2}$. Treat this problem by the method of simplifying assumptions and compare your results with those obtained by computer. Note that Eq. (6A.18) applies equally well to the conjugate bases of any weak diprotic acid. It would even be possible to apply the method to the sodium salts triprotic acids such as Na_3PO_4 or Na_3AsO_4 if we disregarded the third stage of hydrolysis of the $PO_4{}^{3-}$ or $AsO_4{}^{3-}$ as being negligible in comparison to the other two. The 47th edition of the *Handbook of Physics and Chemistry*, page D-87, gives K_a values for H_2S, H_3PO_4, and H_3AsO_4.

You may want to work out the $[H^+]$ for Na_2SO_4 and $NaHSO_4$ at specified concentrations, say, 0.05 M. H_2SO_4 is a *strong* electrolyte with respect to the dissociation $H_2SO_4 \rightarrow H^+ + HSO_4{}^-$, but the $HSO_4{}^-$ ion is a *weak* acid. Use $K_a = 10^{-2}$ for $HSO_4{}^-$.

THE CASE OF SODIUM BICARBONATE (NaHCO₃)

It is interesting to carry out calculations for $NaHCO_3$ similar to those you have already done for Na_2CO_3. With $NaHCO_3$, you will again find six variable species and six simultaneous equations. Five of the six equations are identical for the $NaHCO_3$ and Na_2CO_3 cases. You may let C' be equal to the number of moles of solid $NaHCO_3$ dissolved to make one liter of solution.

It is possible to work out the $NaHCO_3$ case by the method of simplifying assumptions although a considerable degree of ingenuity and chemical know-how is required. An interesting result can be shown. It turns out that $[H^+] = \sqrt{K_{a_1}K_{a_2}}$, regardless of the total concentration C'.

If you combine your six equations for $NaHCO_3$ into a single equation in terms of the $[H^+]$ you get the equation

$$[H^+]^4 + (C' + K_{a_1})[H^+]^3 + (K_{a_1}K_{a_2} - K_w)[H^+]^2 -$$
$$(K_w K_{a_1} + C'K_{a_1}K_{a_2})[H^+] - K_{a_1}K_{a_2}K_w = 0 \qquad (6A.19)$$

A computer solution of this equation gives $[H^+] = 4.87 \times 10^{-9} M$ regardless of the value of C' over a considerable range (say from $0.1\ M$ to $0.01\ M$). *Can you explain from Eqs. (6A.18) and (6A.19) why the $[H^+]$ is dependent on C_b for Na_2CO_3 but independent of the total concentration of $NaHCO_3$?*

Useful Reading

Look up diprotic acids in the following:

J. N. Butler, *Ionic Equilibria, A Mathematical Approach*, Addison-Wesley, Reading, Mass., 1964.
T. R. Hogness and W. C. Johnson, *Qualitative Analysis and Chemical Equilibrium*, Henry Holt, New York, 1964.
E. J. King, *Qualitative Analysis and Electrolytic Solutions*, Harcourt, Brace and Company, New York, 1959.

READINGS AND EXTENSIONS

There are two directions in which the material of the present chapter should be extended. The first is theoretical. Thermodynamics can be developed generally starting with certain fundamental postulates about work and thermal quantities. Several excellent paperback monographs are available which do this for students with only a first introduction to calculus. Some of these are:

J. Waser, *Basic Chemical Thermodynamics*, W. A. Benjamin, New York, 1966.
B. H. Mahan, *Elementary Chemical Thermodynamics*, W. A. Benjamin, New York, 1964.
L. K. Nash, *Elements of Chemical Thermodynamics*. Addison-Wesley, Reading, Mass., 1962.

The second direction of extension is to the calculation of equilibrium concentration and the composition of solutions. This is introduced in the Appendix to Chapter 6 from a point of view that was inspired by J. N. Butler's book *Solubility and pH Calculations*, Addison-Wesley, Reading, Mass., 1964 (paperback). Equilibrium calculations, of course, involve the acquisition of some skills requiring drill work. The subject is probably well suited to programmed instruction. There are some paperback programmed texts available. Two are:

G. Barrow, M. Kenney, J. Lassila, R. Litle, and W. E. Thompson, *Understanding Chemistry 3: Chemical Equilibrium*, W. A. Benjamin, New York, 1967.
H. N. Christensen, *pH and Dissociation*, W. B. Saunders Company, Philadelphia, 1964.

CHAPTER 7

Chemical Families—The Periodic Law and Atomic Structure

7.1 INTRODUCTION

Up to this point, the ancient attitude toward the chemical elements has been an adequate basis for discussion. Elements have been regarded as the simplest constituents of matter, and it has been assumed that it would not be fruitful to ask for any special pattern in their behavior. But, of course, such a point of view is negative. Shouldn't a search be conducted to see if there are regularities in the relationships among elements? If the search is a success, it might become possible to develop a *theory* of the *structure of the atoms* that compose the elements to explain the relationships. Actually, scientific imagination is not easily bridled. From the earliest development of the Daltonian atomic theory there have been efforts to find regularities in the list of the elements. Much of this effort centered around the atomic weights. The earliest proposal is the famous Prout hypothesis of 1815, which proposed that all atomic weights are multiples of the atomic weight of hydrogen and that hydrogen is the *protyle* or first matter out of which all other elements are formed. This idea had intriguing possibilities in a day when atomic weights were still quite imprecise, but it failed in the face of precise atomic-weight determinations which showed clearly that most weights are not integral multiples of hydrogen.

Later in the nineteenth century the question of a connection of properties to atomic weight received more fruitful development. That will be the main subject of this chapter. But the successes had to wait for a considerable extension of the list of elements and the recognition that elements appear to fall into families.

7.2 FAMILIES OF ELEMENTS

The Halogens

Consider the four elements fluorine, chlorine, bromine, and iodine with atomic weights respectively 19.0, 35.5, 79.9, and 126.9. All these elements form similar gaseous molecules F_2, Cl_2, Br_2, and I_2. They are the four elements which are commonly found in nature as simple monatomic anions, X^-. For example, fluorine is often obtained from the mineral fluorspar, CaF_2. Chlorine and bromine are present in important salts of brine (e.g., NaCl and NaBr). It is interesting to note that although salts like NaI and KI containing the I^- ion are easily obtained, iodine is usually derived from a salt that occurs in KNO_3 deposits in nature, namely potassium iodate, KIO_3. All the halogen elements form hydrogen compounds of the binary hydride type, HX. These are all gaseous, molecular substances at room temperature when pure. They dissolve in water to form electrolyte solutions which are of course acidic ($H^+ + X^-$). All are strong electrolytes in dilute aqueous solution except HF, which has an equilibrium constant for dissociation of about 7×10^{-4} (about forty times larger than acetic acid). All the halogens except fluorine can exist in the positive oxidation state, 5+, in a polyatomic anion analogous to the iodate (IO_3^-) mentioned above. This superficial listing of chemical properties suggests strong resemblance, but by no means identity in the chemical behavior of the halogens. What is especially interesting is that many of the differences *evolve* in a regular fashion, if one considers the elements in order of increasing atomic weight.

Consider first the ease of formation, or stability, of the X^- species. We can do this by comparing the standard potentials of the electrode at which the element is reduced to the anion in aqueous solution. The first thing to note about the values which are tabulated in Table 7.1 is that they are all positive.

Table 7.1

Electrode potentials for reduction of halogens

Reaction	$\mathcal{E}^0$, *volts*
$\frac{1}{2}F_2 + e^- \rightleftharpoons F^{-*}$	2.87
$\frac{1}{2}Cl_2 + e^- \rightleftharpoons Cl^-$	1.36
$\frac{1}{2}Br_2 + e^- \rightleftharpoons Br^-$	1.07
$\frac{1}{2}I_2 + e^- \rightleftharpoons I^-$	0.54

* The ions are in aqueous solution solvated by water.

All the halogens will be spontaneously reduced to the *halide* anions in acidic aqueous solution when coupled with oxidation of H_2 to H^+. Looking down the list, we see that the potentials decrease regularly with the increasing atomic weight of the halogen. Fluorine is the most readily reduced, hence the best oxidizing agent, of all the elements. Since the standard electrode potential for the oxidation of water to O_2 and H^+ is 1.23 V, the reaction $F_2 + H_2O = 2HF + O_2$ is favorable to the extent of 1.64 V at standard conditions. No chemical oxidizing agent is strong enough to oxidize fluoride ion to fluorine. Fluorine gas is produced by the *electrolytic* oxidation of a fused salt. In contrast, a number of common oxidizing agents should be able to oxidize chloride ion to chlorine including, for example, the cerium(IV)/cerium(III) couple:

$$Ce^{4+} + e^- \rightleftharpoons Ce^{3+}; \qquad \varepsilon^0 = 1.61 \text{ V.}$$

Even permanganate (MnO_4^-) should do, but the reaction is too slow under most circumstances. Permanganate will oxidize bromide to bromine readily and will oxidize iodide, I^-, beyond iodine, I_2, to the positive oxidation state of five (iodate, IO_3^-).

Problem 7.1   One way to test for the presence of the ions Cl^-, Br^-, I^- in solution is to precipitate them as their very insoluble silver salts, AgX. Unfortunately this test does not distinguish among halides. Consult a table of electrode potentials and suggest oxidizing agents which would permit you to determine whether a mixture of halides in aqueous solution contained one or several of the ions Cl^-, Br^-, and I^-. *Hint:* oxidize some and test for those remaining with Ag^+.

The regular trend in the ease of oxidation as the atomic weight of the halogen increases is reflected in the behavior in positive oxidation states as well as in the ease of oxidation of halide ions. The halogens appear in positive oxidation states most commonly as the central atom of a polyatomic oxyacid or oxyanion. We have mentioned the 5+ state XO_3^-, which is known for $X = Cl, Br, I$. There is also a plus one state represented for these three elements by HOCl, HOBr, and HOI. These three are acids of moderate strength so that the corresponding anions exist in basic solutions. The ease of oxidation of the element to the 1+ oxidation state follows a regular trend in the atomic weights. The electrode potentials are -1.63 V, -1.59 V, and -1.45 V. The less negative standard potential, of course, indicates greater ease of oxidation. In this context, the glaring absence of the positive oxidation states of fluorine may be simply a natural extension of the trend in oxidizability to the lightest element of the family. Chlorine does present some genuinely exceptional behavior in that plus three and plus seven oxidation states occur in the more or less familiar species ClO_2^- and ClO_4^-, but the dominant impression of the chemistry of these elements is of a "family" with a regular progression of behavior as the atomic weights increase.

The Alkaline Earth Metals

As a second example of family relationships in the chemistry of a group of elements, it would be difficult to choose a group less like the halogens than the alkaline earth metals. Let's consider the elements beryllium (Be), magnesium (Mg), calcium (Ca), strontium (Sr), and barium (Ba) with atomic weights 9.01, 24.3, 40.1, 87.6, and 137.3, respectively. All are found in *only one* oxidation state, namely plus two. Commonly they are monatomic cations in salts which are strong electrolytes in aqueous solution. These elements usually occur in nature as these salts of the plus two ions. The pure elements may be obtained by electrolytic reduction of the fused salts or by reduction of oxides with appropriate reducing agents. Two typical reductions are:

$$MgO + C \xrightleftharpoons{\text{heat}} Mg + CO,$$

$$3BaO + 2Al \xrightleftharpoons{\text{heat}} 3Ba + Al_2O_3.$$

The elements all occur as solids at room temperature and are typically *metallic;* that is, they are reasonably good conductors of heat and electricity and are lustrous and malleable.

The degree of similarity within the family is perhaps greater in this case than for the halogens. But, again, an examination of the quantitative aspects of the properties reveals differences that often show a regular progression with increasing atomic weight. For example, the water solubilities of the hydroxides, $M(OH)_2$, and sulfates, MSO_4, are useful for accomplishing the separation of the different alkaline earth metal ions. The hydroxides of the lighter elements are quite insoluble in water, whereas those of the heavier ones are quite soluble. There is an exactly reversed trend for the sulfates. The sulfates of the lighter elements are much more soluble in water than the quite "insoluble" sulfates of strontium and barium. Although all of the alkaline earth metals have standard electrode potentials which are favorable for oxidation to the plus two ions, the oxidation becomes progressively more favored down the series from Be to Ba. This is reflected in the progressively more negative values of standard reduction potentials down the series from Be to Ba as indicated in Table 7.2. The negative reduction potential corresponds to a positive oxidation potential. The ease of oxidation of these elements (or equivalently their ability as reducing agents) is reflected by the fact that they all reduce the relatively inert species, molecular nitrogen, to form salts of the relatively rare N^{3-} ion. The magnesium reaction is

$$3Mg + N_2 \xrightarrow{\text{heat}} Mg_3N_2.$$

Problem 7.2   The ash from burning a magnesium raiiroad flare gives off a gas when it is wetted. Can you guess the nature of the gas and write an equation for the reaction taking place?

Table 7.2

Standard potentials for reduction of alkaline earth metal ions

Reaction	$\mathcal{E}^0$, *volts*
$Be^{2+} + 2e \rightleftharpoons Be$	-1.85
$Mg^{2+} + 2e \rightleftharpoons Mg$	-2.37
$Ca^{2+} + 2e \rightleftharpoons Ca$	-2.87
$Sr^{2+} + 2e \rightleftharpoons Sr$	-2.89
$Ba^{2+} + 2e \rightleftharpoons Ba$	-2.90

In this brief survey, a set of elements, obviously very different from the halogens, are seen to have a family resemblance among themselves. Their chemistries are similar, and the *differences of detail* appear to be related (directly or indirectly) to the atomic weights of the elements.

7.3 THE "EXTENSION" OF THE LIST OF ELEMENTS

Lavoisier's rule that a substance is elemental if it is not broken down into simpler substances when treated chemically led to the development of several general approaches to the preparation and isolation of elements that were exploited vigorously in the early years of the nineteenth century. These included electrolytic decomposition and replacement of other metals by potassium. In the period between 1790 and 1844, 31 new elements were discovered. In 1844 the list of elements had 58 entries. It was not until 1860 that another element (cesium, named for the blue color it produces in a flame) was discovered and this by a new technique that played a very large role in subsequent searches for elements (and later a major role in the study of the internal structure of atoms). The technique is spectroscopy. Its significance will be discussed in Chapter 8. For the present we need only note that the application of the spectroscopic technique led, in addition to the discovery of cesium, to the discovery of thallium (1861), rubidium, named for its red color (1861), indium (1863), gallium (1875),· and the inert gases which were extremely important for the later theories of atomic structure (1895–1898).

The conjunction in the 1860's of a renewed interest in the list of elements with the solution to the atomic-weight problem (Cannizzaro's method was published in 1858—see Chapter 1) set the stage for the successful organization of the list of elements on the basis of atomic weights. This organization emphasized and enriched the notion of family relationships and had two profound results. The first, an important result, is the development of a systematic framework on which to hang discussions of comparative chemical behavior. The other, a result of most fundamental importance, is to suggest *strongly* that atoms have internal structure that might be understood.

7.4 DÖBEREINER'S TRIADS

The work of the 1860's had precursors. It is instructive to approach the climax through these earlier efforts. As early as 1816, J. W. Döbereiner noticed that *his* atomic weight of strontium (50) was the arithmetic mean (or average) of its two close relatives, calcium and barium, which were at that time thought to have atomic weights 27.5 and 72.5, respectively. His initial reaction was (justifiably) to question the independent existence of strontium as an element. But further exploration revealed that chemically similar elements frequently appeared in *triads*, the middle member of which has an atomic weight near the mean of the lighter and heavier. He was convinced of this interpretation by a *prediction* of the atomic weight of bromine from those of chlorine and iodine which was promptly confirmed by a determination made by Berzelius, and he announced a law of triads in 1829. Some of his well-defined triads are shown below.

Li	Ca	Cl	S
Na	Sr	Br	Se
K	Ba	I	Te

Problem 7.3 Check Döbereiner's triad law using modern atomic weights for the triads tabulated. (This idea of atomic weight averaging will reappear subsequently!)

Döbereiner's early idea was one of the clearest and most fruitful of a series of speculations about relationships among atomic weights that appeared prior to 1860. To indicate the extent of activity, we note that numerical classifications were suggested by P. Kremers (Germany), J. H. Gladstone (England), S. P. Cooke (United States), J. B. A. Dumas (France), and A. E. B. de Chancourtois (France).*

No.		*No.*		*No.*		*No.*		*No.*		*No.*		*No.*		*No.*	
H	1	F	8	Cl	15	Co & Ni	22	Br	29	Pd	36	I	42	Pt & Ir	50
Li	2	Na	9	K	16	Cu	23	Rb	30	Ag	37	Cs	44	Os	51
G	3	Mg	10	Ca	17	Zn	24	Sr	31	Cd	38	Ba & V	45	Hg	52
Bo	4	Al	11	Cr	19	Y	25	Ce & La	33	U	40	Ta	46	Tl	53
C	5	Si	12	Ti	18	In	26	Zr	32	Sn	39	W	47	Pb	54
N	6	P	13	Mn	20	As	27	Di & Mo	34	Sb	41	Nb	48	Bi	55
O	7	S	14	Fe	21	Se	28	Ro & Ru	35	Te	43	Au	49	Th	56

Fig. 7.1. Newlands' periodic table of 1866. (From *Chemical News* [1866].)

7.5 NEWLANDS' OCTAVES

In 1866, J. A. R. Newlands read a paper, "The Law of Octaves and the Causes of Numerical Relations Between Atomic Weights," before the members of the Chemical Society of London. Newlands' "law of octaves" suggested that "The eighth element starting from a given one [in the order of increasing atomic weight] is a kind of repetition of the first, like the eighth note in an octave of music." This led to a "periodic table" of the elements showing the repetitions. Newlands' table is Fig. 7.1. Note that the horizontal rows are the similar elements. Many significant family relationships are revealed, but there are certainly

---

*For more details, see A. J. Ihde, *The Development of Modern Chemistry*, Harper & Row, New York, 1964.

some infelicitous matchings (e.g., P and Mn!). The notion of a repetition every eighth element is very close to the idea most fruitfully developed by the Russian chemist, Dmitri Ivanovitch Mendeléeff, but it needs a more cautious formulation than it was given by Newlands. Moreover, Newlands' table of 1866 left no blanks for *as yet undiscovered* elements. This represented a change from a preliminary version he had prepared in 1864. In this change, he missed the feature which made the later contribution of Mendeléeff so compelling; Mendeléeff successfully *predicted new elements*. The response of members of the Chemical Society of London to Newlands' interesting speculations was far from favorable. At the meeting, one Carey Foster, who holds no other claim to fame, rose to ask facetiously if Newlands had ever sought to classify the elements in alphabetical order. The unsympathetic hearing was reported in *Chemical News*, but the full paper was refused publication in the *Journal of the Chemical Society.**

7.6 THE PERIODIC LAW: MENDELÉEFF AND MEYER

The publications of D. I. Mendeléeff and, starting a little later, the publications of the German chemist Julius Lothar Meyer, presented the relationship of properties of elements to atomic weights in convincing fashion. Shall we let Mendeléeff introduce his ideas? Here is an excerpt which explains his thinking in 1869.

The investigations regarding the simple relations of atomic weights have caused many . . . to point out the numerical relations between the atomic weights of those elements that form a group [family]; but, so far as I know, they have not led to a systematic arrangement of *all* known elements . . . When I undertook to write a handbook of chemistry entitled *Foundations of Chemistry* [published 1869–1871], I had to make a decision in favor of some system of elements in order not to be guided in their classification by accidental, or instinctive, reasons but by some exact, definite principle . . .

. . . [Everybody] does understand that in all changes of properties of elements, *something* remains unchanged, and that when elements go into compounds, this material something represents the [common] characteristics of compounds the given element can form. In this regard only a numerical value is known, and this is the atomic weight appropriate to the element. The magnitude of the atomic weight, according to the actual, essential nature of the concept, is a quantity that does not refer to the momentary state of an element but belongs to a material part of it, a part that it has in common with the free element and with all its compounds . . . For this reason I have endeavored to found the system upon the quantity of the atomic weight.

The first attempt I undertook in this direction was the following: I selected the bodies with the smallest atomic weight and ordered them according to the magnitude of their atomic weights. Thereby it appeared that there exists a periodicity of properties and that even according to

---

* The Royal Society (Great Britain) belatedly honored Newlands in 1887 with the award of its Davy medal.

valence, one element follows the other in the order of an arithmetic sequence:

Li = 7	Be = 9.4	B = 11	C = 12	N = 14	O = 16	F = 19
Na = 23	Mg = 24	Al = 27.4	Si = 28	P = 31	S = 32	Cl = 35.3
K = 39	Ca = 40	. . .	Ti = 50	V = 51	*et cetera.*	

In Mendeléeff's table hydrogen is omitted as a somewhat special case. (Note that helium and the other inert gases are missing because that whole family was unknown until 1895.) The first seven elements in the atomic weight order, Li to F, are set down as the first row; the next seven, Na to Cl, as the second. The "periodicity" in chemical behavior has emerged already in such obvious analogs in the vertical column as Li and Na, F and Cl, N and P. In fact, in each column at least the first two elements are members of a "family" in the sense described above (e.g., they both have the same main valence numbers). The addition of the third row starts out emphasizing the success of the analogy. K goes under Na, and Ca belongs to the same family as Mg. The next known element is Ti, which has a maximum valence of *four*, like C and Si, not *three*, like B and Al. Acting on the assumption that all elements are *known* would lead to putting Ti in column 3 and forcing it into the "family" with B and Al. (And of course shifting all subsequent elements one place which might hide relationships.) Mendeléeff left a *space empty* under Al, placed Ti under C and Si, hence: (1) improved the analogy, and (2) *predicted a* new element. A clever view of the significance of this procedure has been given by Holton, Roller, and Roller in their thoughtful work on the development of physics.*

The whole scheme up to this point may be illustrated in terms of a crude analogy. Suppose a librarian were to weigh each of his books individually and then place them on a set of shelves according to increasing weight, and found that on each shelf the first book happened to be on art, the second on biology, the third on classics, the fourth on the drama, and so on. He might not understand in the least the underlying explanation for this astonishing regularity, but if he were now to discover on, say, the third shelf a sequence of books in the order art-classics-drama, he would perhaps be tempted to leave a gap between the books on art and classics, and to suspect that there does exist a biology book of the appropriate weight to fill the gap, even though it had not come to his notice before.

Mendeléeff had no illusions that he understood why the elements could be arranged in this orderly sequence, but he firmly believed that his work would eventually lead to a physical explanation [that is, theory of atomic structure] and that, in the meantime, 'new interest will be awakened for the determination of atomic weights, for the discovery of new elements, and for finding new analogies among the elements.' Later he added, 'Just as without knowing the cause of gravitation it is possible to make use of the law of gravitation, so for the aims of chemistry it is possible to take advantage of the laws discovered by chemistry without being able to explain their causes.'

Figure 7.2 shows Mendeléeff's periodic table in the version published in 1872. Note *two* crucial features. There are many elements indicated as unknown

---

* G. Holton, D. H. D. Roller, and D. Roller, *Foundations of Modern Physical Science*, Addison-Wesley, Reading, Mass., 1948, p. 418.

GROUP→	I	II	III	IV	V	VI	VII	VIII
Higher oxides and hydrides	R_2O	RO	R_2O_3	RO_2 H_4R	R_2O_5 H_3R	RO_3 H_2R	R_2O_7 HR	RO_4
1	H(1)							
2	Li(7)	Be(9.4)	B(11)	C(12)	N(14)	O(16)	F(19)	
3	Na(23)	Mg(24)	Al(27.3)	Si(28)	P(31)	S(32)	Cl(35.5)	
4	K(39)	Ca(40)	—(44)	Ti(48)	V(51)	Cr(52)	Mn(55)	Fe(56), Co(59), Ni(59), Cu(63)
5	[Cu(63)]	Zn(65)	—(68)	—(72)	As(75)	Se(78)	Br(80)	
6	Rb(85)	Sr(87)	?Yt(88)	Zr(90)	Nb(94)	Mo(96)	—(100)	Ru(104),Rh(104), Pd(106), Ag(108)
7	[Ag(108)]	Cd(112)	In(113)	Sn(118)	Sb(122)	Te(125)	I(127)	
8	Cs(133)	Ba(137)	?Di(138)	?Ce(140)	—	—	—	
9	—	—	—	—				
10	—	—	?Er(178)	?La(180)	Ta(182)	W(184)	—	Os(195), Ir(197), Pt(198), Au(199)
11	[Au(199)]	Hg(200)	Tl(204)	Pb(207)	Bi(208)	—	—	
12	—	—	—	Th(231)	—	U(240)		

SERIES

Fig. 7.2. Periodic classification of the elements; Mendeléeff, 1872. (From Holton, Roller, and Roller, *Foundations of Modern Physical Science*, Addison-Wesley, Reading, Mass., 1958, p. 420.)

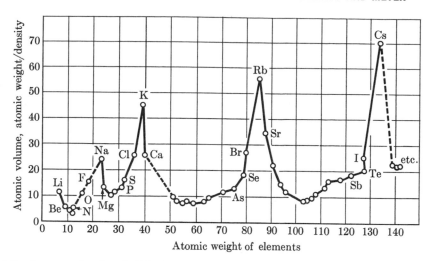

Fig. 7.3. An example of periodicity in the sequence of elements. The atomic volumes vary periodically with the atomic weights. (From Holton, Roller, and Roller, p. 421.)

(horizontal dashes) and there is a *close* following of chemical analogies. Each vertical column is headed with an indication of the formula of the *highest* oxide and hydride of elements in the group.

Mendeléeff enunciated a law at the heart of the construction of this table as follows:

The properties of the elements as well as the forms and properties of their compounds are in periodic dependence on, or (to express ourselves algebraically) form a periodic function of the atomic weights.

This law suggests that the analogies in a family of elements go beyond something as crude as valence numbers (which head the columns in the table). The quantitative richness of the periodic law is best illustrated in the form in which it was presented in the independent formulation of Julius Lothar Meyer. Meyer constructed plots of the value of some physical quantity as a function of atomic weight like the one shown in Fig. 7.3 which plots "atomic volume" (defined as the atomic weight of a substance divided by its density in the liquid or solid phase) versus atomic weight. The graph is a striking demonstration of periodicity. There are five successive sharp maxima. At the peaks we find Li, Na, K, Rb, Cs—the alkali metal family in order of increasing atomic weight. Note, even (see Section 7.2) that the height of the peaks exhibits a regular progression with increasing atomic weight. On the left-hand side of each peak lies a member of the halogen family. On the right-hand side of each peak lies a member of the alkaline earth metal family. Note that if the atomic volume of Br or Rb were *missing*, we could *predict* it with reasonable accuracy from an interpolation (that is, seeing where the curve goes between established points) on this graph!

Problem 7.4

a) Try to restate the "periodic law" in your own words.

b) Using data from the *Handbook of Chemistry and Physics*, or a similar source, plot the melting points of the elements from H to Ba inclusive. Discuss the periodicities that occur.

c) Show how you would go about estimating the melting point of the element which lies below Al if it were still missing as it was in Mendeléeff's periodic table.

Mendeléeff's table was a daring, if perceptive and closely reasoned, innovation. To suggest new and as yet undiscovered elements represented a major departure representing both the strength and the *risk* of the work. Let us let Mendeléeff himself show how predictions may be made employing the periodic law.

Q'	R'	S'
Q''	R''	S''
Q'''	R'''	S'''

Fig. 7.4. Each element such as R'' has properties that are the average of those elements surrounding its position in the periodic table. (From Holton, Roller, and Roller, p. 422.)

If in a certain group there occur elements R', R'', and R''', and if in that series which contains one of these elements, for instance R'', an element of Q'' precedes it and an element S'' succeeds it [Fig. 7.4], then the properties of R'' are determined by the mean of the properties of R', R''', Q'', and S''. Thus, for example, the atomic weight of $R'' = \frac{1}{4}(R' + R''' + Q'' + S'')$. For instance, selenium occurs in the same group as sulfur (at. wt. 32) and tellurium (127) and, in the 5th series, arsenic (75) stands before it and bromine (80) after it. Hence the atomic weight of selenium should be $\frac{1}{4}(32 + 127 + 75 + 80) = 78.5$, which is near the generally accepted value of Se = 79 (there is a possible error in the first decimal, so that 78.5 may be nearer the actual value).

We see the consequence of this at once:

The periodic dependence of the properties on the atomic weights of the elements gives a *new means for determining* . . . the atomic weight or atomicity [valence] of known but imperfectly investigated elements, for which no other means could as yet be applied for determining the true atomic weight.

. . . [For example] *properties** of selenium may also be determined in this manner; for example, arsenic forms H_3As, bromine gives HBr, and it is evident that selenium, which stands between them, should form H_3Se, with properties intermediate between those of H_3As and HBr. Even the physical properties of selenium and its compounds, not to speak of their composition, being determined by the group in which it occurs, may be foreseen with a close approach to reality from the properties of sulfur, tellurium, arsenic, and bromine.

---

* Italics ours.

Mendeléeff then goes on to describe the most important consequence of his approach:

In this manner it is possible to foretell the properties of an element still unknown, especially when it is surrounded by well-known elements. For instance, in the position IV, 5 [that is, in group IV and fifth series] an element is still wanting. These unknown elements may be named after the preceding known element of the same group by adding to the first syllable the prefix *eka-,* which means *one* in Sanskrit. The element IV, 5 follows [on the same side of the column] after IV, 3, and this latter position being occupied by silicon, I named this formerly unknown element ekasilicon and its symbol Es. The following are the properties which this element should have on the basis of the known properties of silicon, tin, zinc, and arsenic. Its atomic weight is nearly 72, it forms a higher oxide EsO_2, a lower oxide EsO, compounds of the general form EsX_4, and chemically unstable lower compounds of the form EsX_2. Es gives volatile organo-metallic compounds; for instance, $Es(CH_3)_4$, $Es(CH_3)_3Cl$, and $Es(C_2H_5)_4$, which boil at about 160°, etc.; also a volatile and liquid chloride, $EsCl_4$, boiling at about 90° and of density about 1.9. EsO_2 will be the anhydride of a feeble colloidal acid, metallic Es will be rather easily obtainable from the oxides and from K_2EsF_6 by reduction, EsS_2 will resemble SnS_2 and SiS_2, and will probably be soluble in ammonium sulphide; the density of Es will be about 5.5, and EsO_2 will have a density of about 4.7, etc. . . .

It was not until 1887 that Mendeléeff's predictions received their crucial test. C. Winckler of Freiberg, Germany, discovered an element in 1887 which was named germanium. Its properties are listed below.

atomic weight,	72.3
density,	5.5 g/cm^3
oxide, GeO_2, of density,	4.7 g/cm^3
chloride, $GeCl_4$	
boiling point,	83°C
density,	1.9 g/cm^3

Here is Mendeléeff's ekasilicon! In similar fashion Mendeléeff did succeed in the prediction of gallium and scandium.

7.7 THE MODERN PERIODIC TABLE

In the 1872 periodic table of Mendeléeff, shown in Fig. 7.2, the members of a given (vertical) group or family of elements are segregated into those placed on the left of the box and those placed on the right of the box. For example, Be, Ca, Sr, Ba are thus distinguished from Mg, Zn, Cd, and Hg in group II. This was done on the thesis, derived from the chemical data, that closer relationships occurred, in general, between elements in alternate (horizontal) periods. Closer inspection of group II shows that this is somewhat of an oversimplification. Mg certainly belongs to the family with Be, Ca, Sr, Ba, but it is true that Zn, Cd, and Hg are different. Once the spectacular successes of Mendeléeff and Meyer (especially the prediction of properties of new elements) had finally silenced critics (recall Carey Foster!) of the whole idea of classification of ele-

IA	IIA	IIIA	IVA	VA	VIA	VIIA	VIII			IB	IIB	IIIB	IVB	VB	VIB	VIIB	
H 1																	He 2
Li 3	Be 4											B 5	C 6	N 7	O 8	F 9	Ne 10
Na 11	Mg 12											Al 13	Si 14	P 15	S 16	Cl 17	Ar 18
K 19	Ca 20	Sc 21	Ti 22	V 23	Cr 24	Mn 25	Fe 26	Co 27	Ni 28	Cu 29	Zn 30	Ga 31	Ge 32	As 33	Se 34	Br 35	Kr 36
Rb 37	Sr 38	Y 39	Zr 40	Nb 41	Mo 42	Tc 43	Ru 44	Rh 45	Pd 46	Ag 47	Cd 48	In 49	Sn 50	Sb 51	Te 52	I 53	Xe 54
Cs 55	Ba 56	* 57–71	Hf 72	Ta 73	W 74	Re 75	Os 76	Ir 77	Pt 78	Au 79	Hg 80	Tl 81	Pb 82	Bi 83	Po 84	At 85	Rn 86
Fr 87	Ra 88	† 89															

*	La 57	Ce 58	Pr 59	Nd 60	Pm 61	Sm 62	Eu 63	Gd 64	Tb 65	Dy 66	Ho 67	Er 68	Tm 69	Yb 70	Lu 71
†	Ac 89	Th 90	Pa 91	U 92	Np 93	Pu 94	Am 95	Cm 96	Bk 97	Cf 98	Es 99	Fm 100	Md 101	102	Lw 103

Fig. 7.5. Long-form periodic table.

ments, it became possible to argue that the pattern was *more complex* than had at first been suggested. Perhaps there were not merely eight groups. And, perhaps *some of the heavier elements had no close analogs among lighter elements.* To give in detail the experimental evidence supporting these two propositions would require a summary of the chemistry of all groups of elements at least as detailed as that given for two *subgroups* in Section 7.2. Let it suffice to say that attempts to represent most faithfully the observed close analogies and regular progressions in the behavior of elements and their compounds led fairly rapidly to the evolution of a more complicated representation of the relationships among elements. The modern periodic table was completed following the discovery of two whole new groups of elements. The inert gases (He, Ne, A, Kr, Xe), which will subsequently be seen to be of *crucial* importance to the theory of chemical bonding, did not create a great problem. They comprised a new vertical group, (designated group 0). The second group, the fourteen rare earth elements, were finally all more or less accommodated in *one* position (box) in group III.

Figure 7.5 shows the modern *"long-form"* periodic table with the so-called *"transition metals"* placed in the middle of the table dividing the periods of *eight* elements formed by the lighter elements and their heavier relatives. The rare earths (or lanthanides) and their analogs, the actinides, are expanded (below the table) out from single boxes in the regular table. The key point about transition elements and rare earths (lanthanides) is the suggestion that *they do not have analogs among the lighter elements,* whereas the rest (that is, the so-called regular elements) include the lightest twenty elements and their analogs.

Now, it is *important* to note that Fig. 7.5 has assigned an order number to each element instead of quoting the atomic weight. This reflects the fact that the *modern* periodic table is based on a property of each atom which is regarded as *more fundamental than the atomic weight.* For the present we may define this *atomic number* as the number designating the position of an element in the periodic table. The hint that arranging the elements in order of increasing atomic weight led to only a *first approximation* of the ordering which revealed the relationships among them was available to Mendeléeff. It was clear that I belonged to the family with F, Cl, and Br. Te (tellurium) however belonged to the family with O, S, and Se. The data indicated that the atomic weight of Te was *greater* than that of I, which would have *reversed* the obvious family relationships if the elements were listed *strictly* in atomic-weight order. Mendeléeff suggested (quite reasonably) that this was only an experimental error and that a more careful determination of atomic weights would show that iodine was heavier than tellurium. Several *decades* of careful experimental work were devoted to attempts to show that Mendeléeff was right, *but the inversion was recalcitrant.* Cobalt and nickel provided a similar and *more subtle* (two-tenths of an atomic-weight unit) reversal of the order of atomic weights. It emerged that the ordering of the elements which brought out the relationships of chemical and physical properties most clearly *was not exactly the order of atomic weight* but something which was only *very nearly parallel* to atomic weight.*

The existence of a periodic table clearly emphasizes a pattern in behavior of atoms which *should* stimulate one to think about a theory of the constitution of atoms (atomic structure). What ideas can *explain* these relationships? A full appreciation of the concept of *atomic number* as opposed to atomic weight must await such a theory. To this theory we shall shortly proceed. But, we shall have to develop it at some length.

Problem 7.5   Find all of the inversions (from the atomic-weight order) in the modern periodic table. *Restate* Mendeléeff's periodic law making allowance for the inversions.

Problem 7.6   In 1871, Mendeléeff predicted that the atomic weight of germanium should be nearly 72. Show the calculations that led to this prediction.

7.8 ATOMIC STRUCTURE AND THE PHYSICISTS

Chemistry, as we noted in Chapter 1, was a late-blooming science. Before Lavoisier's work initiated the modern period in chemistry, much of what we now study in physics was well known. In fact one of the most profound syntheses of natural knowledge, which was able to encompass phenomena as diverse as the behavior of rolling stones, the flight of cannon balls, and the motion of planets, had been developed through the genius of Isaac Newton in the seventeenth century. This is the science known as classical mechanics. Like all great theoretical structures, it employs certain basic models for thought. To put the matter as simply as possible, the basic model behind the arguments of classical mechanics is something like a billiard ball on a smooth surface. Abstract from the picture most of the extension of the billiard ball, and the result is the "mass point." The study of the motion of mass points and the factors which change those motions (forces) is the fundamental task of classical mechanics. Now, Newton believed in atoms, as he stated forcefully, but he is not considered the father of modern atomic theory because his conceptions were not brought to bear on that body of experiment which most clearly showed the rich patterns of phenomena that could be understood in remarkable quantitative detail with the aid of the atomic hypothesis. That opportunity was first exploited by John Dalton. Lavoisier and Dalton would not have recognized our current distinction between chemists and physicists as clearly as we do. The separation of chemistry from physics grew out of attempts to under-

---

* The story just completed drew a comment from Prof. J. A. Young when he read the manuscript of this book. It's well worth quoting. "For all the talk . . . about 'exact' science we make progress lots of times by relying on what we 'know' even though our position is flatly contradicted by facts. I'd go so far as to say that no important contribution ever was first conceived in a manner consistent with what was then known factually—otherwise someone else could have made the contribution earlier. . . In each instance, someone had to make a wild leap—to his credit (since we tend to forget the 'crackpots' who did the same and missed) . . . This . . . attitude should be transmitted. . . This is the poetry of science."

stand the phenomenon of valence which we considered in Chapter 2. The fact that atoms have characteristic *integral* (whole number) combining powers could not be explained using ideas from classical mechanics, even when one's repertoire of forces includes an understanding of electrical ones.

Nineteenth century chemists became convinced that the useful program for chemistry was to explore the idea of valence and the *structure* of chemical substances *without* making an effort to explain what forces held atoms together in molecules or (if it seemed reasonable that electrical forces operated between ions) just why particular ions were formed by particular atoms or groups of atoms. Close relationships to the thinking of physicists appeared only in the area of questions concerning electrical conduction (see Chapter 5) and the energy changes that accompany chemical reaction (see Chapter 6).

With the completion of our account of the origins of the periodic table, the ideas of chemical structure are leading toward the idea of an internal structure of atoms. Toward the end of the nineteenth century physicists interested in electrical phenomena were beginning to probe the internal structure of atoms from a different point of view. In the rest of this chapter, we shall summarize some of the ideas that arose in this research tradition. When we have this material in hand we shall be prepared to describe the incredibly fruitful reunion that occurred between chemistry and physics in our own century, a reunion in which the explorations of nineteenth century chemistry turned out to play an important role in leading to a revolution in physics, which showed that classical mechanics had to be greatly supplemented in order to understand the microscopic world, and which in turn revolutionized chemistry by providing a start toward understanding valence forces. Let us look at several parts of this background of nineteenth and early twentieth century physics now.

7.9 GAS DISCHARGE TUBES—THE ELECTRON AGAIN

Ordinarily, one does not think of molecular gases as conductors of electricity. However, when a tube is evacuated so that the gas pressure is no larger than a few mm of Hg (a few thousandths of an atmosphere), a glow develops between two electrodes attached to an external *high-voltage* source. The apparatus is shown in Fig. 7.6. Now, as the pressure is lowered *still further*, a *dark space* develops to the right of the cathode, and as evacuation continues to pressures as low as 10^{-5} mm of Hg, the dark space fills the entire tube and a greenish glow from fluorescence of the glass itself develops at the end opposite the cathode. The anode (B in Fig. 7.6) may be *displaced* from the line in front of the cathode without changing the location of the glow opposite the cathode, *and it was found that an uncharged plate placed directly in front of the cathode cast a shadow in the glow.* (See Fig. 7.7.) Finally, it was observed that the beam from the cathode could be deflected from its straight line of travel by putting the tube in the field of a magnet. This was significant because it parallels the behavior of a current-carrying wire in a magnetic field. It was also significant

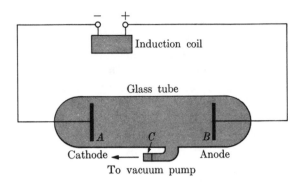

Fig. 7.6. Sketch of apparatus to illustrate phenomena associated with electrical discharge in gases. Metal electrodes A and B are sealed into a glass tube which can be evacuated through outlet C. Electrodes are connected to a high-voltage source such as an induction coil. In Faraday's terminology, electrode A is called the cathode and B the anode.

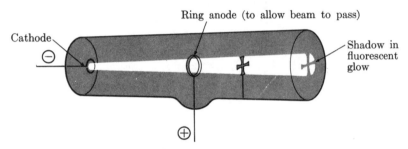

Fig. 7.7. Shadow casting in a gas discharge tube.

that a narrow "cathode beam" producing a small fluorescent region was *only* deflected but not spread out by the magnetic field. These results, largely the outcome of the work of the English physicist, Sir William Crookes, were described by him in 1879 in a language which *assumed that the cathode beam was a beam of particles bearing negative charge.*

Problem 7.7 Adopting Crookes' charged particle model and noting that the *force* exerted on a charged particle is proportional to the intensity of the magnetic field and the *charge on the particle* and that the *acceleration* experienced by any particle acted on by a given force is proportional to the *mass* of the particle, consider the significance of the observation that the deflected beam is *not spread out*. [*Hint:* Suppose the beam were composed of several different kinds of particles.]

The work of Crookes was refined by J. J. Thomson, who succeeded by 1897 in accomplishing the deflection of the cathode beam by an intense *electric* field as well as by a magnetic field. This again provided evidence that the cathode beam was a beam of particles bearing negative charge. As suggested in Problem 7.8, the uniform deflection of all of the beam (no spreading) *implies* that such a beam consists of particles of *uniform* charge, mass, and velocity.

Problem 7.8　The fluorescent screen of a TV tube is one end of a gas discharge tube. Can you account for the changing shape and eventual disappearance of the image seen on a TV screen after the set is turned off?

The most convincing (to most scientists) argument that the cathode beam is a beam of negatively charged particles is to show in *quantitative* detail that the beam behaves just as a beam of negatively charged particles large enough to observe directly would. This is what Thomson did in so far as he could with the deflections of the beam in electric and magnetic fields. According to macroscopic mechanics, the force exerted on a particle by the magnetic field is qHv, where q is the charge on the particle, H is the magnetic field strength, and v is the velocity of the particle. The force exerted by the electric field is qE, where q is the charge and E is the electric-field strength. By appropriate orientation of the electric and magnetic fields, the two forces can be made to balance each other, resulting in no deflection. Then

$$qE = qHv, \qquad v = E/H,$$

and the velocity of the particle beam can be calculated.

How does its velocity arise? The particle was, presumably, accelerated and acquired *kinetic energy* from the electrical work done on it by the potential difference between the cathode and anode in the tube. The electrical work as we have noted before (Chapter 6), is simply the product of the charge times the voltage V. This is converted to the kinetic energy ($\frac{1}{2}mv^2$) of the moving particle

$$qV = \tfrac{1}{2}mv^2.$$

From known E and H, Thomson obtained v. From known v, he calculated q/m, the *ratio* of the *charge* on the particle to its *mass*. Under a *variety* of experimental conditions (including variation of the chemical nature of the gas in the tube), the ratio was *consistently* 1.76×10^8 coul/g. The *charged particle model worked out in quantitative detail*. This negative particle was named *the electron*. In Chapter 6, we saw that there was evidence for an *atom* of charge involved in chemical transformations. It is an immediately *attractive speculation* to identify the cathode-beam particle with that atom and infer that it is a *universal constituent of matter*, a subatomic particle. The full value of that idea was proved on encountering it again in patterns of reasoning applied to other circumstances which unfortunately are beyond the scope of our account.

Problem 7.9　Compare the argument involved in the "discovery of the electron" with the Daltonian "discovery of the atom." How are the patterns similar?

Problem 7.10　Explain the significance of getting the same particle, as recognized by its value of q/m, from chemically different gases.

Let us now call the charge-to-mass ratio for electrons q_e/m_e. It would be most instructive to get q_e and m_e independently so that both could be evaluated.

It was q_e that yielded first to the experimental ingenuity of the American physicist R. A. Millikan. Effectively, Millikan measured charge-to-mass ratios for charged oil droplets that could be seen under a microscope. (Instead of balancing electric and magnetic fields, he balanced the force of gravity with an electric field.) In this case the mass could be taken to be that of the oil drop, but he found that he could detect *very small* changes in the charge on the drops. In his results, there appeared a *smallest change of charge* that could occur. All drops were charged to some multiple of that amount. By reasoning quite analogous to Cannizzaro's solution to the atomic-weight problem, he reasoned that this common factor was the charge on the atom of charge—the electron. It is (by modern values) 1.60×10^{-19} coul! Combining this with q_e/m_e gives $m_e = 9.11 \times 10^{-28}$ g.

Problem 7.11 Using q_e and the number of coulombs in a faraday (a mole of charge —see Chapter 6), calculate Avogadro's number. Does agreement of this value of Avogadro's number with that obtained by other methods (see Chapter 3) support our writing equations in Chapter 6 with electrons and arguing that there are electrons in all matter?

7.10 GAS DISCHARGE TUBES—ISOTOPES

One might also ask about the luminous phenomena which occur in gas discharge tubes at slightly higher pressures. As early as 1886, Goldstein used a *cathode* with *holes* drilled through it. With this apparatus, he discovered *positively* charged luminous rays coming through in the opposite direction to the cathode rays. (See Fig. 7.8.) Later application of Thomson's techniques to the positive rays yielded q/m values thousands of times smaller than those for electrons. If the unit of charge corresponds to an electron, these positive rays must be thousands of times heavier than electrons. W. Wein concluded in 1898 that these were *ionized gas molecules*—the gas phase counterpart of atomic and polyatomic ions in solution. A more complete picture of gas discharge suggests itself. Gas molecules are broken up at the cathode under the influence of the large applied voltage into electrons, which move one way, and residual *positive* ions, which move the other way. There is an intriguing possible chemical appli-

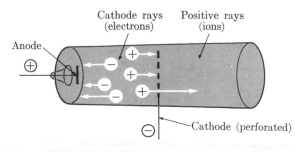

Fig. 7.8. A positive ray tube.

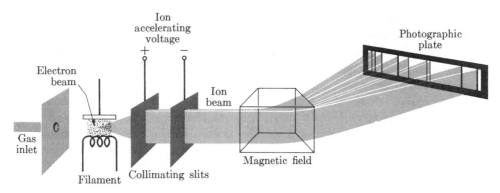

Fig. 7.9. Schematic drawing of a mass spectrograph. (From B. H. Mahan, *University Chemistry*, Addison-Wesley, Reading, Mass., 1965, p. 15.)

cation of this picture. If we can employ electric and magnetic deflection of cathode rays to measure the mass of electrons (once the charge has been determined), why not use deflection of positive ion beams to determine *absolute atomic weights* (once we guess the number of electrons removed from the gaseous atom or molecule and hence the charge it carries). Electromagnetic deflection of positive ions still is an important area of research. The development of techniques in the hands of early workers including J. J. Thomson, F. W. Aston, A. J. Dempster, and K. T. Bainbridge led ultimately to the development of what is now a widely used, highly precise technique. An instrument of the type designed by Bainbridge in 1930 is shown in Fig. 7.9. Such devices are called mass spectrographs. Bainbridge's device produces positive ions by bombardment of gas molecules with an electron beam. All of the ions are then given the *same* kinetic energy ($\frac{1}{2}mv^2$) by accelerating them across a fixed potential difference (voltage). They are then passed through a region of constant magnetic field which causes *different* deflections for *different* q/m values. The *deflections* are measured by allowing the deflected beams to strike a photographic plate which is "exposed" by the incident ions. The most obvious application of the mass spectrograph is direct determination of atomic weights and the solution of problems in molecular structure. A typical mass spectrum, that of molecular sulfur, is shown in Fig. 7.10. It is conventional as in Fig. 7.10 to plot not q/m

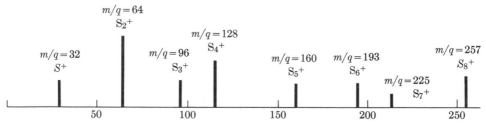

Fig. 7.10. The mass spectrum of sulfur.

but its inverse m/q and to give masses not in absolute units (g) but in masses relative to the standard atom, just as we were *forced* to do in Chapter 1. The heights of lines indicate the relative number of ions at each m/q value (as judged from relative darkness of lines on the photographic plate). As we could determine from freezing-point depression measurements, solid sulfur consists of S_8 molecules. Some of these (ionized!) appear in the mass spectrum of the vapor, but many have broken down into smaller constituents, especially S_2. Much of this fragmentation of molecules occurs in the process of ionization. The *chemical* structure of complex compounds can often be deduced from the pattern of their ionic decomposition products observed in mass spectra. The need for *only extremely* small samples of the compound to be studied makes mass spectra of great use for rare substances such as those extracted from biological origins.

Problem 7.12 Some of the peaks in the mass spectrum of acetone are at $m/q = 58$, 43, and 28. Argue that these can be interpreted as reasonable fragments of a molecule with the structure:

$$\begin{array}{c} O \\ \parallel \\ CH_3-C-CH_3 \end{array}$$

Our digression on the application of mass spectra to current molecular structural research has bypassed the most startling discovery of the early mass spectroscopists. We must return to this matter now. It adds deep insight into our developing theory of atomic structure. In the course of his early experiments on positive rays, Thomson carried out a general survey of chemically different gases. In 1912, while engaged in these investigations, Thomson studied one of the then recently discovered *inert gases*, neon (atomic weight 20.2). Instead of the *single* signal corresponding to a mass of approximately 20 in atomic-weight units, Thomson found *two*, one near 20 and a second (weaker) signal near 22. Why should there be *two* particles of differing mass? Thomson's first inclination was to attribute the second line to NeH_2^+ ions, but the principle characteristic of inert gases is the lack of chemical compounds, and another possible explanation had to be considered. *There might exist different forms of the same element.* That is, there might be *atoms differing in weight* (*mass*) but *identical* in *chemical behavior.* This hypothesis, in the light of subsequent experience, had to be accepted. It became necessary to recognize the existence of what have come to be called *isotopes.*

Problem 7.13 Reconsider the results of Chapter 1. Explain how it is possible for the simple laws leading to *precisely defined atomic weights* to hold if chemically "*identical*" atoms may have *different weights*. That is, explain how the Daltonian model is to accommodate *isotopes*.

7.11 RADIOACTIVITY

In the same period in which the electron was "discovered" and came to be regarded as a constituent part of all matter (and, hence, a subatomic particle), several other fruitful lines of investigation were contributing to the demise of the notion that an atom should be pictured as the ultimate constituent of matter, immutable and indivisible. In 1895, W. K. Röntgen, Professor of Physics at Würzburg, Germany, reported a new ray associated with gas discharge tubes. These rays are detected *outside* the tube. They radiate from the site where the cathode ray strikes the glass and have the ability to pass through matter opaque to light and expose a photographic plate. We talked in Chapter 3 about the consequences of diffracting these rays from crystals. These rays are now also familiar in medical applications (which were begun only a few weeks after the first announcement of the discovery). They are called x-rays. It was interesting that x-rays could discharge an electroscope by making the *air* a conductor of electric current (presumably by ionizing the gases).

The discovery of x-rays proved a tremendous stimulus to research. An early theory of their origin suggested that the phenomenon was related to phosphorescence (emission of light which occurs in some materials after they have been exposed to light). As a consequence, much attention was given to the behavior of phosphorescent materials, and a poor theory led to a great discovery. In 1896, the French physicist Henri Becquerel investigated some phosphorescent minerals containing the element uranium. Almost by accident, he discovered that these minerals emit rays that also pass through certain opaque materials to expose a photographic plate and are able to discharge an electroscope. This radiation was shown to be independent of prior exposure of the material to light and also *independent of the chemical state of the sample*. Thus it must be associated with the atom of the element itself. The study of this "radioactivity" was greatly facilitated by the isolation of two new highly radioactive elements by Marie and Pierre Curie (France), polonium and radium, in 1898.

In 1899, Ernest Rutherford (a name we shall encounter again) working at McGill University in Montreal, Canada, demonstrated that radioactive emission from uranium is of at least two different types which he designated α and β. Villard, in France, soon identified a third component (labeled γ radiation). These experiments were analogous to the experiments of Thomson. The radioactive emission was formed into a beam by passing it along a narrow hole in a lead block. It then passed into a strong magnetic field. The components designated α and β were deflected by the magnetic field just as the "cathode" and "positive" rays in a gas discharge tube. The β-rays, on detailed analysis, proved to have a q/m value consistent with regarding β-rays as a beam of Thomson's electrons. The so-called α-rays were deflected in the opposite direction, showing that they were positively charged. The q/m value obtained indicated that they were either *singly ionized hydrogen molecules*, H_2^+, or *doubly ionized helium atoms*, He^{2+}. Rutherford was ultimately able to show that α-particles, after slowing down, produce *helium* gas. This confirmed the second interpretation of q/m. With this information, it emerged that the "immutable" atom (of uranium) was sufficiently mutable to eject an atomic scale fragment, He^{2+}. [Gamma (γ) rays proved to be a form of radiant energy similar in character, if not in energy, to visible light and x-rays. The nature of such radiation is considered in more detail in Chapter 8.]

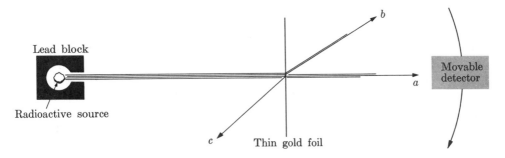

Fig. 7.11. α-particle scattering: *a* indicates the majority of particles; *b*, a few scattered to a wide angle from the initial path; and *c*, a *very* few scattered backward.

7.12 RUTHERFORD AND THE NUCLEAR ATOM

In 1911, Ernest Rutherford, by now in Manchester, England, was associated with colleagues in a study of the scattering of α-particle beams by thin foils of metals. Scattering is a physicist's term for the way in which an initially well-directed beam is modified by passage through a substance. Many workers, including Thomson, were beginning to speculate about the relationships of the *fragments* of an atom that had been discovered (e.g., electrons, alpha particles, etc.) to the *internal* structure of a *normal* atom. These speculations (Thomson's) were directed to placing most of the positive charge in an ill-defined region of space in a smear. As such, it would not present much of a "wall" to resist the passage of α-particles, and the scattering would be expected to be small. Note that this kind of reasoning encourages one to try to use α-particle bombardment as a "probe" for the study of the internal structure of atoms in somewhat the way that a blindfolded man might (in some appropriate state of pique) locate a window by determining through what part of a wall he could throw a baseball. The study of atoms does require obtuse intellectual devices! Rutherford's apparatus is diagrammed in Fig. 7.11. Let's listen to the recollection on the circumstances of the α-particle scattering experiments that Rutherford offered in 1937.*

---

* *Background to Modern Science*, ed. by J. Needham and W. Pagel, Macmillan, New York, 1938.

In the early days, I had observed the scattering of α-particles and Dr. Geiger in my laboratory had examined it in detail. He found in thin pieces [foils] of heavy metal, that the scattering was usually small, of the order of one degree [deflection angle from original beam direction]. One day Geiger came to me and said, 'Don't you think young Marsden, whom I am training in radioactive methods, ought to begin a small research?' Now I had thought that too, so I said, 'Why not let him see if any α-particles can be scattered through a large angle.' I may tell you in confidence, I did not believe that they would be, since we knew that the α-particle was a very fast massive particle with a great deal of energy, and you could show that if the scattering was due to the accumulated effect of a *number of small scatterings* the chance of α-particles being scattered backward [Fig. 7.11(c)] was very small. Then I remember . . . Geiger coming to me in great excitement and saying, 'We have been able to get some of the α-particles coming backwards . . .' . . . It was almost as incredible as if you fired a 15 inch shell at a piece of tissue paper and it came back and hit you. On consideration I realized that the scattering backward must be the result of a *single collision* and when I made calculations I saw that it was impossible to get anything of that order of magnitude unless you took a system in which *the greater part of the mass of the atom was concentrated in a minute nucleus. It was then that I had the idea of an atom with a minute massive center carrying a* [positive] *charge.* [Italics ours.]

Rutherford realized that the deflection of a very small number of α-particles backward, while the majority passed through the foil with little deflection, meant that there were *small* concentrations of mass, positively charged, with which a *few* α-particles collided directly. The idea is illustrated in Fig. 7.12.

The small, dense, positively charged center in the atom is called a *nucleus*. Rutherford suggested that the overall atom is neutral because the nucleus is surrounded (at various distances) by electrons which are orbiting rather like planets orbiting about the sun.

It is possible to analyze the trajectories of all of the α-particles, if we know the mass, radius, and charge on the *nucleus* of the atom, and assume that the force is the Coulomb's law electrostatic repulsion between the positive nucleus and the doubly charged helium atom which is an α-particle. Conversely, a

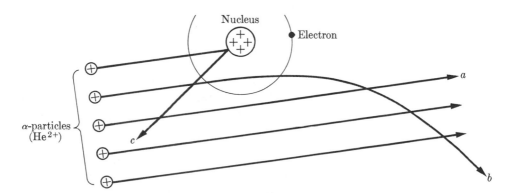

Fig. 7.12. Rutherford's interpretation of Geiger-Marsden experiments: *a,* undeflected beam; *b,* moderately deflected beam that passes near the nucleus and is repelled, hence deflected, by positive charge in the nucleus; *c,* particle in head-on collision is "back-scattered."

careful study of the α-particle trajectories permits inference of the *size* and *charge* of the nucleus. Refinement of Rutherford's scattering experiments demonstrates a very important result: *the charge on a nucleus comes in units of the size of the charge on an electron, the number of which corresponds to the position number of the element in the periodic table!* The order number of an element in the periodic table is called the *atomic number;* it emerges that the atomic number designates the charge on the nucleus (and, of course, for a neutral atom, the number of extranuclear electrons required to balance that charge).*

The problem in Mendeléeff's law is finally understood. It is reasonable that chemical properties do not follow atomic-weight orders exactly. The important variable is atomic number (nuclear charge or number of electrons). The discovery of isotopes shows clearly that atomic weight is not crucial. Atoms of slightly different weight may be chemically the same. It is now attractive to assume that it is numbers of charged particles that determine chemical properties of atoms.

Problem 7.14 Since Rutherford's work it has often been said, "Atoms are mostly empty space." Explain how scattering experiments and earlier results lead to a model of the atom that justifies such a remark.

Problem 7.15 A. J. Ihde's *The Development of Modern Chemistry* calls the twentieth century the century of the electron. Survey what we have learned from electrochemistry, the periodic law, discharge tubes, and scattering experiments. Can you construct an argument for involving electrons as crucial to chemical bonding?

READINGS AND EXTENSIONS

Two excellent paperback books deal with the emergence of the physical theory of atomic structure in the late 19th century with emphasis on the development of understanding and theory. The more chemical in slant is J. J. Lagowski's *The Structure of Atoms*, Houghton Mifflin, Boston, 1964. The more physical in approach is F. L. Friedman and L. Sartori, *The Classical Atom*, Addison-Wesley, Reading, Mass., 1965. This second book requires a good knowledge of calculus.

We must warmly recommend the treatment of the origins of the periodic table which guided our approach in this chapter. It is found in G. Holton, D. H. D. Roller, and D. Roller, *Foundations of Modern Physical Science*, Addison-Wesley, Reading, Mass., 1958. Chapter 24.

---

*It should be noted that prior to the successful refinement of α-scattering experiment, the same conclusion had been reached by Moseley from analysis of x-ray spectra.

CHAPTER 8

Atomic Structure, Waves, and the Periodic Table

8.1 THE WAVE MODEL OF LIGHT

Up to this point, whenever the discussion of atoms, molecules, or ions has turned to physical rather than chemical concepts, the metaphor used has been that of the particle. An atom or even an electron has been conceived of as a minute billiard ball; perhaps as a ball with electric charge, but still a ball; perhaps misshapen, but still of a well-defined shape. When the need has arisen to ask how an atom or an electron moves, we have turned to the physical analysis (Newtonian dynamics—see Appendix 2) which grew out of the study of billiard-ball collisions and planetary motion about the sun. In both these cases, abstraction and simplification of problems proceeds by considering each object of analysis as a *discrete* entity which is to be assigned a mass and a very particular location.

Is this metaphor always useful? Consider riding a wave toward shore at the beach. Is it so simple to separate the wave you are riding as a discrete entity? Probably the answer must be yes and no. The individual wave is recognizable, but the water composing it changes. It is a disturbance in the continuous medium of the water. Close analysis of waves has provided a second area of ordinary experience from which models of subtle physical phenomena may be built. These are phenomena which seem in one way or another to be appropriately comprehensible and explicable as disturbances in continuous media. Wave models notably are central to the theories of sound and light. We shall examine the wave theory of light because of its importance to theories of the structure of matter. Matter and light interact: the observable results of the interactions give us crucial clues to the structure of matter. The discussion will employ basic notions about waves which are reviewed in Appendix 3.

Consider a light beam passing through a slit as shown in Fig. 8.1. If the slit is large, as in Fig. 8.1(a), a sharp shadow is formed and the bright mark is simply the image of the slit. If, however, the slit is made sufficiently small, the bright spot at the middle of the shadow becomes "fuzzy" and wider than

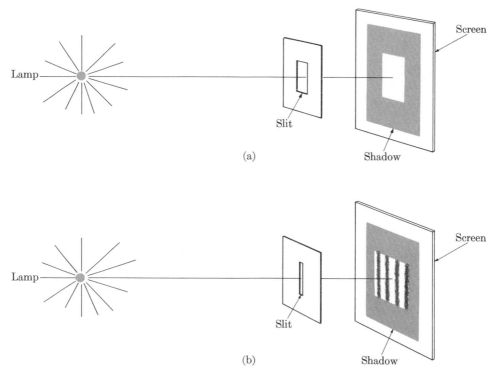

Figure 8.1.

the slit image "should" be. The sharp image could easily be understood if a light beam were taken to be a stream of particles. (If you're willing to accept an implausible analogy, think of the "image" produced on a target by machine gun fired through a piece of heavy armor plate with a slit in it!) But how is the fuzzy image of the narrow slit to be explained? Actually, before jumping into the process of explanation, it's important to describe the "fuzzy" image exactly. You can see the image if you look at a "point light source" (a distant street lamp or a bright star) at night between two fingers. As you squeeze your fingers together, you will see the "fuzzy" image just before you finally succeed in shutting out the light altogether. It will appear as *alternate* bright and dark bands as shown in Fig. 8.2. That such behavior is demonstrable in such simple wave situations as water ripples is shown in Section A3.12 in Appendix 3.

Fig. 8.2. The "fuzzy" image of a slit: a diffraction pattern.

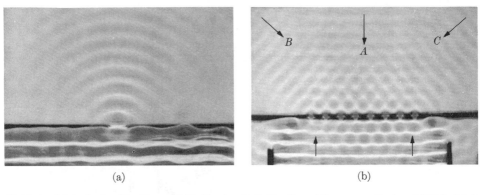

(a)   (b)

Fig. 8.3. (a) A single slit. (From *PSSC Physics*, D. C. Heath and Company, Boston, 1960.) (b) A diffraction grating acting on plane wave ripples on water. (From the film "Interference and Diffraction" produced by Educational Services, Inc.)

Let us jump here from the consideration of a single slit as in Fig. 8.3(a) to a row of slits known as a diffraction grating. Figure 8.3(b) shows what a diffraction grating does to a plane wave on a water surface. In addition to allowing the reconstruction of a wave straight ahead (arrow A in the figure) of the grating, it is also evident that the grating produces new wavefronts coming out at angles from the original direction (B and C). Figure 8.3(b) may be understood as the superposition of a series of events like those shown in Fig. 8.3(a). The new plane waves at angles (B and C in Fig. 8.3b) are produced at angles related to the wavelength of the wave and the spacing of the slits. At the angle such that waves originating in adjacent slits will arrive at a given point exactly in phase (crest to crest *or* trough to trough, etc.), they reinforce each other to form a new wavefront. At other angles, they interfere with each other destructively. The algebra (see Appendix 3) of this situation leads to

$$\sin \theta_1 = \lambda/d, \tag{8.1}$$

where θ_1 is the smallest angle from the direction straight ahead at which a new wavefront is observed (B, C), λ is the wavelength, and d is the distance between slits. In general there will be more (weaker) new wavefronts at other angles given by:

$$\sin \theta_n = n\lambda/d, \qquad n = 0, 1, 2, 3, 4, \ldots, \tag{8.2}$$

where n is an integer indexing the successive angles.

What about diffraction gratings for visible light? They do work well if the slits are sufficiently close together. They are usually produced by ruling lines on a transparent surface. A grating with 14,000 lines per centimeter is quite satisfactory.

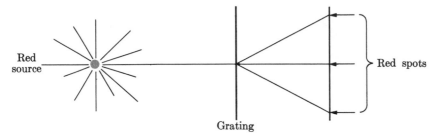

(a) Monochromatic (red) light passing through a grating.

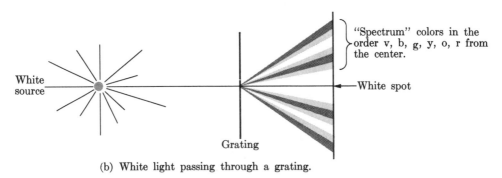

(b) White light passing through a grating.

Fig. 8.4. (a) Monochromatic (red) light passing through a grating. (b) White light passing through a grating.

Problem 8.1   Suppose that the first angle at which a beam is observed to be deviated from its original direction by a grating ruled with 14,000 lines per centimeter is 45°. From Eq. (8.1), determine the wavelength. [Express this wavelength in angstrom units Å (1 Å = 10^{-8} cm). The angstrom unit is a common unit for quoting light wavelengths.]   Why are shadows of objects of *ordinary* size sharp on a bright sunny day?

Let's illustrate two situations. First consider a monochromatic (one-color) light beam passing through a grating (Fig. 8.4a), then a beam of white light passing through the same grating. The monochromatic (say, red) beam produces a series of red spots at fixed angles. The white beam produces a spread-out rainbow of colors—a spectrum. The red is at the same angle as before. Of course, bending different colors through different angles implies that *color* is associated with *wavelength*. Since d in Eq. (8.1) doesn't change with the color, λ must.

Problem 8.2   The spectrum is displayed in the color order blue to red (V, B, G, Y, O, R), reading out from the center. How does the wavelength of blue light compare to that of red?   Consider the frequency of the wave (the number of wave crests passing a given point per second).   If it can be shown that all colors of light travel through empty space (vacuum) at the same velocity, determine the frequency order in cycles per second of the spectrum from highest to lowest. (Recall or learn from Appendix 3 that the velocity of a wave is given by frequency (ν) times wavelength (λ). The velocity of light in a vacuum is 3×10^8 m sec^{-1}.) Consider the beam whose wavelength was calculated in Problem 8.1. What is its frequency?

A word about a subtle question: what is waving? What, in the case of light, is the analog of the water? We cannot give a proper answer here, but we can sketch a suggestive outline of an answer. Electric charges act on each other through space (attractive forces between opposite charges). Suppose that one charge is made to oscillate so that its distance from a second *varies*. The force experienced by the second charge varies as the distance varies. Could the force be said to be waving? And does the second charged object "feel" the movement of the first instantaneously or is there a delay while the "wave" propagates from the first body to the second? These questions lie behind the idea of radio waves. They also prove to lie behind the theory of light waves!

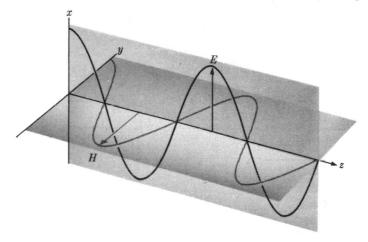

Fig. 8.5. Representation of an electromagnetic wave. The electric field *E* and magnetic field *H* oscillate in perpendicular planes.

One may define the electric field at any point produced by any charged body as the force per unit electric charge that it will produce on another charged body at that point. Clearly as the charged body is oscillated, the field at a point will fluctuate—or, it may wave. Similar language may be developed for *magnetic* interactions. By analysis of this "electromagnetic" situation the great English physicist James Clerk Maxwell (in 1866) was able to derive the velocity at which electromagnetic waves should travel in vacuum and *it proved to be the known velocity of light*. The conclusion was that light waves are fluctuations in values of the electric and magnetic field quantities. Light waves are *electromagnetic* waves differing from radio waves only in wavelength. A sketch of an electromagnetic wave is shown in Fig. 8.5. The category, electromagnetic wave, has been vastly enriched in modern physics by adding (at still other ranges of wavelength or frequency) x-rays, gamma rays, infrared, etc.

8.2 THE PHOTOELECTRIC EFFECT AND THE PHOTON

The analysis of interference and diffraction phenomena in terms of a wave model, followed by the success of Maxwell's electromagnetic wave theory, engenders a feeling that light is well understood. It is imagined as a continuous disturbance in the electric and magnetic fields spread throughout the relevant regions of space. But the slits involved in interference and diffraction experiments, however small, are large scale or macroscopic compared to electrons or atoms. All the phenomena can be seen with the eye aided at most with a simple microscope (to see the rulings of a diffraction grating). There are experiments concerning the interaction of light with matter which lead us down to the scale of the electron. It is found that certain substances emit beams of electrons (that is, beams that are experimentally identical to the cathode rays of Chapter 7) when excited by light of appropriate wavelengths (frequencies). This is known as the *photoelectric effect*. Examination of the photoelectric effect leads to misgivings about the *completeness* of the wave model and to the enrichment of the theory of light in strange and wonderful ways.

Figure 8.6 shows an apparatus which may be used for a systematic study of the photoelectric effect. It is similar to the one constructed by P. Lenard in 1902. The electrode M is made of a metal (e.g., Na, K, Cs) which will emit electrons when illuminated by light passing through the window Q. The electrons emanating from M may strike the plate P, completing an electric circuit and producing a current which is measured by a sensitive galvanometer or microammeter, A. Thus the photocurrent is measurable as the light intensity (or wavelength) is varied. In addition, the apparatus is constructed so that the "slide-wire potentiometer" represented between C and D in the figure can change the voltage between M and P. If the voltage is zero, an electron escaping from

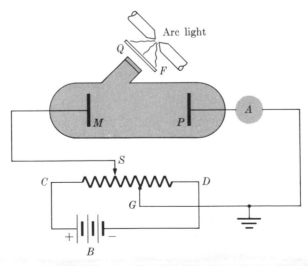

Fig. 8.6. Schematic diagram of tube designed for studying photoelectric current.

M can travel to P unhindered. If, however, M is made electrically positive with respect to P, there is a force retarding the travel of the electron away from M. If the velocity v or kinetic energy $\frac{1}{2}mv^2$ of the electron is insufficient, the force from M may decelerate the electron enough to cause it to fall back into M and *fail* to reach P. Thus it fails to contribute to the "photocurrent." When the voltage between M and P is raised, the photocurrent can be cut off. The voltage which cuts off the photocurrent allows calculation of the *maximum* kinetic energy of the photoelectrons knocked out of M.

The apparatus permits quantitative study of two factors: first, the magnitude of the photocurrent; second, the maximum kinetic energies of photoelectrons. The results of the experiments cannot be simply accommodated by the wave model of light. When the voltage between M and P is zero, the magnitude of the photocurrent depends in a simple way on the intensity of light. The greater the light energy incident on M, the greater the photocurrent. But the kinetic energy of the escaping electrons is *independent* of the *intensity* of light. As more light energy is delivered to the surface the photocurrent, and hence the *number* of photoelectrons, increases. But increasing the energy delivered to the surface *does not increase* the energy of the photoelectrons. If energy is delivered uniformly over the surface of M by wavefronts, one would expect the energy that a given electron could accumulate to be related to the energy per square centimeter delivered to the surface. Instead, the result is that the kinetic energies of photoelectrons remain the same down to *very low* intensities of light as long as the frequency (wavelength) of light remains constant. The only way to vary the kinetic energy is to vary the frequency.

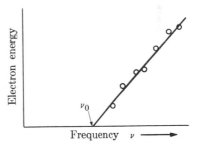

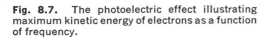

Fig. 8.7. The photoelectric effect illustrating maximum kinetic energy of electrons as a function of frequency.

The maximum kinetic energy of photoelectrons turns out to be proportional to the *frequency* of the light wave. This is shown in Fig. 8.7. Note that electron kinetic energy increases linearly with frequency but that the intercept of the plot is not at frequency $= 0$. There is a minimum frequency below which no photocurrent is produced. The minimum is called ν_0 in Fig. 8.7. Figure 8.8 shows the results of experiments using different metals in the electrode M. It emerges that the proportionality between kinetic energy and frequency is the *same* for all metals. The lines are accurately parallel. But ν_0, the minimum

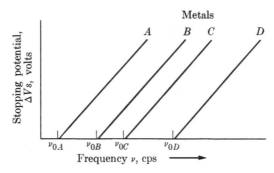

Fig. 8.8. Schematic diagram illustrating Einstein's prediction that stopping potential ΔV_S should be a linear function of the frequency ν of the incident light. The different lines show the behavior of different metals A, B, C, D. The slopes are identical, but the intercepts are properties of each particular metal.

frequency producing a photocurrent, is a function of the metal. How are these results to be understood?

The number of photoelectrons depends on the intensity. The energy of photoelectrons depends on the frequency. These results can be explained by imagining the incident light to be a beam of particles (bundles of energy) rather than a continuous wave disturbance spread uniformly over the surface. Suppose that the light is thought of as a stream of tiny bullets capable of giving their energies up to the electrons on collision with them. An electron may be ejected if hit by a bullet but not if missed. The number of photoelectrons corresponds to the number of hits, hence on the number of bullets, hence on the beam intensity. The kinetic energy of the photoelectrons depends only on the kinetic energy of the bullets which apparently is reflected in the frequency of light. If the kinetic energy of the "bullet" is smaller than the energy required to overcome the force binding the electron in the metal surface, no photoelectrons escape. The observed ν_0 measures that minimum energy.

So far we have regarded light as a continuous wave. Can we reconcile this with bullet-like bundles of energy? There appears to be a contradiction here. But we may have to live with the contradiction. When we make models of submicroscopic "invisible" entities, can we reasonably expect always to accomplish successful explanations with pictures which invoke entities neatly analogous to billiard balls and water ripples? Perhaps we should expect to modify the analogies in surprising ways to get working theories.

The bundles of energy are referred to as *quanta* (from the Latin word quantum meaning how much) or, in the case of light energy, *photons*. The critical test of the need to introduce *photons* comes from study of the timing of photoelectric emission. If the energy required to eject a photoelectron arrives distributed over a wavefront and must be gathered together at a point to eject an electron, the process would seem to require some time. Careful theoretical considerations have led to prediction of the time required for the incident light to eject an electron. Experimentally, exceedingly low light intensities excite detectable photocurrents. Calculations show that if the lowest intensities were spread uniformly through the apparatus they would not stimulate a photo-emission in a time shorter than a year. Yet Lawrence and Beams established experimentally that the time lag between light incidence and photoemission is *no more than* their experimental error of 3×10^{-9} sec even at the lowest light intensities. The wave model fails. We must consider the details of a particle photon model of light first suggested by Albert Einstein in 1905. We shall have to accept a dual theory of light. Waves are required for interference and diffraction; photons are required for photoelectric and related circumstances. Of course, it was not the photoelectric effect alone that led to the quantum theory of light photons. The idea first arose in partial form in Planck's theory of the distribution of wavelengths of light emitted by glowing bodies and has since found many applications.

Let's now turn to the development of Einstein's *photon* equations for the photoelectric effect. Note that in Fig. 8.8 the slope of the plot of the kinetic energy of photoelectrons vs. frequency is the same for all metals. The kinetic energy of the photoelectron is proportional to frequency, and the proportionality constant is the *same* for all metals. This implies that the frequency defines the energy of the photon. The connection is found from the slope of the plots in Fig. 8.8. The constant that indicates the change of kinetic energy per unit change of frequency gives the relation of wavelength to photon energy. The constant is called Planck's constant and given the symbol h; h has the numerical value 6.62×10^{-27} erg secs. The light beam, then, is composed of photons of energy h times the frequency. We write

$$E = h\nu. \tag{8.3}$$

We note that the energy delivered by an incident photon must equal the kinetic energy carried away by the electron plus the energy binding the electron in the surface. This second quantity is called the work function, W_0, of the metal. We write

$$h\nu = W_0 + \tfrac{1}{2}mv^2, \tag{8.4}$$

where W_0 is the work function and $\tfrac{1}{2}mv^2$ is the kinetic energy of the photo-electron. Rearranging, we get for the kinetic energies

$$\tfrac{1}{2}mv^2 = h\nu - W_0. \tag{8.5}$$

Now, expressing W_0 in terms of the minimum frequency required for excitation of photoelectrons we write $W_0 = h\nu_0$. Finally,

$$\tfrac{1}{2}mv^2 = h\nu - h\nu_0. \tag{8.6}$$

Problem 8.3 Show that Eq. (8.6) correctly describes Fig. 8.8.

Problem 8.4 Explain in your own words why no electrons are ejected by light of frequency less than ν_0. Visible blue or violet light causes photoemission from Na or Cs, while ultraviolet light of wavelength less than 2000 Å is required to excite Pt. How do you interpret these data? Is this result of chemical interest?

Perhaps we have generated here an unwarranted enthusiasm for a particle (photon) theory of light. Let's conclude with a wave corrective. It is possible to conduct interference and diffraction experiments using a light intensity low enough so that there is only *one photon at a time* in the apparatus. After a long time exposure the resulting image patterns look just like ordinary high-intensity interference patterns. Or, if you are thinking of a double-slit interference experiment, the photon seems to "know" as it passes through one slit that the other is there! What's going on? Don't expect a straightforward pat answer. The question lies at the heart of "modern" physics. And, modern physics seems to thrive on the contradictions of wave-particle duality!

Problem 8.5

a) Examine the equation $E = h\nu$ and explain why h has the dimensions of energy (ergs) $\times$ time (seconds). (The erg is an energy unit that is 10^{-7} joules. If a process requires 1 kcal per mole, it requires 6.95×10^{-14} ergs per molecule!)

b) When light of wavelength 4500 Å (recall 1 Å $= 10^{-8}$ cm) is incident on a clean sodium metal surface, electrons whose maximum energy is 3.36×10^{-12} ergs are emitted. Given that Planck's constant h is found to be 6.62×10^{-27} erg-sec and the speed of light is 3.0×10^{10} cm sec^{-1}, determine the longest wavelength of light that will excite photoelectrons from a sodium surface.

Problem 8.6 The following are typical wavelengths of several types of "electromagnetic" radiation. Calculate the energy per photon for each in ergs. How many kilocalories are delivered with moles (Avogadro's number, 6.02×10^{23}) of each type of photon?

ultraviolet	2500 Å
visible	4000 Å
infrared	25,000 Å
radio waves	300 cm

8.3 LINE SPECTRA AND ENERGY TERMS

With a fairly satisfactory model of light in hand, it becomes possible to use light as a tool to explore the innards of atoms. After all, some of the light incident on bodies is absorbed. And imparting energy to most substances (for example, by heating) leads in appropriate circumstances to release of energy in the form of light (glowing). Analysis of absorption and emission of photons should help us to build a model of the absorber or emitter.

The study of *individual atoms* (as absorbers or emitters) benefits most from interaction of light with gases. A hot solid body (e.g., the incandescent filament in an incandescent light bulb) emits light which was described as "white" in Section 8.1. If it is passed through a grating or a prism to resolve its wave-

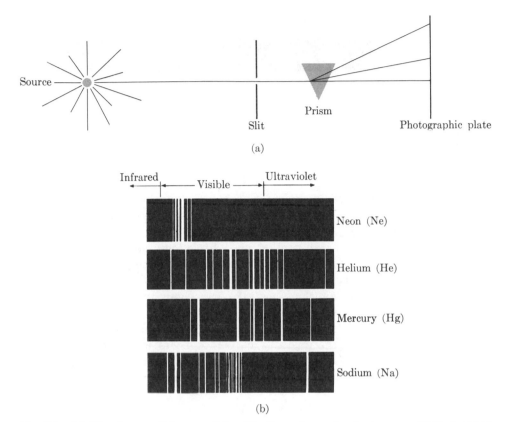

Fig. 8.9. (a) "Spectroscope" for examining "line" spectrum of a glowing gas. (b) Typical line spectra.

lengths (colors), the adjacent colors merge continuously into each other to form a *"continuous spectrum."* The appearance is just as seen in a rainbow. But, luminous gases behave quite differently. When light emanating from a glowing gas (neon signs, etc.) is passed through a prism or grating, it *does not* produce a rainbow-like continuous spectrum. (The rainbow of colors available for "neon" advertising signs requires the use of other gases besides neon!) Rather, only a limited number of definite and *separate* colors appear. This is shown diagramatically in Fig. 8.9(a). Figure 8.9(b) shows the spectra of several luminous gases obtained with an apparatus like the one indicated in Fig. 8.9(a). In Fig. 8.9(b), the spectra are shown as recorded photographically so that *"lines,"* images of the linear slit, lying both at longer wavelength (lower frequency) and shorter wavelength (higher frequency) than visible light may be seen (i.e., infrared and ultraviolet lines are included). These spectra are called *"line spectra."* Interestingly, a very similar phenomenon occurs when "white" light is passed through a gas, then through a grating or prism. The gas absorbs some of the light which passes through, and the wavelengths absorbed may be

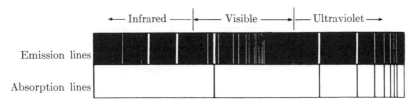

Fig. 8.10. Comparison of the emission and absorption spectra of sodium vapor. Note that the spectra are not identical, in that many of the emission lines do not have counterparts in the absorption spectrum.

identified in the final spectrum from dark areas. Again the dark areas turn out to be *lines*. And *each line in the absorption spectrum matches a line in the emission spectrum*. (See Fig. 8.10.)

What does emission or absorption of light at *only* specific, well-defined, wavelengths (frequencies) imply? In the language of the photon model, it means *only photons of certain energies may be absorbed or emitted* by atoms. Emission of light of frequency ν_1 (wavelength c/ν_1) corresponds to a loss of energy by the atom of exactly the amount $E_1 = h\nu_1$. If there is no emission at slightly higher or lower frequencies, the atom does not emit either slightly more or slightly less energy than the quantity E_1. It appears to be prevented from emitting energies near to but different from E_1. If an atom emits the photon $h\nu$, it loses energy. We may write that the energy of the atom has changed from T_1 to T_2:

$$h\nu_1 = E_1 = T_1 - T_2, \qquad (8.7)$$

where T_1 and T_2 represent the atom energies before and after emission. This last equation is not, of itself, a very profound result. But, connecting it to the existence of line spectra lends it stature. If only a limited set of emission frequencies are observed, there can be only a limited number of T values. That is, there appear to exist only a limited number of possible values of the energy content of the atom! Nothing that has gone before leads us to expect this. By analogy, thinking of the planetary model, note that there are no restrictions on the possible orbits for planets around a star. In the artificial-satellite enterprise, the newly launched satellite may be given any available energy. Each different energy value will correspond to a different orbit, but all are possible.

The T values are called *terms*, and combination of them pairwise leads to reconstruction of the line spectrum. In carrying out this reconstruction, a final economy emerges. The number of terms required to account for the lines in the spectrum is substantially smaller than the number of lines. We are required to postulate a *relatively small* number of possible energy states for the atom and are allowed to claim that it is restricted to these. If it absorbs or emits energy it must jump *discontinuously* from one to another.

Problem 8.7 The following list gives the experimental energies of several lines in the spectrum of gaseous potassium. Find the energy terms required to show that these transitions might be represented by an energy-level diagram similar to the one shown below. "Arbitrarily," the lowest level is taken to be $-22{,}250$ cm^{-1}.

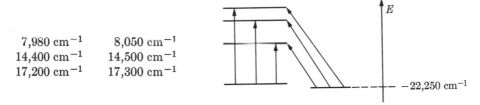

7,980 cm^{-1}	8,050 cm^{-1}
14,400 cm^{-1}	14,500 cm^{-1}
17,200 cm^{-1}	17,300 cm^{-1}

The matter is especially simple in the case of the simplest of all atoms, hydrogen, with only one electron outside a nucleus of charge $+1$ expressed in units of the charge on the electron. Figure 8.11 shows a portion of the line spectrum of hydrogen atoms in a gas discharge tube. A simple numerological relationship among these lines was noticed as early as 1858 by the Swiss school-teacher Johann Balmer. He saw that they could all be represented by the formula

$$\lambda = b \left(\frac{n^2}{n^2 - 2^2} \right), \tag{8.8}$$

where λ is the wavelength, b is an "empirical" constant found to be 3646 Å, and n is one of the integers larger than two; i.e., $n = 3, 4, 5, \ldots$ The series of lines shown in Fig. 8.11 is known, in honor of his insight, as the Balmer series. As more of the spectrum of hydrogen in other regions (far ultraviolet, infrared) was discovered, series very similar in appearance to the Balmer series were found. Balmer had noted in his original paper, apparently with some surprise, that the series of lines for which the 2^2 in his formula should be replaced by 1^2 or 3^2 or 4^2 were missing. He was no doubt anticipating such other series.

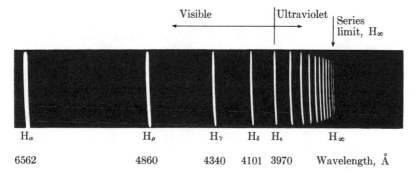

Fig. 8.11. Emission spectrum of atomic hydrogen in the visible and near ultraviolet (the Balmer series).

By 1890 a more complicated formula had been invented by the Swedish physicist Johannes Rydberg which was capable of encompassing all of the known series. Rydberg's formula reduces, in the case of the Balmer series, to

$$\frac{1}{\lambda} = R_\mathrm{H}\left(\frac{1}{2^2} - \frac{1}{n^2}\right), \tag{8.9}$$

where $n = 3, 4, 5, \ldots$, and R_H is 109,678 cm^{-1}. (*Note:* the reciprocal of a wavelength has the units of reciprocal length, cm^{-1}. It is a unit called a wave number, which is frequently used by "spectroscopists.")

Problem 8.8   Show that Eq. (8.9) reduces to (8.8) if $R_\mathrm{H} = 2^2/b$.

Rydberg's formula for the series found in the ultraviolet is

$$\frac{1}{\lambda} = R_\mathrm{H}\left(\frac{1}{1^2} - \frac{1}{n^2}\right), \qquad n = 2, 3, 4 \ldots \tag{8.10}$$

and for the series in the infrared

$$\frac{1}{\lambda} = R_\mathrm{H}\left(\frac{1}{3^2} - \frac{1}{n^2}\right), \qquad n = 4, 5, 6 \ldots \tag{8.11}$$

In general Rydberg wrote

$$\frac{1}{\lambda} = R_\mathrm{H}\left(\frac{1}{n_f^2} - \frac{1}{n_i^2}\right), \tag{8.12}$$

where n_f is a constant characteristic for each series and n_i is an integer indexing the lines in the series running *up* from $n_f + 1$. The constant R_H is the same throughout.

Now, $1/\lambda$ can be reexpressed. If $c = \nu\lambda$ (c is the velocity of light again), then $1/\lambda = \nu/c$. Rydberg's formula can become:

$$\nu = cR_\mathrm{H}\left(\frac{1}{n_f^2} - \frac{1}{n_i^2}\right). \tag{8.13}$$

But $E = h\nu$. Then the energy of a photon absorbed or emitted by a hydrogen atom is:

$$E = h\nu = hcR_\mathrm{H}\left(\frac{1}{n_f^2} - \frac{1}{n_i^2}\right), \tag{8.14a}$$

or

$$E = h\nu = \frac{hcR_\mathrm{H}}{n_f^2} - \frac{hcR_\mathrm{H}}{n_i^2}. \tag{8.14b}$$

Compare this equation to Eq. (8.7). The difference on the right-hand side of

Eq. (8.14b) is the difference between the energy terms. We can thus write a very simple expression for any hydrogen term:

$$E_n = T_n = -\frac{hcR_\mathrm{H}}{n^2}, \qquad n = 1, 2, 3, 4 \ldots \qquad (8.15)$$

Problem 8.9　Look at the Balmer series shown in Fig. 8.11. Note that the lines get closer together as we go to shorter wavelengths. Recalling that $\nu = c/\lambda$, show that shorter wavelengths correspond to higher energies for emitted photons. Do we expect an atom to emit higher-energy photons when it starts from its higher or from its lower energy states? Examine Eq. (8.15). What happens to term values if we insert larger and larger values of n? Is there a limit? Does the series of lines in the Balmer series appear to approach a definite limiting value? Noting carefully the minus sign, what is the value of the highest-energy term described by Eq. (8.15)?

Equation (8.15) has a *minus* sign. When we measure energies, we *always* observe *differences* of energy so we must assign the absolute value of *any* energy arbitrarily. Since the Balmer lines get closer and closer together approaching a definite *limit* at the high-energy end of the scale, we choose the highest possible energy (a fixed point) as the *zero* of energy and make *all* other values *negative*.

Finally, now, *stand back from the algebraic details* and realize that Eq. (8.15) says that we can write *the only possible* energies of a hydrogen atom as a constant $(-hcR_\mathrm{H})$ multiplied by $1/n^2$, where n is an indexing *integer*. It can be 1, 2, or 37, but the energy corresponding to 1.7 or 32.3542 *isn't* allowed; n is called a *quantum number*.

8.4 BOHR'S THEORY OF THE HYDROGEN ATOM

Can an explicit model of the atom be constructed to explain line spectra? The first notable success in this direction was Neils Bohr's theory of the hydrogen atom (1913). Bohr did the calculations for Rutherford's planetary picture with the addition of the *quantum* restriction. He imagined hydrogen as being composed of an electron orbiting in a circle about a singly positive nucleus. If one knows the force pulling an orbiting particle toward the center of its orbit (the centripetal force), it is a simple matter of elementary physics to calculate the velocity with which the particle must be orbiting in order for it to maintain a constant distance from the center and neither fly off nor fall in. Given these two restrictions it is also straightforward to calculate the energy of the orbiting particle.

Bohr argued that the *force F* was simply the well-known *coulombic* force of attraction between oppositely electrically charged particles:

$$F = \frac{(+e)(-e)}{r^2}, \qquad (8.16)$$

where $+e$ represents the charge (one electron unit positive) on the nucleus, $-e$ the charge on the electron, and r the separation of the electron from the

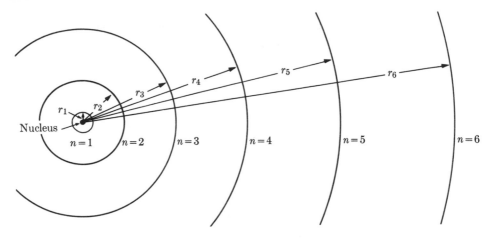

Fig. 8.12. Radii of the first few allowed orbits of the hydrogen atom.

nucleus (the radius of the orbit). The next question was how to restrict the model to *only* those orbits matching the allowed energies of Eq. (8.15). Here Bohr resorted to an ingenious and subtle analogy. He argued that for large orbits a formula with a *quantum* restriction should give the *same result* as older formulas constructed for larger than atomic-scale phenomena that had no quantum restrictions but had *worked well* on the large scale. The angular momentum of an orbiting particle of mass m moving in a circle is mvr, where v is the velocity (tangential to the orbit) and r is the radius of the orbit. Bohr showed that the condition on angular momentum,

$$mvr = n(h/2\pi),   (8.17)$$

would serve his needs, where n is an *integer* ($n = 1, 2, 3, 4, \ldots$) and h is Planck's constant. Thus allowed orbits were restricted to those for which the radius was $r_n = nh/2\pi mv$. When this is coupled with restrictions on v to produce stable orbits, the problem can be completely solved. The energy expression is

$$E_n = -\frac{2\pi^2 me^4}{n^2 h^2},   n = 1, 2, 3 \ldots   (8.18)$$

This expression involves the charge on the electron, e, the mass of the electron, m, and Planck's constant, h, all known from other lines of investigation. When we compare this to Eq. (8.15), it appears that we should be able to calculate R_H, the Rydberg constant. The success of this calculation was a triumph whose excitement is difficult to recapture fifty-six years later!

Bohr's atom is pictured in Fig. 8.12. As the electron moves *out* to allowed orbits of larger radius, it moves to higher (less negative) energies. The reason for higher energy at larger radius is easily understood in terms of potential-energy changes. As a negative particle is pulled away from a positive center,

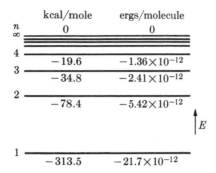

Fig. 8.13. An energy-level diagram for the hydrogen atom. The spacing between energy levels is not drawn to scale.

work must be invested against the attractive force. Energy must be put in. This simple idea is qualitatively correct for the total energy (sum of potential and kinetic energy). The diagram showing the energy levels for hydrogen atoms is shown in Fig. 8.13.

Problem 8.10 Assume that a hydrogen atom is normally found in its lowest energy state. From Fig. 8.13, determine how much energy is required to remove the electron (to $r = \infty$, i.e., $n = \infty$). Removing the electron corresponds to ionizing hydrogen. Show that this ionization energy can be calculated from Eq. (8.15). (We shall find ionization energies or "ionization potentials" of great chemical interest!)

Enthusiasm for Bohr's theory *must* be tempered. Although it answered many questions, the quantum restrictions are merely *incorporated*, not explained. Also, it turned out to *fail* the test of quantitative extension to other atoms and was of very little help in understanding molecule formation. (The first calculations based on Bohr's theory suggested that H_2 would *not* be stable!) However, Bohr took a crucial step toward our present understanding in that the key role of Coulomb's electrostatic law is carried over into all later theories. After all, it explains the hydrogen spectrum—it must be doing something right! The electron as planet about a nuclear sun will not, however, fare too well in the "new wave" to follow.

8.5 MATTER WAVES

According to "classical" (nineteenth century) physics, matter has mass and comes in particles, light is massless and is an electromagnetic wave disturbance. This simple division has already fallen with the recognition that light has particle characteristics in addition to wave behavior. Is it possible that the completion of this chapter in the story of the study of matter should be the extension of the wave model to material particles? Such an extension was suggested in 1924 by a French theoretical physicist, Louis de Broglie, as an elaboration of ideas implicit in the famous theory of relativity of Albert Einstein.

In particular, the line of reasoning was based on the Einsteinian idea of mass-energy equivalence. It led de Broglie to suggest an equation relating the *momentum mv* (where *v* is velocity) of a freely traveling particle to a wavelength, λ:

$$mv = h/\lambda. \tag{8.19}$$

What would count as an experimental test of de Broglie's speculation? If there is a wave, there should be a grating that would produce a diffraction pattern. Not quite with the intention to do so, Davisson and Germer (at the Bell Telephone Laboratories) discovered the appropriate grating in 1927.* In a study of the surfaces, they reflected electron beams (cathode rays) off nickel crystals. To their surprise, the reflected beams were *not* exactly in the expected directions. They were spread out over various directions with regions of minimum and maximum intensity, precisely like the pattern from light passing a diffraction grating. It appeared that the spaces between atoms in the crystal were serving as the grating. A photograph of the result of passing a beam of electrons of circular cross section through a thin foil of gold is shown in Fig. 8.14. Wavelengths calculated from such experiments confirm de Broglie's equation.

Fig. 8.14. Electron diffraction pattern of gold. (From G. P. Thomson, *Wave Mechanics of the Free Electron*, McGraw-Hill, New York, 1930).

Problem 8.11 An electron accelerated between electrodes with a voltage of 100 V between them has a velocity of about 5.9×10^8 cm sec^{-1}. What is its wavelength? What kind of substance would function as an appropriate diffraction grating?

These studies restore a certain symmetry to our models. The wave-particle duality applies to material particles as well as to light.

Problem 8.12 It is not difficult to drive a car weighing about a ton into a garage at 20 mph while ignoring diffraction effects. Does this mean de Broglie is wrong? Calculate the "associated wavelength" of such a moving car and look again at the question at the end of Problem 8.1.

---

* Davisson and Germer were trying to study the role played by Ni crystal surfaces in catalyzing chemical reactions. But they were aware of de Broglie's idea and didn't miss the significance of their results.

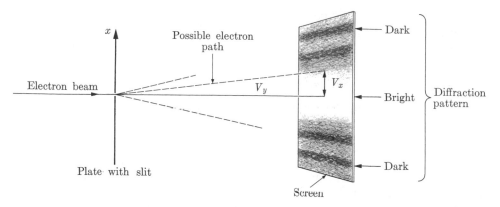

Fig. 8.15. Device for observing electrons. If the slit is small, the screen shows diffraction pattern.

8.6 THE UNCERTAINTY PRINCIPLE

Developments like the Rutherford-Bohr atom model, which visualizes the electron as a planet orbiting a nuclear sun, involve making statements, which are in principle precise, about the *position* and *velocity* (v) or *momentum* (mv) of a particle as small as an electron. With the idea of matter waves in mind, perhaps we should worry about this.

Of course, the way to worry usefully is to ask how one would *observe* (measure) the position and velocity of something like an electron. We must have something to see. A good possibility is the flash of light an electron makes when it hits a fluorescent screen (as in a TV tube). Consider the hypothetical experiment in Fig. 8.15. Let's define an x-coordinate along the length of the plate. If we make the slit very small, we know the value of the x-position of the electron when it went through the slit fairly precisely. We have an uncertainty about its position of only Δx, the width of the slit. Now, if we see a sharp bright spot directly in front of the slit, we know what the velocity of the electron was in the x-direction when it went through the slit: zero. But, if the slit is *small* so that Δx is *small*, the wave nature of electrons tells us that the pattern on the screen will be smeared out. From a particle point of view this means that v_x, the velocity in the x-direction wasn't zero. In fact, Δv_x, our *uncertainty* in the value of velocity in the x-direction, is directly related to the spread of the diffraction pattern. That is, *the smaller we make* Δx, the *larger* Δv_x *becomes*. The *smaller we make* Δv_x (by enlarging the slit), the larger Δx becomes. We are caught in a trap by the wave nature of the electron. The more accurately we try to observe its position, the less accurately we can know its velocity. The more accurately we try to learn its velocity, the less accurately we can pin down its precision. In 1927 W. Heisenberg examined this general problem of observation of microscopic entities and announced his uncertainty principle. The result of the "new physics" shows in general that there is a fundamental limit on precision of measurement which *cannot* be overcome by any instrument maker's art. Quantitatively it is expressed in terms of the uncertainty in measurement of *position*, Δx, times the uncertainty in measurement of *momentum*, Δp $(p = mv)$ as

$$\Delta x \cdot \Delta p \cong h. \tag{8.20}$$

That is, the uncertainty of position times the uncertainty of momentum can be *no smaller* than of the order of Planck's constant!

Problem 8.13 According to Bohr's theory of hydrogen atoms the *radius* of the first allowed orbit is about 0.5 Å. Although the theory exactly specifies the value, let's accept that we would allow an uncertainty of 10%. Δr would then be about 5×10^{-10} cm. Find Δp. The mass of an electron is 9×10^{-29} g. What is the uncertainty in its velocity in the r-direction? How far *may* it be from the nucleus 0.001 sec after it was found at $r = 0.5 \pm 0.05$ Å?

Let's return for a moment to Fig. 8.15. We would like to ask, as we did for light, *what is waving* in matter waves. If we imagine the intensity of the electron beam reduced to the point that only one electron passes through the apparatus at a time, do we see a smear on the screen or a localized flash indicating a particle hitting? This circumstance gives the *particle* result: a localized flash. But, if we allow flashes to accumulate, they form *statistically* the diffraction pattern. The best interpretation we can give is to suggest that the *motion* of the electron is *governed* by a *mathematical expression* of wavelike character, but when the observation is made, the flash is found at a *point* as a particle model suggests. The trouble is that we can't *predict* the point of impact precisely in advance. The mathematical wave expression describes the *probability* that the particle will be observed at a given point. In modern theory, the behavior of the electron is described by a *wave function* usually called ψ (*psi* pronounced "sigh") which has a value at each point (x, y, z) in the space of the system. The probability that a "particle" will be found (in a single observation) at a given point is governed by the square of the wave function, ψ^2. Note for future reference that the wave function, ψ, may have both positive and negative values (corresponding to crest and trough of a simple wave), but that ψ^2 will be always *positive* as the probability of observing the "particle" must be. Two parts of waves may arrive at a point crest to trough (the positive and negative contributions cancelling). This will result in low ψ^2, and low value of the probability of observing a "particle" at that point. This is the source of diffraction patterns for material "particles." To answer the question which headed this paragraph: the waves are, so to speak, probability waves. (This answer may be more picturesque than informative!)

8.7 THE SCHRÖDINGER MODEL FOR HYDROGEN ATOMS

We are now faced with the problem of understanding an atom which is in some senses successfully modeled by the "planetary-system" approach of Rutherford and Bohr but whose electrons exhibit the crucial wave characteristics of interference and diffraction. We have suggested that the motion of an electron is not to be described as an "orbit" but rather by a wavelike mathematical function of positions, the square of which specifies the probability of finding an electron at a particular position. What sort of a wave is it to be?

Consider a "stretched string" (a piece of laboratory rubber tubing will do) which is attached to a fixed support at one end and held in the hand at the other. If it is given a shake, a ripple travels down to the fixed end and is *reflected* and travels back. Now, suppose a second pulse is started from the hand while the first is coming back. They will cross in the middle somewhere. As they cross, they superpose (add together). If at a point at a given instant both are pulling the string above its rest position, the upward deflection will be greater than produced by either pulse alone. If one is pulling down while the other is pulling up, the effects tend to cancel. Suppose, now, that the shaking is made continuous and regular. There will be a continuous superposition of waves moving down the string and those moving back after reflection from the fixed end. The net result is a *sensitive function* of the frequency of the shaking. For most frequencies, an irregular and unstable pattern is observed on the string, but there exists a set of frequencies for which the string vibrates *regularly* with the appearance of something not traveling back and forth but standing.

A few moments' reflection will convince you that there is one very important aspect to the *standing-wave* phenomenon. As noted above it is not possible to set up a standing wave with just any *arbitrary shaking rate*. Only the rates of propagating a wave down the string that lead to an appropriate matching of "phase" with the reflected wave lead to standing waves. An analogy with a crucial and mysterious aspect of atomic behavior is clear. Standing waves are "quantized." There is a *discrete* set of possible standing waves determined by the length of the string and the velocity with which a wave disturbance travels along it. The properties of the string determine a *discrete set* of possible standing or stationary wave patterns which are *temporally stable*.

The most familiar example of generating such standing waves is in music. Choose a particular fingering on a flute. Blow into it with variable air pressures and frequencies. You can sound only one note, or at best that note's octave. The flute air column will vibrate at only certain well-defined frequencies. It won't sound the intermediate note unless you change the fingering to make a different vibrating system.

If we insist that an electron in a given state in an atom (which persists in time, a stable state) displays its wave properties in some form of standing wave, we shall have a *natural explanation of quantum numbers and discrete states*. Recall that these had to be introduced rather arbitrarily in the Bohr model.

In 1927, Erwin Schrödinger developed a theory of the hydrogen atom* making full use of the wave approach to the behavior of the electron. In his theory, the discrete orbits of the Bohr theory are replaced by mathematical

---

* We shall attempt to "describe" this theory. Even to write down the equations of Schrödinger would be an effort to impress rather than to inform at our mathematical level.

functions describing standing waves. The functions (standing waves) describe the probability of finding the electron at various positions about the nucleus, or, in a convenient "shorthand" phrase, describe the motion of the electron. Recall that standing waves arise in a stretched string out of the superposition of a wave traveling down the string and one *reflected* back from the end. The fact that the string is not infinitely long and that the wave disturbance is "bound" within a certain region is crucial to the origin of standing waves. What is the binding factor in the case of an electron wave about the proton which forms the hydrogen nucleus? Here Schrödinger's theory carries over an important idea from Bohr's theory and its predecessors. An electron is held in the neighborhood of a nucleus by the electrostatic force between the positive charge on the nucleus and the negative charge on the electron. In fact, Schrödinger took the *potential energy* of the electron to be specified, just as it had been earlier, as proportional to $1/r$, where r is the distance between the electron and the nucleus. The *kinetic energy* is described in terms of a more strictly wavelike mathematical formalism with one familiar particle feature remaining: the electron *mass* remains important.

Since Bohr's theory had correctly interpreted the Rydberg formula for the spectrum of the hydrogen atom (i.e., properly calculated all the *differences* in energy between the various allowed states), it is clear that the energies calculated by Schrödinger for an electron whose motion was governed by the various possible standing waves generated in the case of an electron bound at the center by the electrostatic force would have to be very close to Bohr's expression. It is, in fact, identical. Once again, it was found that the energy depends upon a *single quantum number n*

$$E = -\frac{2\pi^2 m e^4}{n^2 h^2} = \frac{E \text{ (for } n = 1)}{n^2}, \tag{8.21}$$

where all symbols have their previous meanings. The only difference is that it is now possible to *understand* n. We see why only certain levels are allowed. They correspond to standing waves. Intermediate possibilities cannot give the time-independent waves required for a stable state.

Problem 8.14 Itemize ideas in the Schrödinger wave theory that are carried over from earlier theories of atomic structure.

8.8 THE SIMPLEST HYDROGEN ORBITALS

For our subsequent purposes it will be important to have a feeling for the appearance of the standing waves which arise as the solutions to Schrödinger's equation for the hydrogen atom. These are called (by analogy with the orbits in Bohr's theory) *orbitals*. They are mathematical functions of the spatial coordinates. That is, they assign a number (either positive or negative) to each point in space, the *square* of which (clearly always positive) is proportional to the probability of finding an electron in that orbital (moving according to that mathematical function) at the point in space in question.

Obviously, the hydrogen standing waves are somewhat different from those on a stretched string. An electron moves in three dimensions and is bound only at the center. Actually, there is a useful analog in simple wave situations. There is some similarity between the vibrations of an electron bound to a nucleus and the vibrations of air in a spherical cavity. The analogy would be *quite close* if the electron could be thought of as a particle confined in a spherical box instead of being tied down from the center. The comparison of binding at the center to enclosing in a ball is shown to be a reasonable analogy by Fig. 8.16, which shows the *potential energy V* of an electron as a function of its distance from the nucleus in the $1/r$ case of electrostatic force and for a spherical box.

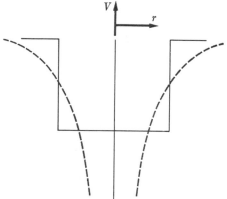

Fig. 8.16. Potential energy functions (potential energy V plotted vs. distance from one center r) for an electron in a spherical box (solid line) and an electron in a hydrogen atom (dashed line).

Let us try to visualize the standing waves in a spherical box. First there are a number of spherically symmetric modes of vibration in which the motion of the air is simply in or out radially (along the spokes of the wheel, to use a two-dimensional analog). The simplest of these is a simple "breathing" sort of motion in which all of the air at any instant is either *all* moving outward or *all* moving inward. Because the wave is *spreading* out from the center or *converging* on the center, the intensity of the wave disturbance decreases from the center outward. This vibration is shown in Fig. 8.17. This vibration is

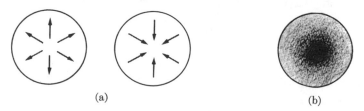

(a)   (b)

Fig. 8.17. (a) Fundamental spherically symmetric standing vibration of an air-filled sphere. (b) Intensity of wave motion at various points in the sphere for this mode.

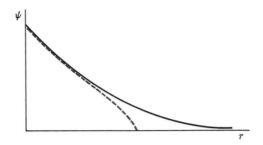

Fig. 8.18. The ground state hydrogen wave function (solid line) compared to the simple breathing mode of a spherical air cavity (dashed line).

Fig. 8.19. (a) Second mode of the vibrating air in a cavity. (b) Intensity of the wave disturbance at various points.

very closely related to the ground (lowest energy) state wave function for the hydrogen atom. The key difference is that the ground-state wave function does not cut off sharply at an outer wall but "tails off to" or asymptotically approaches zero at infinite radial distance from the nucleus. Figure 8.18 shows a plot of the wave function ψ vs. the distance from the nucleus. Remember that the wave function determines the wave amplitude and the wave function squared determines the probability of finding the electron at the distance r from the nucleus. The hydrogen wave function is compared to the wave function for the simple breathing mode of air in a cavity.

A second mode of the air in the spherical cavity is shown in Fig. 8.19. In this case, air in the inner and outer parts of the sphere are moving in *opposite* directions. Note that where the *direction* of the motion *changes*, there is a node or quiet ring which remains undisturbed, like the *node* in the middle of the second standing wave on the string. This corresponds to a sphere in the second hydrogen wave function where there is a *node* in the function and a *zero* probability of finding the electron. Figure 8.20 shows a plot of the wave function ψ vs. r and the square of the wave function for this second standing wave. The hydrogen functions are again compared to vibrations in the air cavity. Note that the change of direction in the motion of the air corresponds to a change in *sign* of ψ. There is a comparable phenomenon in the wave function for hydrogen. Note, of course, that the wave function may be positive or negative

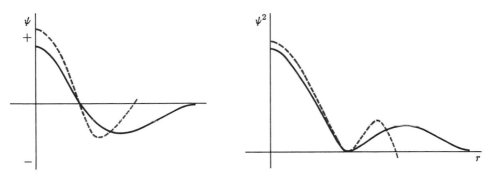

Fig. 8.20. (a) Second wave functions of hydrogen atom and air cavity. (b) Square of wave functions for hydrogen (solid line) and air cavity (dashed line).

designating the relative *phase* (direction) of the vibration but that the square is necessarily positive and thus the probability of finding the electron at any point is positive or zero (as it should be).

Problem 8.15 (A question to think over.) It is often asked, "How does the electron get past the node where it can never be found?" How does the air wave get past the node?

One more point needs to be made about the standing waves we have been describing. The square of the wave function, ψ^2, *determines* the probability of finding an electron in a given distance from the nucleus but we have not yet completely specified that probability. There is a geometrical factor. Obviously, at a geometrical point of *no extension*, there is no probability of finding the electron. The probability P is the product of the value of ψ^2 in some volume in space times that volume. Equation (8.22) may be written

$$P = \psi^2 \cdot V, \tag{8.22}$$

if the volume V is small enough that the value of ψ does not change significantly in it. Let us now consider the probability of finding an electron at a given distance r from the nucleus, whose motion is described by one of the standing waves we have discussed. This is the question of the probability that the electron is located in a thin spherical shell of thickness Δr a distance r away from the nucleus. The volume of such a shell is $4\pi r^2\,\Delta r$. The probability of finding the electron in the shell is $\psi^2 4\pi r^2\,\Delta r$. With this result we can construct a plot of the probability of finding the electron at any value of r. This is shown in Fig. 8.21 for the two standing waves we have described. The maximum in this plot gives the *most probable* distance from the nucleus. It is labelled a_0 for the first standing wave. Interestingly enough, a_0 turns out to be the same as the radius of the *first orbit* in the Bohr theory.

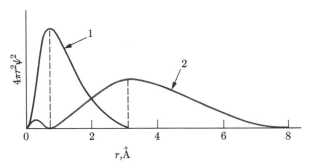

Fig. 8.21. Probability of finding an electron at various distances from a hydrogen nucleus. (1) First standing wave. (2) Second standing wave. (Compare with Figs. 8.18 and 8.20.)

An electron with the energy to be moving as described by the second orbital is not strictly farther from the nucleus than one in the first orbital. Sometimes it may be the other way around. But, *on the average*, the one following the second orbital is farther out. That is, it is at *higher potential* energy and, as suggested above, at higher total energy.

Problem 8.16 Recapitulate in your own words an argument to show that increasing distance from the nucleus leads to increasing energy to the electron. Hence, argue that the second standing wave described must be for an "excited" state of the hydrogen atom. According to the Schrödinger model, describe what is happening to the motion of the electron when a hydrogen atom absorbs a photon of light energy such that it jumps from its lowest energy state ($n = 1$) to the next higher state ($n = 2$).

8.9 OTHER HYDROGEN ORBITALS

The standing wave shown in Fig. 8.20, which we have just been discussing, corresponds to the hydrogen-atom energy quantum number $n = 2$. Spectroscopic experiments show that there must be at least one standing wave according to which an electron may move and have the *energy* corresponding to $n = 2$ in Eq. (8.21). This is *not* to say that there *can't be more than one*. In fact, there are *four* different Schrödinger standing waves or orbitals for which $n = 2$ and the *energy is the same*. Orbitals of the same energy are said to be *degenerate* (no pejorative connotation). Physically, the different energetically degenerate orbitals differ with respect to *angular momentum*. Recall that for an object moving on a circular path the angular momentum is mvr, where m is the mass, v the velocity tangential to the path, and r the radius. This describes the magnitude, *but* angular momentum also has a direction. In the language of vector algebra, angular-momentum expressions may be given for any arbitrary path.

It is well enough to *assert* that there are different solutions of Schrödinger's problem that differ in angular momentum, but what experimental consequence does this have? Eventually, we shall see that indirectly it provides the flexi-

(a)   (b)   (c)

Fig. 8.22. Examples of line splitting in a magnetic field (Zeeman effect). (a) *Normal* Zeeman triplet of the Cd line 6438 Å ($^1P - {}^1D$ transition). Above, the exposure was made so that only light polarized parallel to the field direction could reach the plate (single component at the position of the original line). Below, the components were polarized perpendicular to the field; they lie symmetrical to the original line. (b) *Anomalous* Zeeman splitting of the two D lines of Na, 5889 Å and 5895 Å ($^2S - {}^2P$ transition). Above, with magnetic field. Below, without magnetic field. (c) *Anomalous* splitting of the Zn line 4722 Å ($^3P_1 - {}^3S_1$ transition). (Normal Zeeman splitting is closely related to orbital angular momentum. Anomalous Zeeman splitting is related to the "spin" angular momentum of electrons.) (After Back and Landé, *Zeeman-Effekt und Multiplettstruktur der Spektrallinien*, Springer, Berlin, 1925. Reproduced in G. Herzberg, *Atomic Spectra and Atomic Structure*, 2nd ed. Dover, New York, 1944.)

bility in the Schrödinger model required for the explanation of a wide range of chemical phenomena (e.g., the periodic table), but fortunately there is a much more immediate confirmation of the significance of angular momentum. As a simple coil of wire wound into an electromagnet reminds us, moving electric charge produces a magnetic field. Charge moving on a *closed loop* of nonzero angular momentum will act as a *magnet*. Suppose that a small magnet is placed in the field of a large one. If it lines up *with the field* of the large magnet, it will be *in a state of lower potential* energy than if it lines up against the field. How is this detected on the *atomic* scale? By spectroscopy again! If there are different energies for different orientations, then there will be changes in the energy of *photons* absorbed. Again, it turns out with electron waves everything is still quantized. There are only certain values of angular momentum possible. And, there are only certain *allowed orientations of the submicroscopic atomic magnets!* Figure 8.22 shows an atomic spectrum recorded when the gaseous atoms are enclosed between the pole faces of a large magnet. Note the splitting of lines produced by the magnetic field. (A warning: so far, what there is to be said about the effects of magnetic fields remains incomplete. The most important source of "angular momentum" hasn't yet been mentioned.)

In the absence of "fields," the Schrödinger model for hydrogen suggests that *one* quantum number, n, is sufficient to specify the energy of the hydrogen electron. But for n larger than one, the Schrödinger model allows more than one *orbital* for each n value. The additional orbitals may be indexed according to two other quantum numbers which are *physically* most conveniently associated with the *total* angular momentum of the electron and with its components in *one direction* (e.g., the direction of an applied magnetic field). These two new quantum numbers are called l and m respectively. Their values are not independent of n. Rules for their values may be summarized:

$$n = 1, 2, 3, 4, \ldots,$$
$$l = n - 1, n - 2, \ldots, 0.$$
$$m = +l, (l - 1), \ldots, 0, \ldots, -(l - 1), -l.$$

This becomes, for the lowest values of n:

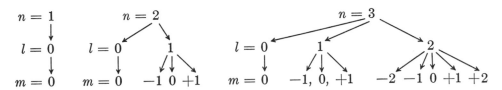

Looking at the "m" line, we see that for $n = 1$ there is only *one* orbital. This one was illustrated in Fig. 8.18. For $n = 2$, there are four of which the $l = 0$ ($m = 0$) one was illustrated in Fig. 8.20. In the case of $n = 3$, there are *nine* degenerate orbitals for the hydrogen atom, $(5 + 3 + 1)$. That is, an atom with energy corresponding to the third (starting from the lowest) energy state may actually have its electron "moving" according to *any one* of nine distinct standing waves.

The new quantum numbers l and m have been introduced in the context of the physically measurable quantity, angular momentum. This approach is important, but it should not be allowed to obscure the fact that there is a simple reason for three quantum numbers. It takes *one* quantum number (which counts the number of nodes) to specify the sequence of possible standing waves of a *one-dimensional* string. When an object moves in the three independent coordinates of a three-dimensional space, three quantum numbers will be required. The n, l, m set is connected with treating the three-dimensional H-atom problem in *spherical polar coordinates* (the distance from the center and two angles). It is not surprising (remember Coulomb's law) that the *energy* of the H-atom depends on *only one* of these quantum numbers (n), the one associated with the distance from the center.

We can draw standing-wave pictures of some of these additional orbitals. Figure 8.23 shows the *analog* of air vibrating in a sphere for the case of $n = 2$, $l = 1$ and *one* of the three of different m values. Note that what really distinguishes this vibration from the one in Fig. 8.19 is that the *node*, or no vibration surface, is now a plane cut through the middle of the sphere instead of a shell. This nodal plane would include the H-atom nucleus so that the probability of finding the electron whose motion is governed by this function at the nucleus would be *zero*. This becomes important when we turn to atoms heavier than H that contain more than one electron.

The three waves differing only in m values differ only in their orientation. They could all be derived from Fig. 8.23. The other two are simply those having

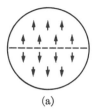 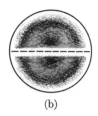 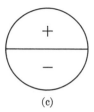

(a)   (b)   (c)

Fig. 8.23. Longitudinal vibrations of air in a sphere. The analog of a *"2p"* orbital of hydrogen. (a) Direction of air motion. (b) Intensity of wave disturbance. (c) Algebraic sign of the wave function.

the nodal plane in the two planes *perpendicular* to the one shown. The similarity of the three functions of the same l but different m leads to a common notation for H-atom orbitals which designates only n and l values. The n values are still identified by number but the l values are given lower-case letter names as follows:

$$l \text{ value} \quad 0 \quad 1 \quad 2 \quad 3,$$
$$\text{name} \quad s \quad p \quad d \quad f.$$

Thus we speak of the hydrogen orbitals as:

$1s$	only one orbital,
$2s$	only one orbital,
$2p$	three orbitals,
$3s$	one orbital,
$3p$	three orbitals,
$3d$	five orbitals,
$4s$	one orbital,
$4p$	three orbitals,
$4d$	five orbitals,
$4f$	seven orbitals.

The number of different orbitals and the designations listed above are simply derived from the rules just above and on p. 250. Roughly speaking, the same l designation means the same shape for all n values.

Problem 8.17 Derive these designations and the number of orbitals under each, using the rules on page 250.

8.10 CONTOUR DIAGRAMS

The drawings which have been used to emphasize the similarity of Schrödinger's model of the H-atom to standing waves in a sphere full of air are not always the best way to indicate the essential "shapes" of orbitals, and these "shapes"

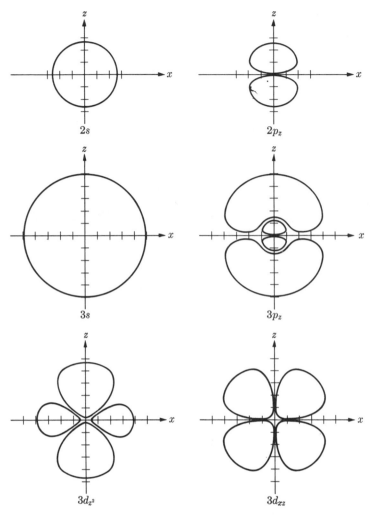

Fig. 8.24. Contour diagrams of hydrogen orbitals showing contours of constant probability which enclose the region of space where the probability of finding the electron is 0.99. (The diagram is a cross section in the z, x-plane.)

will prove to be of great importance to the theory of chemical structure. As a result, an alternative picture, a contour diagram, is the most commonly encountered orbital representation. The contour diagrams of Fig. 8.24 are constructed to show *lines along which the value of ψ^2 is constant*. Moreover, they are drawn so that the probability of finding the electron within a sphere just enclosing the contour line is 0.99. A three-dimensional (solid) picture of these contours is attempted in Fig. 8.25. These "three-dimensional" versions are generated by rotating drawings like those of Fig. 8.24 about the appropriate axes to sweep out the space.

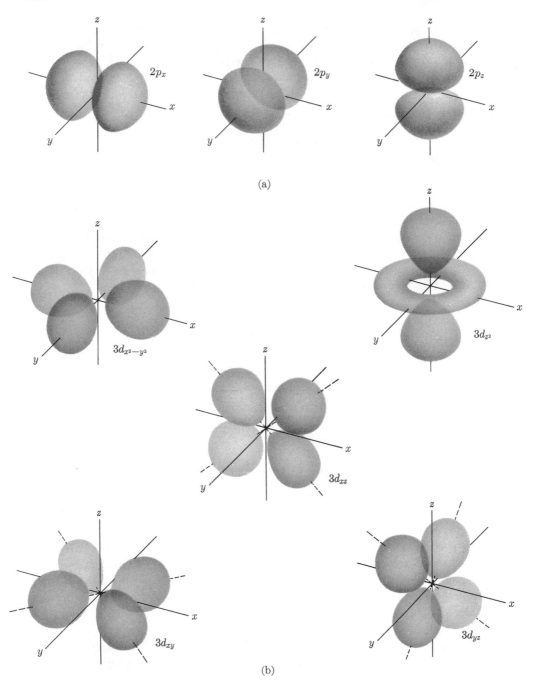

Fig. 8.25. (a) The 2p-orbitals of the hydrogen atom. (b) The 3d-orbitals of the hydrogen atom. Note the relation between the labeling of the d-orbitals and their orientations in space. (Adapted from K. B. Harvey and G. B. Porter, *An Introduction to Physical Inorganic Chemistry,* Addison-Wesley, Reading, Mass., 1963.)

These diagrams very clearly reveal the directional properties of the various hydrogen orbitals. The subscripts (e.g., x or y on $2p_x$ or $2p_y$) are related to the differing values of m and show that m differences are mostly a matter of the coordinate directions along which the orbitals are preferentially oriented.

8.11 THE ELECTRON SPIN QUANTUM NUMBER

The main experiments against which models of the atom must be tested are spectroscopic. Strong evidence in favor of a theory follows from successful predictions of the number and energies of "lines" in the line spectrum. It has been suggested that the hydrogen-atom spectrum is fairly simple and that is true unless one undertakes a search for subtleties. One way of searching for experimental subtleties to test the details of the Schrödinger model is to "perturb" with electric and *magnetic* fields the atoms whose spectra are under observation. It was suggested above that such a procedure might reveal the importance of "angular momentum" and give a clue about the significance of the "other" quantum numbers, l and m, of the Schrödinger model.

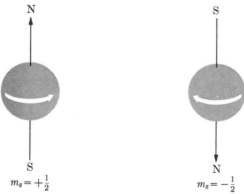

Fig. 8.26. The spinning electron. (N and S designate the poles of the micromagnet.)

It was also noted that this sort of experiment reveals even more. Magnetic fields can lead to more elaborate "splitting up" of lines in the hydrogen spectrum into components than the Schrödinger model *predicts!* The remaining features were first rationalized by Goudschmit and Ulenbeck in 1925 by suggesting that the electron, of itself and as a particle, might form a current loop. The essence of the suggestion was that a charged sphere rotating on its own axis would form a current loop that would behave like a small magnet in the same way that an electron in orbit may. This is usually called *electron spin*. A fourth quantum number is introduced, the *spin quantum number*, m_s. To properly account for the observations, it was necessary to allow this number to have only *two* values. For reasons connected with the quantity of angular momentum involved, the values are given as

$$m_s = +\tfrac{1}{2}, \qquad m_s = -\tfrac{1}{2}.$$

In terms of a spinning charged spherical particle, which the elegant theoretical work of Dirac (1927) has finally shown to be somewhat naive, these two values may be thought of as the electron spinning either clockwise or counterclockwise at the same rate (Fig. 8.26). Adding the spin quantum number to the three that we cataloged before, we recognize the need for the specification of four quantum numbers to specify the state of an electron in a hydrogen atom. The first three describe the shape and orientation of the probability cloud describing the motion of the electron in the usual three spatial dimensions, and the fourth specifies the spin (or angular momentum) intrinsic to the electron.

8.12 HEAVIER ATOMS, THE PAULI PRINCIPLE, AND THE PERIODIC TABLE

Almost as soon as the first Bohr theory of the hydrogen atom appeared with its many orbits for electrons, attempts were made to account for the behavior of atoms containing more than one electron by assigning the several electrons to several Bohr orbits. The first step in the development of such theories had to be a decision as to how to "fill" orbits. Could each Bohr orbit be used by only one electron at a time? Could each be occupied by two, or three, or many electrons at once? We can ask the same questions about the Schrödinger *orbitals*, and we shall wish to do so because the theory of many-electron atoms still rests on the use of the answers to the hydrogen problem as a starting point. The H problem is extended with the suggestion that the several electrons may be described by assigning them in an appropriate way to orbitals of the type derived in Schrödinger's model of hydrogen.

Before describing such a theory of many-electron atoms, the warning should be recorded that it is *only approximate*. Schrödinger's formulation of the problem of electrons moving as described by standing waves determined by the electrostatic forces made it clear that one electron should influence the behavior of another (both are charged). Using the hydrogen standing waves to describe several electrons amounts, according to Schrödinger's theory, to an initial *neglect* of the repulsive forces between the like charged electrons. It does emerge that the repulsive forces may be reintroduced in an approximate fashion at the last stage of the approximate theory, and we shall have to be satisfied with this approach because the *exact* and *rigorously correct* (no approximations) solution of the problem of standing waves of many-electron atoms remains to this date beyond the range of available mathematical tools. (To chemists, this means additionally that almost all *molecules* are accessible to only approximate description in the language of the wave theory, but that is a problem for the next chapter.)

Two of the most obvious aspects of the "behavior" of many-electron atoms that a theory of their structure should account for are the *energy terms* in their line spectra and the *periodicity* of their chemical properties. The first was probably historically the aspect that received the most attention but it is a fairly complicated story. We can appreciate some of the flavor of arguments made by concentrating on arguments designed to explain the periodic table, and this will also be a major accomplishment of importance to chemistry.

The first thing that the periodic table suggests is that there is something repetitive about atomic structure. There is also something repetitive about hydrogen orbitals: the ones of the same l and m values but different n values have similar shape. That is, as we go to higher-energy hydrogen orbitals of higher n value, the basic shapes are repetitive. Perhaps the repetitive chemical behavior reflected in the periodic table is to be explained by putting the electrons into hydrogenlike orbitals in order of increasing energy and expecting similar behavior when similar groups of orbitals of l and m values recur for successive n values. This idea can be reexpressed by saying that the electrons behave as if they were organized in successive *shells* where all orbitals of a given n value form a shell and that the chemical properties of the atom are determined by the distribution of the electrons in the outermost shell according to l and m values. The theory begins, then, with the following two postulates:

1. Many-electron atoms may be modeled by feeding their electrons into orbitals like those found in the hydrogen problem. The order of filling is that of increasing energy.

2. The orbitals of a given n value may be thought of as forming a shell, and the chemical properties of the atom depend principally on the population of the outermost shell.

But these postulates are insufficient for the complete theory. The question of how many electrons per orbital appears. Perhaps the simplest answer to try would be *one*. If that worked, it would follow that for $n = 1$ there would be only one atom (hydrogen) which employed the $1s$ orbital. The two-electron atom (helium) should then use orbitals $1s$, $2s$ and, according to postulate 2, be *similar* to hydrogen in chemical behavior. The suggested similarity is obvious nonsense, since He is chemically inert, so the rule must be faulty. In fact, the first element that could plausibly be said to be similar to hydrogen is lithium, which has the atomic number three and, therefore, three electrons.

In 1925, working still in the general framework of Bohr's theory, W. Pauli had examined the structure of the periodic table *and of atomic spectra* and suggested that the correct number of electrons per orbital was *two*. The number *two* does not appear entirely arbitrary after a moment's reflection. The two electrons whose positions in space are to be described by the same orbital may *differ* in the value of their spin quantum number so that each electron may be thought of as unique. In fact this is just what Pauli suggested. He enunciated a final postulate for the theory of many-electron atoms which may be stated:

No two electrons in a system (e.g., atom) may have the same values of all *four* quantum numbers n, l, m, and m_s.

Only one problem stands in the way of now developing a complete theory of the periodic table. As yet, the order of filling orbitals of differing l and m within a given shell has not been specified. According to the solution of the hydrogen-atom problem, all are of the same energy.

At this point, it becomes essential to consider electron-electron repulsion, which leads to differences between different values of l. Consider an atom with a pair of electrons of opposite spin occupying the 1s orbital. Figures 8.17 through 8.20 suggest that this electron distribution would resemble a spherical cloud of negative charge. (The physical resemblance of the probability distribution to a cloud of charge was established in a famous theorem of Hellman and Feynman.) The next electron in this atom goes into an orbital of $n = 2$. It has a choice between the 2s and one of the three similarly shaped 2p orbitals. Both 2s and 2p orbitals represent electron distributions "centered" farther from the nucleus than the 1s so that the third electron will be largely located farther from the nucleus than the first two. It may be said to lie mostly *outside* them. As a result, the repulsions of the first two electrons for the third *screen* it from the nucleus and contribute to an increase of its energy over and above that resulting from the simple circumstance that it is on the average farther out. If it can be shown that the inner electrons screen one type of the $n = 2$ orbitals more than the other, it would follow that that type of orbital lies at higher energy than the other. This is in fact the case for the 2p orbitals because of their "hour-glass figures." The 2p orbitals have their *nodes* at the nucleus, whereas the 2s orbital has its node on a sphere removed from the nucleus. A detailed analysis of the s and p types of probability distributions shows that the s type corresponds to probability for the electron to lie at distances from the nucleus shorter than those for electrons of lower n more often than p type. It is said that an s orbital *penetrates* the inner shell better than a p type and is therefore less shielded from nuclear attractions and at lower energy. The order of energies corresponding to different l values which is the reverse of the order of shielding is:

$$s < p < d < f.$$

This effect is even more pronounced in the case of d and f type orbitals. It becomes so large that the d orbitals of a given n (say 3d) turn out to have higher energy than the s orbitals of the next higher n value (4s in the particular case). The order of filling of levels of differing l value which are sometimes called *subshells* is shown for uncharged atoms* in Fig. 8.27.

---

* This order is somewhat modified in the case of ions.

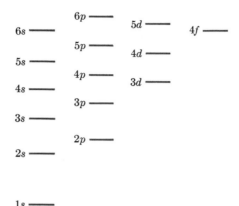

Fig. 8.27. Order of filling of shells and subshells of *neutral* atoms. This corresponds to the order of increasing energy. Penetration effects are taken into consideration. (The figure is not to scale with respect to energy.) Each orbital is represented by a dashed line. It has the capacity to contain two electrons of opposite m_s values. Penetration effects change if an electron is added or removed! This diagram does not apply to ions!

The periodic table may be analyzed using Fig. 8.27 as a guide. (In the following discussion we shall introduce notation in which a superscript number next to a designation of an orbital type will specify the number of electrons in that type of orbital. For example, $3d^2$ means two electrons in $3d$ type orbitals.) Starting from atomic number one, we designate H as $1s^1$ and He as $1s^2$ with the two electrons of opposite m_s in the $1s$ orbital. At atomic number three, Li is $1s^2 2s^1$ and should be somewhat analogous to H (e.g., the oxidation state $+1$). The atoms of the next seven atomic numbers, Be to Ne, run from $1s^2 2s^2$ to $1s^2 2s^2 2p^6$. The last one, Ne with the $2p$ orbitals filled to capacity, is chemically analogous to He, which is reasonable since it has a filled shell. Next comes element eleven, Na; with the three-shell beginning at $3s^1$, it is clearly an analog of Li, which has $2s^1$ in its outermost shell. The next seven elements are analogous to the corresponding two-shell elements. Examples are: Mg, $3s^2$ analogous to Be, $2s^2$; Si, $3s^2 3p^2$ analogous to C, $2s^2 2p^2$; Cl, $3s^2 3p^5$ analogous to F, $2s^2 2p^5$. At the end of this group comes Ar, $3s^2 3p^6$, which is analogous, in that it is chemically inert, to Ne, $2s^2 2p^6$ and He, $1s^2$. It is interesting that the least reactive elements all correspond to the configuration $ns^2 np^6$ with the single exception of He. It is not in general a requirement for this inertness that the whole set of orbitals of a given n value including d and f be occupied.

The next element, number 19, is K with the configuration $1s^2 2s^2 2p^6 3s^2 3p^6 4s^1$ as indicated by its chemical similarity to Li and Na. Number 20 is Ca with $4s^2$ which is analogous to Mg. The next element, number 21, Sc, has *no analogs earlier in the periodic table.* As Fig. 8.27 suggests, it is the element with the configuration $4s^2 3d^1$. The next nine elements are similarly unprecedented and are the elements described in terms of the filling of the first d subshell. They do have analogs later in the periodic table when the $4d$ shell is filling. The next set of unprecedented elements arise starting with element 57 and occupy 14 places usually shown at the bottom of the periodic table diagram. They are the elements explained as arising during the filling of the first f type subshell, the $4f$. Since there are seven m values in this case, the orbitals can accommodate 14 electrons. The theory begins to make comprehensible the complications in the idea of periodicity which could only be characterized in Chapter 7 by saying that a "long form" was needed to properly represent the family relationships among the elements.

In a modern periodic table, the first two families on the left can be called the "s-block." The first column is ns^1 elements and the second ns^2 elements. Next we find columns of three extending for ten places. These are the "d-block" or "transition elements." At the right are six columns corresponding to the "p-block" running from the family B, Al, Ga, In, Tl, which is the np^1, to the chemically unreactive gases Ne, Ar, Kr, Xe, Rn, which are represented by np^6. The two known "f-block" rows, the lanthanides and actinides, are grouped at the bottom.

Problem 8.18 Write the configuration for each of the following elements identified by their atomic numbers:

N(7),   P(15),   Cr(24),   Ni(28),   As(33),
Sr(38),   Ag(47),   Te(52),   Eu(63).

An example would be O(8): $1s^2 2s^2 2p^4$.

8.13 MANY-ELECTRON ATOMS: ENERGY LEVELS

We have agreed that electron configurations explain the periodic table. This implied that the outer-shell configuration determines the bonding (ionic or molecular) of an atom and the chemical reactions in which it will participate. In Chapter 6 we saw certain energy changes connected with reaction. If we are to prepare to find an electron theory of bonds and reactions, we will now have to face a complex task. We must understand energy levels in many-electron atoms clearly and *completely*. The task is difficult because we must consider carefully the balance between electron-nuclear attraction and inter-electronic repulsion not only for neutral atoms but also for positive and negative ions.

It might be supposed that electrons filling orbitals above the one used in the ground state of hydrogen would be higher in energy. This would imply that electrons would be more easily removed from all other atoms than from hydrogen. What is overlooked by such a naive supposition is that atoms of atomic number greater than hydrogen have larger nuclear charge. A greater nuclear charge will mean a lower electron energy at any given distance from the nucleus. In fact, the orbitals of the hydrogen type must be thought of as "contracted" when an electron "moves" under the influence of a larger positive charge. The electron remains closer to the nucleus on the average. In many-electron atoms, then, electrons will be occupying orbitals associated with higher quantum numbers which would be progressively *higher-energy orbitals* in hydrogen. But this "rise" does not apply simply to many-electron atoms because the nuclear charge is larger.

Problem 8.19   Calculate the force between an electron and a proton when they are (a) 1 cm apart and (b) 10^{-8} cm apart. Express the charges in "electrostatic" units such that Coulomb's law has the form $F = q_1 q_2 / r^2$ with no constant. In this unit system the electronic charge is -4.80×10^{-10} esu. If the distance r is in centimeters, the force is given in *dynes* which is the force *accelerating* an object of mass 1 g by 1 cm/sec^2. What is the effect of doubling the nuclear charge?

There is another effect which is very important for the energy of the electron assigned to the outermost orbital of a many-electron atom. This effect is the effect of *repulsions* among electrons. The outermost electron does not "see" simply the charge of the nucleus. It "sees," approximately, the charge of the nucleus *minus* the charges of electrons in inner shells of lower principle quantum number.* The inner shells screen the nucleus from the outermost-shell electron. The result is that the first electron to occupy the shell of a given principle quantum number n sees the atom "inside" almost as if it added up to a net charge of *one unit*. That is, the first electron in a shell approaches the behavior of an electron in *that shell in hydrogen*. However, as a given shell fills, the subsequent electrons which are all at approximately the same distance from the nucleus do not shield each other from the nucleus so well. The additional electrons in a given shell have an opportunity to "feel" the increasing nuclear charge. Of course, increasing nuclear charge, unshielded, means decreasing orbital energy!

The situation may be explored experimentally. Earlier (Problem 8.10) we considered the significance of the *limiting term* that appears in a Rydberg series. Then it was argued that the convergence of lines to a definite limit was associated with promotion of the electron from a given level to the highest possible "level" which is, of course, the condition of being *free*, removed from the atom. If series of lines, similar to Rydberg series, which converge on definite limits can be found in spectra of other atoms, then energies for removal of electrons from atoms can be calculated. Correct assignment of a series of lines as starting from the normal ground state can lead to calculation of the energy required to remove one electron for the normal ground state or to "ionize" the atom. Such energies are called *ionization potentials*. Spectral data on Na appear in Fig. 8.28.

Note in the figure that levels of higher energy get closer and closer together. This corresponds to the lines in the spectrum coming closer together because an electron "coming" from a lower level has two very closely spaced "choices"

---

* A theorem of electrostatics proves that a charge outside a charged sphere is affected just as it would be if all the charge on the sphere were concentrated at the center.

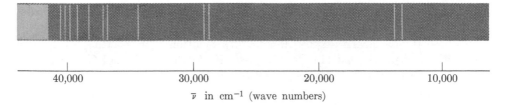

$$\bar{\nu} \text{ in cm}^{-1} \text{ (wave numbers)}$$

Fig. 8.28. The "principal" series in the spectrum of gaseous sodium.

of energy level. The limit on which the spectral lines converge in Fig. 8.28 is the zero of energy. The difference between the lowest level and the zero is the energy required to remove an electron from Na or the *ionization potential* of Na. (Note that the process we have described is the formation of Na^+. This will clearly be important for an understanding of compounds like NaCl.)

8.14 TRENDS IN IONIZATION POTENTIALS

Figure 8.29 shows the variation of *ionization potential* with atomic number. This is an example of a typical "periodic" relationship. The maximum ionization potentials for each row occur at the "inert" gases with the electron configuration ns^2np^6 which corresponds to effective "completion" of a shell. The next highest values of ionization potential are for the *halogens* (F, Cl, Br, I)

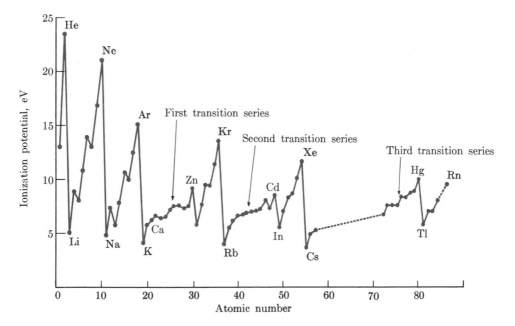

Fig. 8.29. Ionization potentials of neutral atoms as a function of atomic number. The energy unit is electron volts; 1 eV/atom equals approximately 23 kcal/mole.

which have the configuration ns^2np^5. The lowest values of ionization potential are associated with the alkali metals (Li, Na, K, Rb, Cs).

These trends clearly expose the factors mentioned in Section 8.13. The successive low points, ns^1 (Li, $2s^1$; Na, $3s^1$; K, $4s^1$; Rb, $5s^1$; and Cs, $6s^1$) show that as one electron is put into a shell of new n, that electron is more easily removed than the one from the row before. That is, when the shielding of inner shells makes the last electron "see" nearly only a $+1$ charge at the center (nucleus), the energies of the orbitals rise as n is increased and correspondingly the ionization potentials decrease. Going across any row staying within the same

Table 8.1
Some values of ionization potentials

Z	Atom	Orbital electronic configuration	Ionization potential (I), electron volts	Ionization potential (II), electron volts
1	H	$1s$	13.60	–
2	He	$1s^2$	24.58	54.40
3	Li	$(He)2s$	5.39	75.62
4	Be	$(He)2s^2$	9.32	18.21
5	B	$(He)2s^22p$	8.30	25.15
6	C	$(He)2s^22p^2$	11.26	24.38
7	N	$(He)2s^22p^3$	14.54	29.61
8	O	$(He)2s^22p^4$	13.61	35.15
9	F	$(He)2s^22p^5$	17.42	34.98
10	Ne	$(He)2s^22p^6$	21.56	41.07
11	Na	$(Ne)3s$	5.14	47.29
12	Mg	$(Ne)3s^2$	7.64	15.03
13	Al	$(Ne)3s^23p$	5.98	18.82
14	Si	$(Ne)3s^23p^2$	8.15	16.34
15	P	$(Ne)3s^23p^3$	11.0	19.65
16	S	$(Ne)3s^23p^4$	10.36	23.4
17	Cl	$(Ne)3s^23p^5$	13.01	23.80
18	Ar	$(Ne)3s^23p^6$	15.76	27.62
19	K	$(Ar)4s$	4.34	31.81
20	Ca	$(Ar)4s^2$	6.11	11.87
21	Sc	$(Ar)4s^23d$	6.56	12.89
22	Ti	$(Ar)4s^23d^2$	6.83	13.63
23	V	$(Ar)4s^23d^3$	6.74	14.2
24	Cr	$(Ar)4s\ 3d^5$	6.76	16.6
25	Mn	$(Ar)4s^23d^5$	7.43	15.7
26	Fe	$(Ar)4s^23d^6$	7.90	16.16
27	Co	$(Ar)4s^23d^7$	7.86	17.3
28	Ni	$(Ar)4s^23d^8$	7.63	18.2
29	Cu	$(Ar)4s\ 3d^{10}$	7.72	20.34
30	Zn	$(Ar)4s^23d^{10}$	9.39	17.89

n, we find that increases in nuclear charge overweigh increased shielding and ionization potentials rise (admittedly a bit irregularly but we overlook small effects here and talk of general trends). The most "stable" electron configuration is ns^2np^6. For a given configuration the lighter element displays the larger ionization potential. It is clear that inner-shell shielding does fairly effectively counter increased nuclear charge and that electrons in higher n orbitals are farther out. This should be reflected in size effects which we shall discuss in Chapter 9.

Table 8.1 gives ionization potentials for the first 30 elements in detail. The value I is the work required for the process $M \rightarrow M^+ + e^-$. The value II, or second ionization potential, is the value for the process $M^+ \rightarrow M^{2+} + e^-$, where M represents any atom. Close inspection of ionization potentials higher than one reveals that effects dependent upon *shielding* (interelectronic repulsion) become less pronounced. This is reasonable. When the species has lost an electron, nuclear charge is relatively more important. An electron is missing and interelectronic repulsion must be reduced. An *important* special case of this last effect is encountered in the transition metals. These atoms come after the ns^2 elements because the $(n-1)d$ shell is better shielded (less penetration) and falls at higher energy. For example, $3d$ is filled after $4s$ but before $4p$. In positive ions, shielding becomes less important since there are fewer electrons. The situation becomes more like the one-electron H atom. Thus the first two electrons lost by the transition metals are usually the ns pair. This accounts for the common $2+$ oxidation state represented by M^{2+} ions. For example, Fe (neutral) is $4s^23d^6$ but Fe^{2+} is $3d^6$. The reduced interelectronic repulsion allows $3d$ to reapproach the $3s$ and $3p$ orbitals and come *below* $4s$ as it does in H atom.

8.15 ELECTRON AFFINITY AND ELECTRONEGATIVITY

The energy required to remove an electron from a neutral atom is of interest in the discussion of formation of positive ions. The energy charge accompanying negative-ion formation is also important. In most cases it does not *require* energy to put a *free* electron onto a neutral atom in the gas phase. Rather, it takes work to remove an electron from a negative ion to give a neutral atom. That is, a negative ion has a *positive* ionization potential just like a neutral atom. This ionization potential may also be measured from analysis of negative-ion line spectra, although the experiment is usually difficult.

Despite the similarity of *anion* and *neutral atom* ionization potentials, the quantity usually tabulated is the energy *released* in the process

$$M + e \rightarrow M^-.$$

This is clearly just the *negative* of an ionization potential. It is called the *electron affinity*. Table 8.2 lists some values of electron affinities. It is not complete because of experimental difficulties. For some elements the best available values

Table 8.2

Atomic electron affinities (values in parentheses are estimated)

H							
0.747							
Li	Be		B	C	N	O	F
(0.54)	(−0.6)		(0.2)	1.25	(−0.1)	1.47	3.45
Na	Mg		Al	Si	P	S	Cl
(0.74)	(−0.3)		(0.6)	(1.63)	(0.7)	2.07	3.16
		Zn	Ga	Ge	As	Se	Br
		(−0.9)	(0.18)	(1.2)	(0.6)	(1.7)	2.96
		Cd	In			Te	I
		(−0.6)	(0.2)			(2.2)	2.66

of electron affinities are not much more than "informed guesses." The table includes only fairly secure values.

Electron affinities are smallest for elements having only a few electrons in the outermost shell and greatest for elements of nearly filled shells. These reflect again the balance between nuclear charge and screening effects. Again the special stability of ns^2np^6 configurations is emphasized by the high electron affinity of the halogen atoms. Since anions have electrons in excess over positive nuclear charge, outer-electron repulsions are relatively more important than they are in determining ionization-potential trends.

It is worth remarking that the family of elements of highest electron affinity is the halogen family. In this family addition of an electron gives the configuration ns^2np^6. The lowest ionization potentials are associated with the alkali metal atoms of configuration ns^1 which on losing the electron give ions $(n − 1)s^2(n − 1)p^6$. But there is an imbalance in values. The ionization potentials of the alkali metal atoms are larger than the electron affinities of the halogens. This means there is *not* a net energy release associated with the transfer of an electron from metal atom to halogen atom to form two ions free in the gas phase. It's important to remember this energy deficit when trying to explain formation of ionic compounds.

Ionization potentials and electron affinities are useful in discussing ionization. Not all compounds are ionic. What if a neutral molecule is formed? Can we then hope to "locate" electrons? It is sometimes necessary to try to assess the "affinity" of an atom for electrons when it is clearly not an ion (e.g., when it is imbedded in a neutral molecule). A number of proposals have been made as to ways of defining an appropriate atomic characteristic. Most of them have been called *electronegativity* (a term first introduced by the American theoretical chemist and Nobel laureate, Linus Pauling). We follow Mulliken (another American theoretical chemist and Nobel laureate) in suggesting that an appropriate quantity to consider as a measure of the affinity of neutral atoms for electrons is the average of the first ionization potential (IP) and electron affinity (EA). Thus electronegativity χ (chi) may be defined as:

$$\chi = k \left(\frac{\text{EA} + \text{IP}}{2} \right),$$

where k is a constant chosen to put electronegativities on a convenient scale. The scale is so chosen that the largest value, that for F, is 4.0. Table 8.3 shows the electronegativities of the elements according to one set of estimates. Note that trends are similar to those for ionization potentials and electron affinities for obvious reasons. Electronegativity increases across the periodic table from left to right and up the table from bottom to top.

Problem 8.20 Make a plot of second ionization potentials of the elements He through Ca. Explain the differences between this graph and Fig. 8.29.

Problem 8.21 It has been suggested that the "filled-shell" configuration ns^2np^6 has unusual stability. Can you find any evidence from values of ionization potentials and electron affinities that a filled subshell (e.g., ns^2, nd^{10}) or a half-filled shell (e.g., np^3, nd^5) is a stable configuration?

A summary restatement of our main conclusions about atomic energetics is now in order. It will be important to have these ideas clearly in mind in order to approach the problem of the factors that lead to chemical bonding that we shall examine in the next chapter. In developing a theory to account for the formation of ions and molecules it will be necessary to have the following main points in mind.

1. In the H atom (or any other one-electron system) all orbitals of the same n have the same energy (e.g., $3s = 3p = 3d$).

2. As interelectronic *repulsion* becomes important in many-electron atoms, the levels of different l separate with the lowest l corresponding to lowest energy because it is best able to "penetrate" the inner electrons and get close to the nucleus (e.g., $3s < 3p < 3d$ and $4s$ is actually below $3d$ in *neutral* atoms).

3. In positive ions, the comparative scarcity of electrons leads to reduced importance of interelectronic repulsion and there is a return toward the H atom (one-electron) orbital energy order (e.g., $3d < 4s$).

4. As we move down the periodic table, screening effects (repulsions from inner electrons) become more important. Ionization potentials *and* electron affinities become smaller in magnitude with increasing atomic number in a given family.

5. As we move left to right across a row of the periodic table electrons are being added to orbitals at about the same average distance from the nucleus. Nuclear charge is increasing faster than interelectronic repulsions. Ionization potentials and electron affinities are increasing in magnitude.

Table 8.3

Atomic electronegativities

H 2.20																	
Li 0.98	Be 1.57											B 2.04	C 2.55	N 3.04	O 3.44	F 3.98	
Na 0.93	Mg 1.31											Al 1.61	Si 1.90	P 2.19	S 2.58	Cl 3.16	
K 0.82	Ca 1.00	Sc 1.36	Ti 1.54	V 1.63	Cr 1.66	Mn 1.55	Fe 1.83	Co 1.88	Ni 1.91	Cu 1.90	Zn 1.65	Ga 1.81	Ge 2.01	As 2.18	Se 2.55	Br 2.96	
Rb 0.82	Sr 0.95	Y 1.22	Zr 1.33		Mo 2.16			Rh 2.28	Pd 2.20	Ag 1.93	Cd 1.69	In 1.78	Sn 1.96	Sb 2.05		I 2.66	
Cs 0.79	Ba 0.89	La 1.10			W 2.36			Ir 2.20	Pt 2.28	Au 2.54	Hg 2.00	Tl 2.04	Pb 2.33	Bi 2.02			

6. The rough overall trends in the periodic table of ionization potential, electron affinity, or *electronegativity* may be shown as indicated in the sketch below.

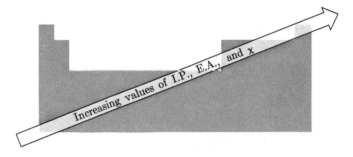

READINGS AND EXTENSIONS

Part of the treatment of wave mechanics given in this chapter was inspired by a somewhat more mathematically sophisticated treatment given in a paperback text by M. J. S. Dewar, *An Introduction to Modern Chemistry*, Oxford University Press, New York, 1965. Chapters 5 and 6. A good book dealing in some detail with the electronic structure and spectra of atoms at a level consistent with this book is R. M. Hochstrasser, *Behavior of Electrons in Atoms*, W. A. Benjamin, New York, 1964. Incidentally, the books cited at the end of the next chapter also deal with the questions of electrons in atoms. It is not really possible to talk about the electronic theory of molecules without discussing atoms.

Seminar: A Wave Treatment of Diatomic Molecule Rotation

Consider a particle moving on a circular *ring* about some center. This resembles a stone being whirled on the end of a rope or a planet orbiting the sun. Suppose there is a standing wave associated with the particle. Try to convince yourself that standing waves on a ring would be required to satisfy the condition

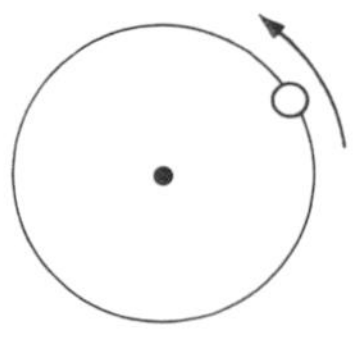

$$n\lambda = 2\pi r, \tag{1}$$

where n is an *integer* (quantum number), λ is the wavelength of the standing wave, and r is the radius of the ring.

Make sketches representing the first three standing waves. This relates possible λ's to the radius r.

According to de Broglie's suggestion, λ is related to the momentum $(m\mathbf{v})$ of the particle by the equation $\lambda = h/m\mathbf{v}$. Given that kinetic energy, KE, is $\frac{1}{2}m\mathbf{v}^2$, prove that

$$KE = h^2/2m\lambda^2. \tag{2}$$

Now find the kinetic energy of rotation as a *function of n* and r [eliminate λ from (2)].

There is a theorem of elementary physics that establishes that a system of two particles rotating about some center, as shown in the figure, may be treated as a problem equivalent to that of a single particle rotating about a point (our case above) if the mass of the hypothetical single particle is taken as the so-called reduced mass, μ, of the two,

$$\mu = \frac{m_1 m_2}{m_1 + m_2},$$

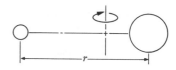

and the distance between the two, r, is taken as the radius of rotation of the hypothetical particle of mass μ.

Now, the diagram of the two particles on a line rotating about an axis between them looks like a fair model of a diatomic molecule such as HCl. We might undertake to calculate the quantized rotational energy levels of HCl if we knew r (the bond length in HCl) and then attempt to find corresponding line spectra with "light" (electromagnetic radiation) of appropriate energy. We can get the needed masses from known atomic weights and N_0, Avogadro's number. Conversely, if we can find the right sort of line spectra, we might use them to find r (the HCl bond length).

Note that your final expression for the rotational energy can be written as

$$KE = n^2 B, \tag{3}$$

where B depends only on μ, r, and universal constants. That is, B is a constant for a given molecule. This suggests a series of energy levels as shown in the figure.

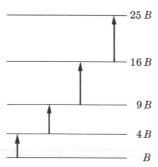

Consider the series of energy level transitions indicated by the arrows in the preceding sketch.

When diatomic molecules are irradiated with electromagnetic radiation in the *far infrared* or *microwave* (radio frequency) region of the spectrum they absorb energy at characteristic frequencies to give line spectra. The interesting thing is that these lines are very nearly *equally* spaced. For example, the HCl molecule has lines at the frequencies (in wave numbers, cm^{-1})

$$83.03, \quad 104.1, \quad 124.3, \quad 145.03, \quad 165.51, \quad 185.86, \quad 206.38, \quad 226.50.$$

Show that these lines are interpretable in terms of the wave analysis above. Find the "bond length" in the HCl molecule containing the mass 35 isotope of Cl (not 35.5). Does the value seem reasonable?*

As an interesting sidelight to this analysis, the data given above are for HCl composed exclusively of the mass 35 isotope of Cl. Replacement of ^{35}Cl with ^{37}Cl shifts the spectrum (because μ is changed), but analysis yields the same value of bond length! What do you take to be the significance, with respect to the theory of chemical bonding, of this nuclear mass independence?

Useful Reading

H. B. Dunford, *Elements of Diatomic Molecular Spectra*, Addison-Wesley, Reading, Mass., 1968. Chapters 1 and 2.

G. M. Barrow, *The Structure of Molecules* W. A. Benjamin, New York, 1964. Chapters 1 and 2.

W. S. Brey, *Physical Methods for Determining Molecular Geometry*, Reinhold, New York, 1965. Chapter 6.

---

* A footnote is *necessary*. The details of modern quantum theory would modify the above treatment. Our quantum number n would be replaced by J with energy dependent upon $J(J + 1)$ instead of n^2.

CHAPTER 9

The Electronic Theory of Valence

9.1 INTRODUCTION

The discussion in Chapters 2 through 7 developed three main ideas about chemical structure. It was shown that the fundamental unit of many substances may be regarded as a *molecule* composed of a small number of atoms arranged according to certain *bond* patterns, which determine shape and energy. Other substances do not occur in molecule units but as aggregates of charged particles or *ions*. Finally, these alternatives are realized by particular elements in ways that suggest definite patterns of relationship among atoms. In Chapter 8, this "pattern" of relationship, the *periodic table*, was explained in terms of the arrangement of the extranuclear electrons. Since it was argued there that it is the disposition of *electrons* in the outermost orbitals, what we shall call the *valence* shell, that determines the chemical family to which the element belongs, it is clearly implied that chemical combination depends critically on the outer or valence-shell electrons. We must tackle now our two early problems. *What* is a bond in a molecule, and *what* leads to the formation of ions? These questions demand a *theory* formulated with due regard for the full sophistication of the contemporary wave-mechanical theory of the structure of atoms, but also sensitive to its present limitations.

9.2 LEWIS-KOSSEL THEORY

As a first step in theory building, it is important to notice a set of fairly simple numerical rules. Consider the elements of the second and third rows of the periodic table. Typical compounds of their *highest* and *lowest* oxidation states are listed below.

$\underline{Li}Cl$	$Be\underline{Cl}_2$	$\underline{B}F_3$	$\underline{C}H_4$	$\underline{N}H_3$	$H_2\underline{O}$	$H\underline{F}$
			$\underline{C}O_2$	$\underline{N}O_3{}^-$		
$Na\underline{F}$	$\underline{Mg}Br_2$	$\underline{Al}Cl_3$	$\underline{Si}H_4$	$\underline{P}H_3$	$H_2\underline{S}$	$H\underline{Cl}$
			$\underline{Si}Cl_4$	$\underline{P}_2O_5$	$\underline{S}O_3$	$\underline{Cl}O_4{}^-$

Consider first that compounds of C, Si, N, P, S, and Cl display *both* positive and negative oxidation states. The *sum*, in each case, of the highest and lowest oxidation number is (neglecting sign) eight. Recall that the number eight is the cycle-generating number for the original version of the periodic table and in the modern form is associated with the "stable" electron configuration ns^2np^6.

The problem is to relate the "rule of eight" in oxidation states to electron configuration. A useful first model was introduced by G. N. Lewis in 1916 (for molecular substances), and its complement was developed in the same year by W. Kossel (for ionic substances). Lewis' idea may be phrased as follows (in anachronistically modernized terms):

1. The s^2p^6 or octet is an unusually favorable outer-shell electronic configuration; witness the *slight* chemical reactivity (and the high ionization potentials) of the inert gas elements, He, Ne, A, Xe, Kr.

2. Other atoms form *molecules* in ways which achieve approximate s^2p^6 configurations in the outer shell about each atom.

3. These "octet" configurations are achieved by *sharing* of electrons among atoms in the molecule.

4. The sharing occurs in pairs.

This is best elaborated by an illustration. Consider the F_2 molecule. The electron configuration of a free F atom is $1s^22s^22p^5$. If we represent the nucleus *plus* inner shell by the symbol F and the outer or "valence"-shell (so-called because it determines bonding) electrons by dots, a diagram of an F atom becomes:

$$: \overset{..}{\underset{..}{F}} \cdot$$

Two F atoms, each with *seven* outer or "valence"-shell electrons have between them 14 electrons. If they *share* one pair, the result may be diagrammed as:

$$: \overset{..}{\underset{..}{F}} : \overset{..}{\underset{..}{F}} :$$

There are 14 electrons in the molecule but *each* atom has a *share* in *eight* (six owned outright and two shared), which is the appropriate number for the ns^2np^6 configuration. The one *shared pair*, according to Lewis, represents the familiar single-bond line we draw in an F_2 molecule:

$$F—F.$$

This scheme can easily be extended to other molecules with the addition of one idea. For the H atom, only *one* pair of electrons would complete the configuration of the inert gas He ($1s^2$), so pairs instead of octets of electrons are crucial for H. Consider the water molecule. Each H atom contributes one valence electron (its $1s$ electron). The oxygen valence-shell configuration is $2s^2 2p^4$, or *six* electrons. There are, then, *eight* valence-shell electrons in a water molecule. If two pairs are assigned to oxygen, unshared, and one pair is shared between oxygen and each hydrogen, we have

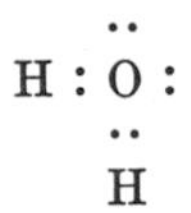

The oxygen has at least a share of eight (for $s^2 p^6$), and each hydrogen has a share of two (for $1s^2$). The two shared pairs correspond to the two single bonds of the usual diagram of the water molecule:

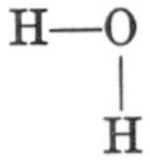

$$H\!-\!O$$
$$\;\;\;\;\;\;|$$
$$\;\;\;\;\;\;H$$

Carbon has the valence-shell configuration $2s^2 2p^2$ or *four* electrons. The molecule C_2H_4 (ethylene) then has $4 + 4 + 1 + 1 + 1 + 1 = 12$ valence electrons. Assigning one pair to be shared between a carbon and each hydrogen leaves *four* to be shared between the two carbons.

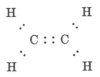

Note that each carbon has a share of eight electrons, each hydrogen has a share of *two*, and the sharing of *two* pairs between the two carbons corresponds to the double bond of ethylene.

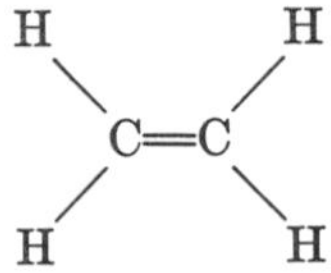

Problem 9.1   Construct Lewis diagrams similar to those just given showing patterns of electron sharing in accord with the octet rule for NH_3, CH_4, HCl, N_2 (nitrogen is usually represented with a triple bond), CO, and $H\!-\!C\!\equiv\!C\!-\!H$.

After practice on Problem 9.1, some more sophisticated examples may be in order. The nitrite ion (NO_2^-) is one good one. Nitrogen ($2s^2 2p^3$) contributes *five* electrons, each oxygen ($2s^2 2p^4$) contributes *six*, and the overall *negative* charge on the ion means there is *one* more electron, or $5 + 6 + 6 + 1 = 18$.

Note that a structure with single bonds only,

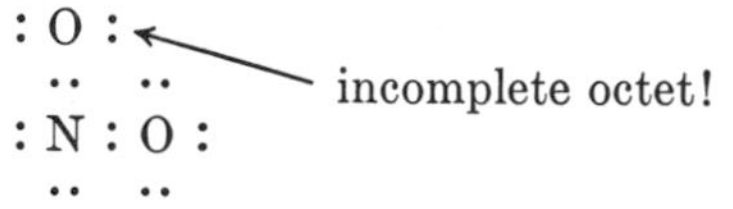

$$: \mathrm{O} : \longleftarrow$$
$$: \mathrm{N} : \mathrm{O} : \qquad \text{incomplete octet!}$$

would leave one oxygen short a pair of electrons, if we exclude the ring structure which does not fit the geometry. Shifting one of the lone pairs on N to share with oxygen to make a "double" bond gives:

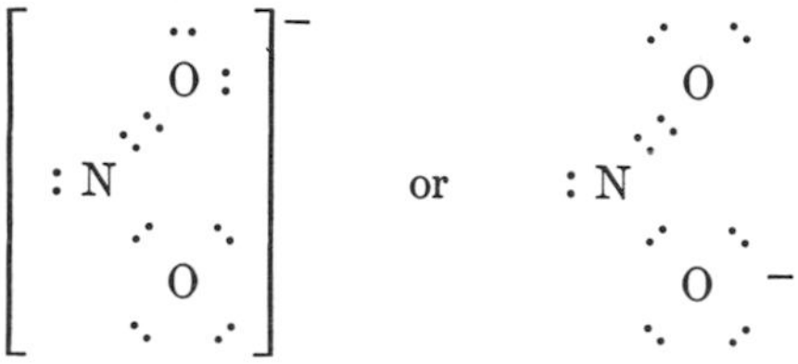

$$\left[\begin{array}{c} \cdot\cdot \\ \mathrm{O} : \\ \\ : \mathrm{N} \\ \\ \mathrm{O} \end{array} \right]^{-} \qquad \text{or} \qquad \begin{array}{c} \mathrm{O} \\ \\ : \mathrm{N} \\ \\ \mathrm{O} \ - \end{array}$$

All atoms now have a share in an octet. The bond diagram for this ion would be

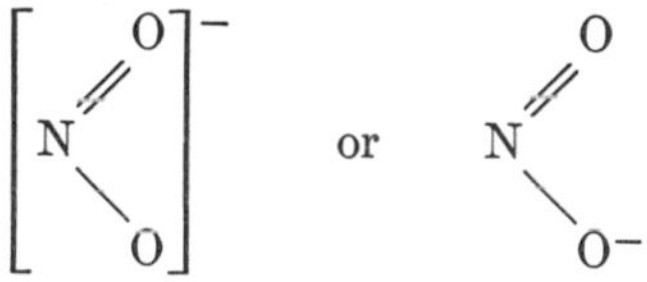

$$\left[\begin{array}{c} \mathrm{O} \\ \| \\ \mathrm{N} \\ \diagdown \\ \mathrm{O} \end{array} \right]^{-} \qquad \text{or} \qquad \begin{array}{c} \mathrm{O} \\ \| \\ \mathrm{N} \\ \diagdown \\ \mathrm{O}^{-} \end{array}$$

It is sometimes conventional to assign the negative charge to the oxygen with the single bond, as shown in the second version of the diagrams. The argument for this assignment is that a shared pair may be thought of as *belonging* half to each atom and an unshared pair as *all* belonging to the atom on which it is shown. On this basis, the single bonded oxygen has *six* electrons plus *half* the pair in the bond for a total of $6 + \frac{1}{2}(2) = 7$. This is *one more* than a neutral oxygen atom $(2s^2 2p^4 = 6)$, so the charge is minus one. A similar calculation for the doubly bonded oxygen gives $4 + \frac{1}{2}(2) + \frac{1}{2}(2) = 6$. This is the number for a *neutral* oxygen. These charges are called *formal* charges because they are based on the simplest but not necessarily the best assumptions about how to divide the shared electrons. (For example, electronegativity differences are ignored.)

Problem 9.2 Find the formal charge on N in the NO_2^- ion.

Problem 9.3 Show a Lewis structure for NO_3^- [*Hint:* there is a single bond from N to O involving the pair that is unshared in NO_2^-.] Assign formal charges to each atom. Do the same for the carbonate ion, CO_3^{2-}. *Note:* carbonate is said to be *isoelectronic* to NO_3^- because there is the same number of valence-shell electrons in each. Similarly, N_2 and CO (see Problem 9.1) are isoelectronic.

Experimentally, NO_2^- does not have distinct oxygens. The two are equivalent. One is not bonded differently from the other. After encountering benzene in Chapter 2, we should not be surprised. Again, the distinct ways of drawing the molecule, which are in all ways equivalent except for the distribution of bonds,

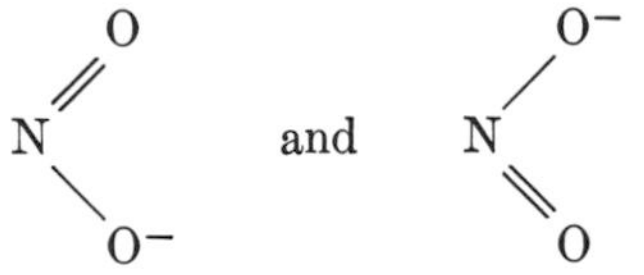

must both be regarded as appropriate but insufficient for the description of the nitrite ion. *Resonance* is involved in the description of NO_2^-:

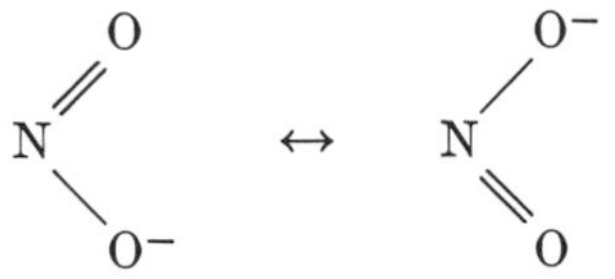

There is no difficulty in the incorporation of resonance into Lewis' theory:

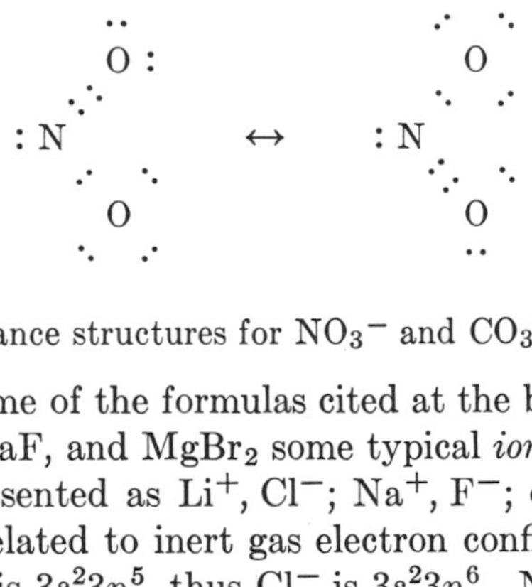

Problem 9.4 Draw resonance structures for NO_3^- and CO_3^{2-} using Lewis diagrams.

Looking back at some of the formulas cited at the beginning of this section we recognize in LiCl, NaF, and $MgBr_2$ some typical *ionic* salts which probably should have been represented as Li^+, Cl^-; Na^+, F^-; or Mg^{2+}, $2Br^-$. These formulations are also related to inert gas electron configurations. Li is $1s^2 2s^1$ so Li^+ is $1s^2$ (He). Cl is $3s^2 3p^5$, thus Cl^- is $3s^2 3p^6$. Na^+ is $2s^2 2p^6$, as is F^-. Mg^{2+} is *isoelectronic* with Na^+ (has the *same* valence-shell electron configuration). Br^- has the configuration $4s^2 4p^6$. All of these are *octet* configurations. As Kossel pointed out in 1916, the majority of stable simple ions are formed by *gain* or *loss* of electrons to the extent of formation of *octet* configurations.

Problem 9.5 Identify the valence-shell electronic configuration of the monatomic ions: O^{2-}, Cl^-, Cs^+, Sr^{2+}, Al^{3+}, S^{2-}. Which of these are "isoelectronic"?

9.3 ELECTRONEGATIVITY AND ELECTRON SHARING

In Section 9.2, the scheme of assigning "formal" charges to atoms was introduced. It was based on the assumption that sharing of electron pairs was even. Since Chapter 5, we have been forced to recognize that some molecules (such as those composing the solvents that dissolve electrolytes) are *polar*, even though they are electrically neutral overall. This means, clearly, that the question of distribution of charge in molecules is not so simple a matter as the "formal"-charge scheme suggests. Possibly, the *electronegativity* parameter, which was defined in a way to suggest that it would correlate with the tendency of an atom to attract electrons in a molecule, will help.

Consider the gaseous HCl molecule. It is clearly *covalent*, but it is *dipolar* with the H end positive and the Cl end negative, and it readily reacts with water to form $H^+(aq)$ and $Cl^-(aq)$. Its Lewis structure is

$$H : \overset{\displaystyle ..}{\underset{\displaystyle ..}{Cl}} :$$

According to formal-charge notation both H and Cl "own" the appropriate number of electrons to be neutral. But the electronegativity of Cl (3.2) is greater than that of H (2.2). Thus an improved Lewis diagram might be:

$$\overset{\delta+}{H} : \overset{\delta-}{\overset{\displaystyle ..}{\underset{\displaystyle ..}{Cl}}} :$$

implying a stronger hold by Cl on the pair of electrons shared with H. The symbols δ^+ and δ^- signify charges of a part of an electron unit at each atom. This model does account for the dipolar nature of HCl. It provides one explanation for the behavior described in Chapter 5, where it was seen that an HCl molecule will align itself in an electric field as if it were two small charges δ^+ and δ^- separated by some distance r usually taken to be the length of the bond. If r is taken to be the length of the bond, the dipole moment could be used to find δ^+ and δ^-, which are perhaps more nearly realistic charge values for the atoms than are formal charges.

In general, we should argue that we expect *polar* molecules to be formed when atoms of significantly different electronegativity form a molecule. This is a good first approximation if we are appropriately cautious about the role of molecular geometry. For example, it is necessary to be cautious about the interpretation of the measured value of *zero* for the dipole moment of CO_2. The electronegativity of C is 2.55 and that of O is 3.44, but if the molecule has a *linear* geometry, O=C=O, there would be dipoles associated with each bond which would be oppositely directed and would cancel. The zero value of the dipole moment is best viewed as evidence for the linear structure of CO_2.

Problem 9.6   The electronegativity of C is 2.55, that of Cl is 3.16. The results of Chapter 3 suggest that CCl_4 has tetrahedral geometry. Make a guess as to what the dipole moment of CCl_4 would be.

Problem 9.7   There are two isomers of the complex $Pt(NH_3)_2Cl_2$, usually assumed to have Pt surrounded by four groups in a planar arrangement. One has a dipole moment; the other doesn't. Sketch structure diagrams for the two isomers.

Problem 9.8   Consult the table of electronegativities and order the following molecules in order of increasing values expected for the dipole moments: HF, IBr, NO, Cl_2.

The extreme example of unequal electron sharing must be admitted to be those compounds whose structures are best represented not as molecules but as collections of charged particles or ions. On comparison of the exemplary compounds tabulated in Section 9.1 with the table of electronegativities, it emerges that *ionic* compounds arise in those cases where the difference in atomic electronegativities is largest.

If the rules concerning electron transfer or sharing to form ns^2np^6 configurations are supplemented by electronegativity arguments, a fairly good scheme for the prediction of the nature of the interaction between a pair of atoms is available. It would be a good start toward molecular theory if the geometry of the molecules could be predicted.

9.4 VALENCE-SHELL ELECTRON-PAIR REPULSION THEORY

A bond has now been pictured as a covalent sharing of a pair of electrons between a pair of nuclei. There is an implication that two electrons (perhaps of opposite spin property!) will stay together in the internuclear region. Electrons should repel each other, since they are all negatively charged, the potential energy associated with this repulsion varying as $1/r$, where r is the distance separating the electrons. If the pair can be taken for granted, the pairs may be expected to avoid each other at least on the average. In 1920, G. N. Lewis reproduced a sketch he had made in 1902 (when he was first starting to speculate about electron pairs), which showed a suggested arrangement for the four pairs comprising the "octet" of electrons, namely on the alternate corners of a cube. This *maximizes* the separation of the four pairs of electrons as one might expect from Coulomb's law. Note that a *tetrahedron* inscribed in a cube touches the four alternate corners where the electron pairs sit in Lewis' picture! This static picture of the electrons "sitting" at a particular position was hard to reconcile with the orbits of Bohr, which were popular in 1916 and 1920 when Lewis published his theory, so he did not pursue the consequences of his diagram to the development of a complete theory of molecular geometry. That was left for the English inorganic and theoretical chemist N. V. Sidgwick in the 1930's and for later workers. Since the 1950's a "valence-shell electron-pair repulsion" theory (VSEPR theory) of molecular geometry has matured.

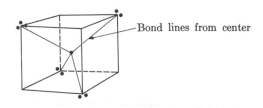

Bond lines from center

The essence of this theory is the idea that the pairs of electrons involved in bonds tend to remain as well separated as possible. Thus all *octets* of electrons should prefer the *tetrahedral* distribution implied by Lewis' drawing, which may be redrawn as below.

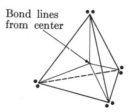

Bond lines from center

If the four hydrogens of CH_4 share a pair of these electrons, then CH_4 should have the well-known tetrahedral geometry!

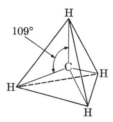

The H—C—H *angle* is about 109°. This is very near the bond angle in most hydrocarbon compounds, e.g., 109.3 in CH_3OH, 109.3 in C_2H_6, or 110.5 in CH_3Cl.

Water has the Lewis formula

$$H : \overset{\cdot\cdot}{\underset{\cdot\cdot}{O}} : H$$

If this structure is based on the tetrahedral distribution of the pairs, the angle H—O—H is again predicted to be 109°. This prediction isn't quite so good; the angle is 105°.

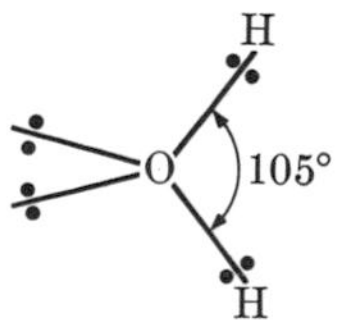

Presumably, the angle between the "unshared pairs" or "lone pairs" on the oxygen is larger than the 109° angle by an amount compensating the smaller HOH angle. The VSEPR theory suggests that this difference of angle between CH_4 and H_2O is not *anomalous* but *expected*. The lone (unshared) electron pairs are probably *less constrained* than those shared between two nuclei. If they are free to move, they are free to interfere with each other. The lone pairs, then, should repel each other *more* than the bond pairs, and the angle between them should be *larger*. As a result, the angle between the bonding pairs is forced to be smaller. The VSEPR theory adopts the following four postulates:

1. geometry of molecules is principally determined by electron-pair repulsion,

2. repulsions are greatest between *unshared* (lone) pairs,

3. next in importance are repulsions between *unshared* pairs and *bond* pairs,

4. finally, last in importance are repulsions between *two bond pairs*.

Given these postulates, the geometry of H_2O becomes entirely comprehensible. The HOH angle is *smaller* than the tetrahedral 109° because the bond-pair–bond-pair *repulsion* of the electron pairs in the O—H bonds are *smaller* than the repulsions between the lone pairs or the repulsions between lone pairs and bonding pairs.

We really should pause now to reflect on our accomplishments. They are imposing. We have now started to talk *predictively* and *confidently* about intimate details of the structure of the incredibly small molecules, such as their bond angles. We have done this with only a small number of physical concepts about the behavior of electrons *added* to the concepts of bonding which *first emerged* without any reference to electrons in the discussions in Chapter 2. We have reached the high point (to date) of a story of great intellectual achievement beginning with Kekulé and continuing through van't Hoff and Le Bel, Lewis, and the developers of valence-shell electron-pair repulsion theory. In the later sections of this chapter we shall look at the recent efforts to explain this success in terms of the new ideas about electron behavior as governed by standing probability waves introduced in Chapter 8, but it is worth noting that even the simpler phase in evolution of structural ideas is not closed. J. W. Linnett has added new ideas to it in this decade.* His paper is recommended to the interested reader.

Problem 9.9 Ammonia (NH_3) has an H—N—H bond angle of 107.3°, less than 109° but more than 105°. Account for this angle using the valence-shell electron-pair repulsions theory. Would you expect ammonia to have a dipole moment?

Problem 9.10 A general theory for molecular structures is now available, at least for the lighter elements. For CH_3Cl, C_2H_4, HF, HOOH, and NCl_3

a) draw Lewis diagrams,
b) predict geometries on the basis of valence-shell electron-pair repulsions, and
c) decide the question of polarity.

---

* *Journal of the American Chemical Society,* Vol. **83**, p. 2643 (1961).

9.5 HEAVIER ATOMS

Let's look now at the application of the octet rule to heavier atoms. The sulfate ion is an interesting first case. A Lewis structure could be:

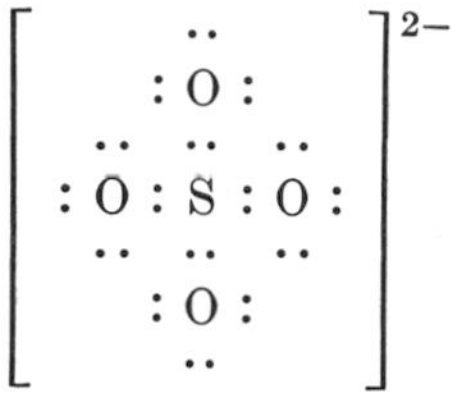

$$
\left[
\begin{array}{c}
\ddot{\,}\vphantom{O} \\
: O : \\
\ddot{\,}\ \ \ddot{\,}\ \ \ddot{\,} \\
: O : S : O : \\
\ddot{\,}\ \ \ddot{\,}\ \ \ddot{\,} \\
: O : \\
\ddot{\,}\vphantom{O}
\end{array}
\right]^{2-}
$$

But a formal-charge calculation shows that each oxygen has a *formal* charge of 1−, and sulfur has, simultaneously, a formal charge of 2+. The formal charge on sulfur seems rather too large for an atom of its electronegativity. An alternative theory has the following diagram as one of several resonance forms:

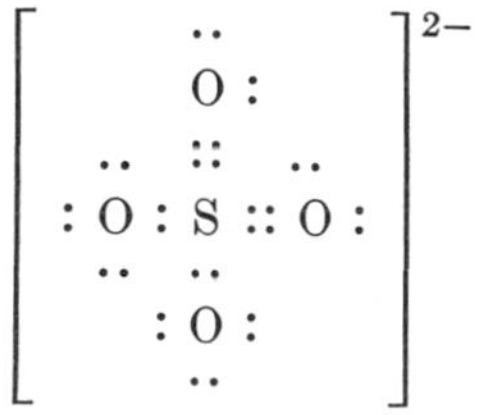

$$
\left[
\begin{array}{c}
\ddot{\,}\vphantom{O} \\
O : \\
\ddot{\,}\ \ ::\ \ \ddot{\,} \\
: O : S :: O : \\
\ddot{\,}\ \ \ddot{\,} \\
: O : \\
\ddot{\,}\vphantom{O}
\end{array}
\right]^{2-}
$$

Here two oxygen lone pairs have been moved to participation in double bonds with sulfur at the *price of increasing the number of electrons around sulfur to 12.* The octet has been expanded. Since the "octet" at sulfur is the $3s^2 3p^6$ configuration, this new structure is not entirely unreasonable. It simply requires use of the $3d$ orbitals. If the idea that the sulfur octet can be expanded is correct, there should be predictable consequences. F is a good oxidizing agent; perhaps many F's will bond to an "expandable" S. Indeed, we find SF_6:

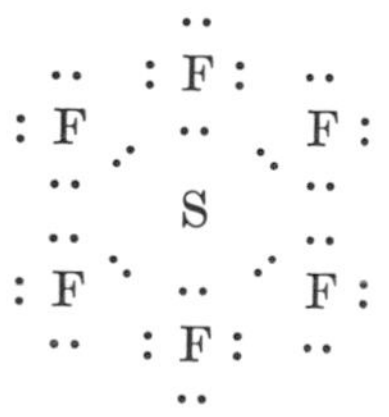

$$
\begin{array}{ccc}
 & \ddot{\,} & \\
\ddot{\,}\ \ : F : \ \ \ddot{\,} & & \\
: F \ \ \ddot{\,}\ \ F : & & \\
\ddot{\,}\qquad S\qquad \ddot{\,} & & \\
: F \ \ \ddot{\,}\ \ F : & & \\
\ddot{\,}\ \ : F : \ \ \ddot{\,} & & \\
 & \ddot{\,} &
\end{array}
$$

which requires twelve electrons around the sulfur atom. Similar evidence for the occurrence of "octet expansion" when the $3d$ orbitals are available is seen in the existence of PF_5:

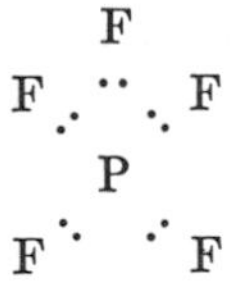

$$
\begin{array}{c}
\text{F} \\
\text{F} \cdot \overset{\cdot\cdot}{} \cdot \text{F} \\
\text{P} \\
\text{F} \cdot \overset{\cdot\cdot}{} \cdot \text{F}
\end{array}
$$

Incidentally, the geometry of SF_6 or PF_5 can also be understood via the valence-shell electron-pair repulsion theory. For minimum repulsions six pairs should be directed toward *six* points of an octahedron and *five* in a trigonal bipyramidal arrangement. These are diagrammed as follows:

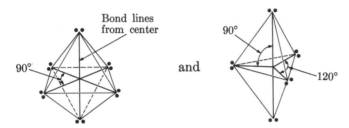

The model suggests F—S—F angles of 90°, which is in good agreement with experiment, and *two* different F—P—F angles of 90° and 120°, respectively.

With a similar flexibility, Lewis theory can be extended to the explanation of complex ions such as $Co(NH_3)_6{}^{3+}$. Six ammonia molecules are bound octahedrally to cobalt in this species:

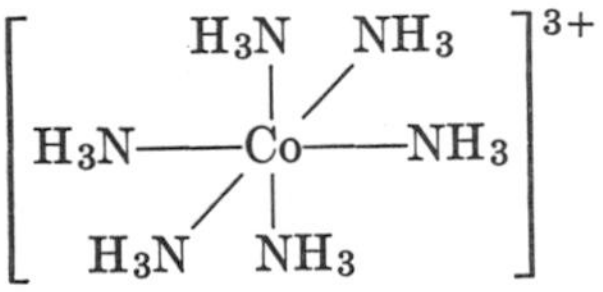

$$
\left[
\begin{array}{c}
\text{H}_3\text{N} \quad \text{NH}_3 \\
\backslash \quad / \\
\text{H}_3\text{N} \text{---} \text{Co} \text{---} \text{NH}_3 \\
/ \quad \backslash \\
\text{H}_3\text{N} \quad \text{NH}_3
\end{array}
\right]^{3+}
$$

A Co^{3+} ion without any coordinated groups has the electron configuration $[Ar]3d^6$, that is, six $3d$ electrons outside the electron shell similar to the inert gas argon. If the six bonds are taken to be shared pairs *supplied* by the "lone" pair on ammonia

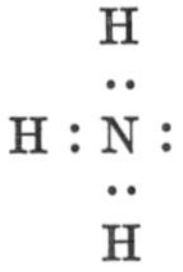

$$
\begin{array}{c}
\text{H} \\
\cdot\cdot \\
\text{H} : \text{N} : \\
\cdot\cdot \\
\text{H}
\end{array}
$$

the Co^{3+} has gained a *share* in twelve additional electrons. This would give an "effective" configuration $[Ar]3d^{10}4s^24p^6$ *or* the configuration of the "inert gas" Kr. This configuration involves $12 + 6$ or 18 electrons in the valence shell but is clearly the sort of configuration that Lewis theory would suggest for compounds of transition metals with partly filled shells. It strongly suggests an analogy between the reactions

$$
\begin{array}{ccc}
\text{H} & & \text{H}^+ \\
\ddot{} & & \ddot{} \\
\text{H} : \text{N} : \; + \text{H}^+ & \rightleftharpoons & \text{H} : \text{N} : \text{H} \\
\ddot{} & & \ddot{} \\
\text{H} & & \text{H}
\end{array}
$$

and

$$
\begin{array}{ccc}
\text{H} & & \text{H} \\
\ddot{} & & \ddot{} \\
\text{H} : \text{N} : \; + \text{Co}^{3+} & \rightleftharpoons & \text{H} : \text{N} : \text{Co}^{3+}. \\
\ddot{} & & \ddot{} \\
\text{H} & & \text{H}
\end{array}
$$

The first is a simple reaction of a base with H^+, in the Brønsted sense of base. The second is sometimes called a *Lewis* acid-base reaction. Co^{3+} becomes an *acid* from this point of view.

Problem 9.11 Draw Lewis structures to characterize the structure of the phosphate ion ($PO_4{}^{3-}$) and the perchlorate ion ($ClO_4{}^-$).

Problem 9.12 Write Lewis structures of SO_2. Why is SO_2 a *polar* molecule, whereas CO_2 is *not*? (Consider the geometry from the VSEPR theory.)

Problem 9.13 Consider a possible Lewis diagram for the complex ion $[PtCl_4]^{2-}$. Show that the Pt atom achieves a share in an electronic configuration like that of the "inert gas" Rn.

9.6 LIMITATIONS OF THE LEWIS THEORY

The last examples, $Co(NH_3)_6{}^{3+}$ and $PtCl_4{}^{2-}$, lead directly to consideration of failures of Lewis theory. There is another ion, $Co(NH_3)_6{}^{2+}$, which is quite stable. If the *plus-three* ion has just the number of valence-shell electrons of the next inert gas, the *plus-two* ion with one more electron *must have one too many*. A number of transition metal complex compounds, similar to the example cited, are known, which have more valence-shell electrons than is appropriate for the configuration of the next "inert gas." You might check for yourself the "electron count" on:

$$
\begin{bmatrix}
 & \text{H}_2\text{O} \quad \text{OH}_2 & \\
 & \big| \diagup & \\
\text{H}_2\text{O}\!-\!\!-\!\!-\!\text{Ni}\!-\!\!-\!\!-\text{OH}_2 & & \\
 & \diagup \big| & \\
 & \text{H}_2\text{O} \quad \text{OH}_2 &
\end{bmatrix}^{2+}
$$

On the other hand, there are some compounds that do not reach a complete octet (or other inert gas configuration). Perhaps the simplest is $BeCl_2$. Beryllium (atomic number 4) has two valence electrons. The structure

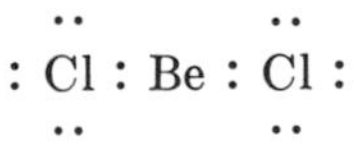

$$
\overset{\cdot\cdot}{}\quad\overset{\cdot\cdot}{}
$$
$$
:\text{Cl}:\text{Be}:\text{Cl}:
$$
$$
\underset{\cdot\cdot}{}\quad\underset{\cdot\cdot}{}
$$

gives complete octets around Cl but puts only four electrons at Be. According
to valence-shell electron-pair repulsion theory, two pairs should be 180° apart,
and the molecule should be linear, which it is found to be, in agreement with
the representation shown. Another example is BF_3, boron trifluoride, which
may be represented:

$$
\text{F}
$$
$$
\overset{\cdot\cdot}{\text{B}}:\text{F}
$$
$$
\underset{\cdot\cdot}{}
$$
$$
\text{F}
$$

The three pairs in the valence shell in BF_3 suggest the planar trigonal geometry

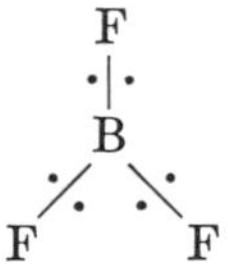

which is, in fact, observed.

It is especially interesting that boron and hydrogen do not form BH_3.
The simplest boron hydride is

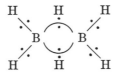

where two of the hydrogens are represented as associated with "two electron
bonds" bridging between the two borons so that three nuclei are bonded by one
electron pair. This is the simplest of a large family of complicated boron hydride
structures all of which are characterized by too few electrons for each pair of
atoms to be "properly" bound by an electron pair. Such molecules have been
called "electron deficient." Of course, the real "deficiency" is in the Lewis theory.

It is clear that the Lewis theory is a useful yet still fallible tool for predicting
formulas. If a better one is to be developed, it will be necessary to raise a
question that has so far been ignored, but should not be if deep insight into
chemical bonding is to be obtained. The question is: what are the energetic
factors that make a particular aggregate of atoms chemically stable? As was
argued in Chapter 6, a spontaneous process is one which is "exothermic" in the

sense that it can be harnessed to do *work*. When atoms react to form a particular aggregate that is stable, where does the free energy decrease originate?

Our next stage of development of a semiquantitative theory of chemical combination will not be deep enough to show exactly how the free-energy changes (ΔG) arise, but we shall explore some changes of *potential energy* which are suggestive. First, ion aggregation is treated, then the more subtle case of covalent bond formation is examined.

9.7 FORMATION OF IONIC LATTICES

The reaction

$$2\text{Na(solid)} + \text{Cl}_2(\text{g}) \rightarrow 2\text{NaCl(solid)}$$

is spontaneous with a large negative ΔG. If we are to understand the stability of ionic lattices, we must see why this reaction is so favorable. It is a complicated process in which a sodium metal lattice and Cl_2 molecules are broken down, electrons are transferred, and an ionic crystal lattice forms. Chapter 6 has suggested, however, that the stages may be treated separately. That is, ΔG_{IV} of Fig. 9.1 may be written as $\Delta G_{\text{I}} + \Delta G_{\text{II}} + \Delta G_{\text{III}}$. We shall try to make estimates of these three and calculate ΔG_{IV}. If this can be accomplished, it should provide substantial insight into the factors which make ΔG_{IV} quite *negative* (favorable reaction). We won't get ΔG values precisely from a simple model but we'll talk about them to keep in focus that it is only ΔG that tells when a reaction can go.

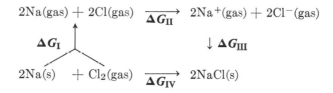

Fig. 9.1. Alternative paths for formation of solid sodium chloride.

1. ΔG_{I}. This process involves conversion of a metallic solid, Na, and a gaseous molecule, Cl_2, into *gaseous atoms*. We shall take experimental values for these, since we are not, just now, asking about the energies of covalent bonds. Temporarily assume that $\Delta G = \Delta H$ (we shall postpone discussion of ΔS) for these processes,

$$
\begin{array}{lll}
2\text{Na(s)} \rightarrow 2\text{Na(g)}, & \Delta H^0 = +\ 51.9\ \text{kcal/2 moles,} \\
\text{Cl}_2(\text{g}) \rightarrow 2\text{Cl(g)}, & \underline{\Delta H^0 = +\ 58.0\ \text{kcal/mole,}} \\
\Delta G_{\text{I}}\ \text{(very approximately)} & {+109.9\ \text{kcal.}}
\end{array}
$$

Note that this is *unfavorable* by about 110 kcal.

2. ΔG_{II}. This process is the ionization in the gas phase. The work "available" might well be estimated as the work required to ionize 2 moles of Na atoms or twice Avogadro's number times the *ionization potential* plus the work available from formation of 2 moles of Cl^- ions or twice Avogadro's number times the *electron affinity* of Cl. This gives:

$$
\begin{array}{lll}
2Na(g) & \rightarrow 2Na^+(g), & +235.6 \text{ kcal/2 moles,} \\
2Cl(g) & \rightarrow 2Cl^-(g), & -176.5 \text{ kcal/2 moles,} \\
\Delta G_{\mathrm{II}} \text{ (approximately)} & & +59.1 \text{ kcal.}
\end{array}
$$

Again this is a *positive* or *unfavorable* contribution. This result is significant by itself. Na^+ and Cl^- are ions of the "octet" or "inert gas" configurations that figure in Lewis theory. This result shows that there is *no spontaneous tendency for atoms to reach such configurations* when they are *isolated* in the gas phase. Reasons for the common occurrence of the octet configuration must be more *subtle*.

3. ΔG_{III}. The process corresponding to ΔG_{III} is apparently no more than the aggregation of charged particles (ions) from the gas phase where they are effectively infinitely separated, into a crystal where they are separated by a constant lattice distance, which is known from x-ray crystallographic study. If we presume that they interact via Coulomb's law, we should be able to calculate the change of *potential energy* accompanying the aggregation process. Since bringing together oppositely charged particles lowers potential energy, we might expect this process to release energy and be exothermic. Let's explore the consequences of assuming that ΔG_{III} is equal to the potential energy change in question.

Two ions of opposite charge located close together exert the force given by Coulomb's law on each other.

$$
F = \frac{Z_+ Z_- e^2}{r^2} ,
$$

where Z_+ and Z_- are the signed number of charge units on the ion and e is the electronic charge. To pull them apart, it is necessary to *do work*, and this work is stored as increased potential energy. As we have previously noted, we choose to set the zero of potential energy when the particles are infinitely far apart so that potential energy is *negative* at all finite distances. In fact, it is (remember that Z_+ and Z_- are of opposite sign):

$$
\mathrm{PE} = \frac{Z_+ Z_- e^2}{r}
$$

at any distance r so that allowing for *two* ions to come together from infinite separation to the separation r_{eq} (eq for equilibrium) observed by x-ray in ionic

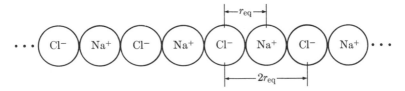

Fig. 9.2. Hypothetical linear NaCl crystal.

crystals leads to lowering the potential energy from zero to:

$$\text{PE} = \frac{Z_+Z_-e^2}{r_{\text{eq}}}.$$

Of course, energy release is the consequence of this decrease.

Now, a sodium chloride lattice is not composed of a *pair* of ions but an "almost infinite" cubic array of alternating positive and negative ions. Consider first a linear row taken from an NaCl crystal to form a hypothetical linear NaCl crystal (see Fig. 9.2). For the pair of nearest neighbors to a given ion, we do have the contribution to potential energy $-|2Z_+Z_-/r_{\text{eq}}|$ (the bars are absolute value symbols), but the next nearest neighbors have *like* charge and repel the ion. This makes a contribution $|2Z_+Z_+e^2/2r_{\text{eq}}|$. For linear NaCl where all the Z's are *one*, we write for the potential energy of a linear crystal:

$$\text{PE} = -\frac{2e^2}{r_{\text{eq}}} + \frac{2e^2}{2r_{\text{eq}}} - \frac{2e^2}{3r_{\text{eq}}} + \frac{2e^2}{4r_{\text{eq}}} - \cdots,$$

$$\text{PE} = \frac{2e^2}{r_{\text{eq}}}\left(1 - \frac{1}{2} + \frac{1}{3} - \frac{1}{4} + \frac{1}{5} - \cdots\right).$$

The term in parentheses is a pure number and a known infinite series which has a definite sum. If we knew that sum, we could evaluate the potential energy of the linear crystal. The problem for the three-dimensional array is still more complicated but a purely mathematical problem of summing an infinite series. The infinite series arising depends only on the geometry of the crystal lattice. The appropriate sum is called the Madelung constant and has the value 1.75 for the NaCl cubic lattice. That is, the potential energy is given by

$$\text{PE} = -1.75(e^2/r_{\text{eq}}).$$

Since x-ray data give r_{eq} as 2.80 Å for NaCl, we get

$$\text{PE} = -1.75\frac{(4.80 \times 10^{-10}\ \text{esu})^2}{(2.80 \times 10^{-8})} = -1.44 \times 10^{-11}\ \text{ergs/molecule}$$

$$= -206\ \text{kcal/mole} \quad (-412\ \text{kcal/2 moles}).$$

This value will be taken to approximate ΔG_{III}, although we should recognize that the argument is a bit oversimplified. We talked as though the two ions simply approached until they touched, then stopped. But ions, like atoms, don't have fixed boundaries, only electron "clouds." The distance apart at which they stop approaching must be determined by repulsions between the overlapping electron clouds, and this factor could (and should) have been put into the expression for potential energy. However, a more sophisticated model which derives a repulsive potential from study of crystal compressibility only changes the result by *about 10%*. We shall see that the error we make is only about this magnitude.

Note that, at least, we have found a factor which leads to a *negative* (favorable) contribution to the ΔG for formation of NaCl from Na and Cl_2.

Let's now "compute" ΔG_{IV}.

$$\Delta G_{IV} = \Delta G_I + \Delta G_{II} + \Delta G_{III}.$$

$$\Delta G_{IV} = +110 + 59 - 412 = -243 \text{ kcal/2 moles}.$$

This suggests a value of about -122 kcal/mole for $\Delta G^0{}_f$, the free energy of formation, of one mole of NaCl(s). The experimental value is -92 kcal/mole at 25°C. A part of the difference results from our casual identification of potential energies and ΔH with ΔG. The "lattice energy" of -206 kcal/mole which was calculated is in fact directly comparable to appropriate experiment, which gives it as -186 kcal/mole. (The error is, significantly, about 10%.)

We must now attend to the factors which allowed us to "get away" with neglect of ΔS for the steps. There are *two* steps in the cycle for which we could *anticipate* large positive values of ΔS. The first is the one in which the ordered Na crystal is converted to gaseous Na atoms:

$$\text{Na(s)} \rightarrow \text{Na(g)}.$$

This should make a *large* positive contribution. A smaller positive contribution arises in the dissociation of Cl_2 molecules into free Cl atoms:

$$Cl_2(g) \rightarrow 2Cl(g).$$

There is one large *negative* ΔS contribution when $Na^+(g)$ and $Cl^-(g)$ are converted from their disordered state into the ordered NaCl solid crystal:

$$Na^+(g) + Cl^-(g) \rightarrow \text{NaCl(s)}.$$

Other steps should have small ΔS values, and our success must follow from ΔS cancellation (approximately) between these steps. Figure 9.3 summarizes the argument.

The calculation shows that ionic lattices are formed when the energy released on forming the lattice is sufficient to *compensate* the energetic *cost* of forming

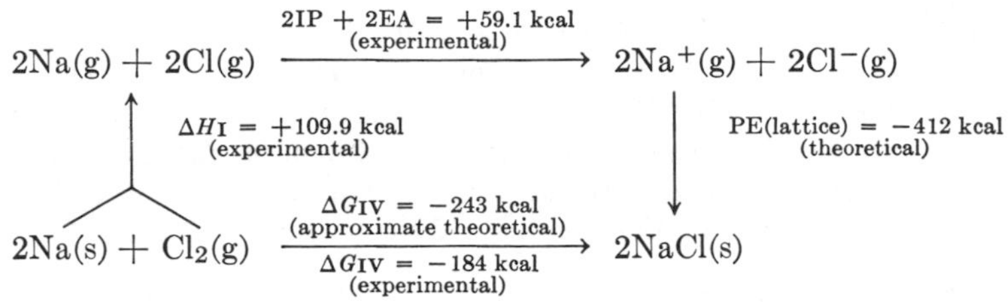

$$2\text{Na(g)} + 2\text{Cl(g)} \xrightarrow{\substack{2\text{IP} + 2\text{EA} = +59.1 \text{ kcal} \\ \text{(experimental)}}} 2\text{Na}^+\text{(g)} + 2\text{Cl}^-\text{(g)}$$

$$\Delta H_{\text{I}} = +109.9 \text{ kcal} \quad \text{(experimental)}$$

$$\text{PE(lattice)} = -412 \text{ kcal} \quad \text{(theoretical)}$$

$$2\text{Na(s)} + \text{Cl}_2\text{(g)} \xrightarrow[\substack{\Delta G_{\text{IV}} = -184 \text{ kcal} \\ \text{(experimental)}}]{\substack{\Delta G_{\text{IV}} = -243 \text{ kcal} \\ \text{(approximate theoretical)}}} 2\text{NaCl(s)}$$

Fig. 9.3. Theoretical and experimental energy values for the "cycle" of Fig. 9.1. (IP means ionization potential; EA, electron affinity.)

the ions themselves. This energy release will increase with increasing *ionic charge* and increase with decreasing r_{eq}. That is, the most stable lattices will be formed by "small" ions of "large" charge. Table 9.1 gives some values of calculated and experimental "lattice energies" for several salts. The calculated values are obtained with the expression for potential energy that includes repulsion explicitly. Values of the ionization energies required to produce the metal positive ions are also included to give some idea of what must be "overcome." Note the *hypothetical* compound of the "K^{2+}" ion, KF_2, which violates the rule that stable ions have "inert gas" electron configurations. Formation of the lattice is quite favorable but the compound does not form because the price to ionize K *twice*, removing an electron from the stable $[Ne]3s^2 3p^6$ configuration, is too high. Note, also, the change in "lattice energy" with changing ion charge and r_{eq}.

Problem 9.14 Considering the series NaF, NaCl, NaBr, explain the trend in "lattice energy." Explain the similar trend in the series NaCl, KCl, CsCl.

Problem 9.15 Formation of the Ca^+ ion is achieved with an ionization potential of only 141 kcal/mole. Based on the assumption that the Ca^+ ion would be very similar

Table 9.1

Salt	Metal ionization energy, kcal/mole	Potential energy of lattice, kcal/mole	
		Calculated	Experimental
NaF	118	−217	−219
NaCl	118	−184	−186
NaBr	118	−175	−177
KCl	100	−167	−169
CsCl	90	−150	−156
CaF_2	414*	−618	−624
"KF_2"	834*	(−618)†	

* First ionization potential plus second ionization potential.
† Theoretical estimate assuming the hypothetical K^{2+} ion to resemble Ca^{2+}.

to the K^+ ion it is possible to calculate a lattice energy for CaCl and hence to reach the conclusion that the free energy of formation of CaCl would be *negative* in the amount of a few kcal/mole. Can you suggest why it is that *only* $CaCl_2$ and *not* CaCl is a known compound? (The free energy of formation of $CaCl_2$ is -179 kcal/mole.) [*Hint:* In attempting to formulate an answer you should try to estimate the energetics of the reaction which consumes the hypothetical CaCl as follows: $2CaCl \rightarrow Ca + CaCl_2$.]

Problem 9.16 Develop an argument from the "lattice energies" of Table 9.1 which allows you to predict the order of the *melting points* of the salts NaF, NaCl, NaBr.

9.8 IONIC RADII AND IONS IN SOLUTION

Table 9.1 shows that atoms of higher atomic number form ions which form less stable lattices. This is fairly obviously a result of the larger size of these ions, which leads to larger values of r_{eq} and less negative lattice energies. It would be interesting to inquire about the size (i.e., radius) of each ion *individually*. That such an approach is reasonable is shown by nearly *constant difference* between the values of r_{eq} for pairs of crystals like NaCl, KCl and NaBr, KBr. This distance relationship is a close analog of Kohlrausch's law (see Chapter 5), and just as Kohlrausch's law suggests regarding conductivities as a sum of cation and anion contributions, so the present result suggests consideration of r_{eq} as the sum of a *cation radius* and an *anion radius*. The only problem is that of dividing r_{eq}. What part of the distance between nuclei belongs to the *cation* and what part to the *anion*? The simplest (and first) approach is the suggestion that in LiI the cation Li^+ is *so small* that the spacing in this crystal lattice is determined by the *closest packing* of *iodides* so that r_{eq} would be related only to the radius of an iodide ion (see Fig. 9.4). Subsequently, a number of more sophisticated procedures of a more theoretical character have been developed for division of r_{eq} and analysis of ion sizes. The prominent American theoretician Linus Pauling has played an important role in this development. Figure 9.5 shows a plot of ionic radii according to Pauling.

Problem 9.17 Considering factors of *nuclear attraction* vs. *interelectronic repulsion*, attempt to explain why cations are usually much smaller than anions. (Refer to Chapter 8.)

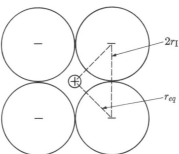

Fig. 9.4. Proposed relationship of r_{eq} in LiI to the radius of $I^-(r_I)$.

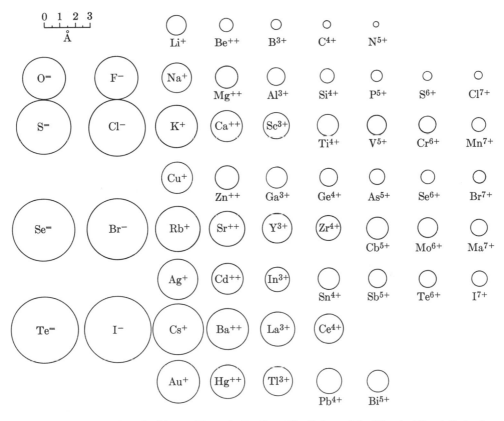

Fig. 9.5. The crystal radii of ions. (From L. Pauling, *The Nature of the Chemical Bond,* 3rd ed. Cornell University Press, Ithaca, New York, 1960. Reproduced by permission.)

Now, attention must be given to an urgent problem. The argument has established that ionic crystals should exist. What about ions in solution? If lattice energies are crucial to ion formation, how do ions form in solution? The answer to these questions is to be found in the answer (see Chapter 5) to the question of the nature of good solvents for ionic materials. The *polar molecules* of a solvent of high dielectric constant align around ions and function just as the neighboring ions in a crystal lattice to lower the potential energy (see Fig. 9.6). This lowering will, clearly, depend on two factors:

1. the dipole moment of the solvent molecule,
2. the radius of the ion.

Note that the *smallest* ions will be most strongly solvated.

With an account of ions in solution we can regard our energetic examination as complete, at least for a first look.

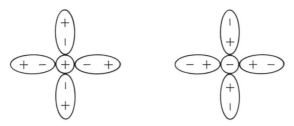

Fig. 9.6. Lowering of ion potential energy by ion solvation.

Problem 9.18   Try to describe the energy changes associated with dissolving an ionic crystal in water.   Why are ΔH values for solution sometimes positive, sometimes negative, and usually small?

9.9   THE ENERGETICS OF ELECTRON SHARING

We have developed a theory of ion formation, but are still faced with the assumption that electrons are shared in molecules.   The key question must now be faced.   What is the advantage of sharing electrons?   (And, why is the sharing pairwise?)   Unfortunately, the Schrödinger theory does not permit a simple straightforward answer to these questions.   The relevant mathematics of the Schrödinger equation are too complicated.   It is relatively easy to write the correct form of the Schrödinger equation for a molecule containing several electrons, but mathematical difficulties of a more or less fundamental sort prevent a simple determination of the wave functions (ψ) which "satisfy" or "solve" the equation.   As in many electron atoms, it will be necessary, finally, to rely on astute approximations based on *the one* well-known set of wave functions, those found in the solution of the hydrogen atom problem.

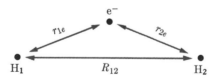

Fig. 9.7. An electron in the internuclear region of an H_2 molecule.

How should such "astute approximations" be developed?   Let's explore the effect of a large value for the probability of finding an electron in the *internuclear* region of a diatomic molecule.   To make it the simplest case take the H_2 molecule.   In Fig. 9.7 an electron is shown in this region.   The potential energy of this arrangement will be determined by three terms.   There will be a negative term $-e^2/r_{1e}$, representing the attraction of nucleus H_1 for the electron.   A similar term $-e^2/r_{2e}$ arises, representing the attraction of nucleus H_2 for the electron.   The third contribution arises from the *repulsion* of nucleus

H_1 for nucleus H_2, $+e^2/R_{12}$. When the electron is in the *internuclear* region, *both* favorable (negative) terms will be larger than the unfavorable (positive) term. The total potential energy is lower in the molecular situation imagined than it would be if the second nucleus H_2 were removed from the scene. This is because the distance r_{2e} is *smaller* than R_{12} so that the negative term $-e^2/r_{2e}$ is more important than $+e^2/R_{12}$. Of course, if the molecular wave function should lead to a *relatively low probability that the electron will be found in the internuclear region*, the result would be opposite. Then, $+e^2/R_{12}$ would be *greater* than $-e^2/r_{2e}$, and potential energy would be *lower* if H_2 separated. The point is that it is *high* probability of finding the electron in the *internuclear* region that gives a *lower* potential energy to the *molecule* than to the separated atoms! And, the lower potential energy of the molecule provides a source for the energy to make molecule formation *exothermic* (ΔG negative). Again, we can get no more than a rough idea of the origins of ΔG in a reaction like $H + H \rightarrow H_2$ from a simple model, but the rough idea will be very helpful.

9.10 SIMPLE MOLECULAR ORBITALS

The last section points to the development of a description of shared electrons that leads to a high probability for finding shared electrons in the region directly between the sharing nuclei. Can we construct orbitals which will "contain" pairs of electrons and localize them to some extent in the bond region? Of course, we shall attempt to construct only *approximate* orbitals using the hydrogen atom orbitals as a basis.

The new theory of *molecular orbitals* begins as does the theory of many-electron atoms by overlooking repulsion effects among electrons and asking the question, how does a single electron move or distribute itself under the influence of the nuclei? If an electron in a diatomic H_2 molecule is *near* nucleus H_1, then H_1 influences the electron much more than H_2 does, and the electron *should* be distributed as it is in a free atom, namely in an atomic orbital centered on H_1. Similarly, near H_2, the second nucleus, the electron distribution must resemble an *atomic* orbital at H_2, the nucleus. If the electron is in a *low* energy state, the atomic orbital to choose is the one corresponding to the normal or ground state of hydrogen, namely the $1s$. Figure 9.8(a) suggests the distribution

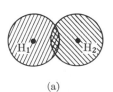 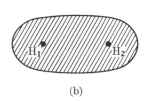 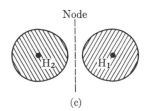

(a)   (b)   (c)

Fig. 9.8. (a) Two overlapping $1s$ atomic orbitals on the nuclei of an H_2 molecule, (b) $1s_{H_1} + 1s_{H_2}$, (c) $1s_{H_1} - 1s_{H_2}$.

of an electron in the 1s type of orbital around each nucleus in a hydrogen molecule. The interesting question concerns the region between the two nuclei. The two 1s orbitals *overlap* in the internuclear region. What is the nature of the electron behavior there? It would be convenient if we could represent it approximately using the atomic orbitals that *must be* nearly correct in the region near one nucleus.

Again, we can exploit an analogy between the *wave* aspect of electron motion and simple wave phenomena to guide the development of our theory. If we consider a vibrating system, we have already pointed out that it has a series of "quantized" modes of vibration, its standing waves, and its fundamental "note" and harmonics. There was a subtlety about these standing waves which could be safely ignored in Chapter 8 but which forms the basis of the "molecular orbital" theory of molecules. The vibrations which we described are a *complete* set to characterize the possible standing waves, but are not a *unique* set. As we noted, standing waves arise from the *superposition* of traveling waves. When two waves travel through the same region of space, the *observed* vibration is the sum of the two waves. This superposition can, of course, enhance or increase the observed vibration depending on whether the waves meet in phase or out of phase. Meeting in phase or out of phase is represented arithmetically by considering not only sums (addition) but also differences (subtraction). There is an entirely analogous property of standing waves. They may be *superposed.* Sums and differences of standing waves are also standing waves.

In Chapter 8, this point was not important because the standing waves of *different* energy have the property of being "orthogonal." That is, if you summed their superpositions in all regions, the in-phase waves would be exactly canceled by the out-of-phase waves, and the resulting superposition is not different from its components. Shown immediately below is a sketch of the superposition of contour diagrams of an s- and p-orbital on the same atom which may clarify this point:

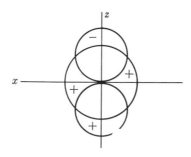

Note that where p is positive so is s, but where p is negative s is still positive and the two contributions cancel. This is an example of "orthogonality." A second sketch shows that putting the two orbitals on *different* nuclei will spoil the orthogonality.

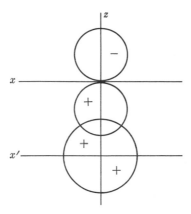

We shall, therefore, want to consider combinations of atomic orbitals as being as good "candidates" for molecular orbitals as the atomic orbitals themselves are. With these ideas let's return to the H_2 molecule.

The problem is to construct an orbital that will "hold" the two electrons of hydrogen. As a first guess try an atomic $1s$ orbital on one hydrogen nucleus:

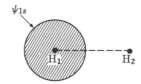

This arrangement is good for a description of the motion *near* H_1, but the function is correct for a situation in which the *potential energy* becomes high as the electron moves away from H_1. In the direction of H_2, the potential is not getting higher in the molecule because the electron is approaching a second nucleus, where the potential should get low. The alternative

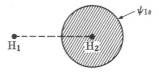

would be better for the H_2 nucleus neighborhood but defective near H_1. But if we *may* consider each as a possible *approximate* molecular orbital, then we may consider two *combinations:* $\psi_{1s_1} + \psi_{1s_2}$ and $\psi_{1s_1} - \psi_{1s_2}$. These are the *sum* and the *difference* of atomic orbitals located on nucleus **1** and nucleus **2**. These are illustrated in Figs. 9.8(b) and 9.8(c). In the case of taking the *sum*, the superposition of two waves of the *same phase* in the internuclear region (the sign always positive means the same phase; see Chapter 8) leads to constructive interference and reinforcement. In the case of taking the difference, the overlapping waves interfere destructively, tending to cancel each other where they overlap in the internuclear region.

Both the waves represented by the contour diagrams Figs. 9.8(b) and 9.8(c) are to be expected to be *better* approximations to the "molecular orbital" than either atomic orbital alone because they are symmetrical with respect to the two nuclei. As they have been written, they need only one modification. The sum of two one-electron orbitals is $\psi_{1s_1} + \psi_{1s_2}$. We need to multiply by a constant to "scale it down" so that when it is "occupied" by *one* electron, the total probability of finding an electron adds up to only *one*. This is accomplished by writing $N(\psi_{1s_1} + \psi_{1s_2})$, where N is called a normalizing constant and can always be found by mathematically straightforward procedures. Having recognized the need to "normalize," we shall not explore the process in detail because the mathematics does not affect a "pictorial" description.

The next question is which of the molecular orbitals formed by "linear combination" of the atomic orbitals is the lowest energy one, and does it correspond to *bonding* between the two H atoms? Consider the linear combination $(\psi_{H_1} + \psi_{H_2})$ which is diagrammed in Fig. 9.8(b). This leads to a concentration of electron density in the region *between* the two nuclei. An electron found between the two nuclei, recall, feels strong attractive forces from both nuclei. This is a *low* potential energy situation and, as we have seen, one which leads to *bonding*. This *energy* analysis can be shown in an *energy-level* diagram. Figure 9.9 shows such a diagram for H_2. The energy levels of the *atomic orbitals* for the *separated* atoms are shown on the left and right. These will always be crucial to determine the energy of molecular orbitals because the electron must display atomic orbital-like behavior *near* a nucleus. In the center the *energy* of the two linear combinations, i.e., the molecular orbitals, is shown. The bonding molecular orbital, of course, is lowest. Note that the $\psi_{H_1} - \psi_{H_2}$ combination orbital lies *higher* than the orbitals of the separated atom. This is to be expected from Fig. 9.8(c), where we see that the probability of finding the electron in the internuclear region is small and no favorable interactions compensate the repulsion between the two nuclei. Electron occupancy of this orbital *will not contribute to bonding*. In fact, it will favor the separated atoms. However, the hydrogen molecule has only two electrons and both may occupy the *bonding* level, $N(\psi_1 + \psi_2)$.

Some useful terminology is now worth recording. The method of constructing approximations to molecular orbitals is called "linear combination of atomic orbitals" (LCAO for short). We call an atomic (hydrogenlike) orbital an AO and a molecular orbital, whether grossly approximate or more exact, an MO. Bonding orbitals are distinguished from ones whose occupancy would favor separation of the constituents of the molecule to free atoms by the terms *bonding* and *antibonding*. Antibonding orbitals (like $\psi_1 - \psi_2$) are often designated with an asterisk (*). Molecular orbitals like those shown in Fig. 9.8(b) and 9.8(c), which *lack* a *node* along the internuclear axis, are called *sigma* (σ) molecular orbitals. The simplest orbitals which do change sign across the internuclear axis and have a node containing it are called *pi* (π) orbitals. These designations are shown in Fig. 9.9.

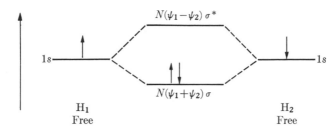

Fig. 9.9. Energy-level diagram for H_2 molecule orbitals. The H_2 molecule contains only two electrons which occupy the lowest level with their "spins" paired.

Problem 9.19 Suppose that an H_2 molecule is irradiated with a photon (light) of the correct energy to promote one electron from the σ_b orbital to the σ^* (antibonding) orbital. What would you expect to be the consequence for molecular stability?

Problem 9.20 On the basis of the molecular orbital energy diagram can you suggest a reason why He_2 (4 electrons) doesn't form? Electronically excited He does form He_2. Is this reasonable?

Let's now examine molecular orbitals for the diatomic molecules formed by atoms of the second row of the periodic table, those with partly filled $2s$ and $2p$ valence shells. This group includes N_2, F_2, and O_2 molecules. We shall make one further approximation. We shall assume that only the orbitals in the *valence shell* are affected when two atoms interact so that only the linear combinations of the *valence-shell* atomic orbitals need be considered. The similarity of chemical behavior of elements with the same valence-shell electron configuration (periodicity) provides good chemical evidence for this approximation.

If we take the lowest energy atomic orbitals in the $n = 2$ shell first, the $2s$ orbitals, they will yield linear combinations very similar to the two derived from the $1s$ orbitals of H. A positive linear combination yields a σ_b (bonding) and a σ^* (antibonding) negative linear combination. The new features emerge when the p atomic orbitals are introduced. Figure 9.10 suggests the appearance of the linear combinations derived from the three p-orbitals, p_x, p_y, p_z. The *internuclear* axis is taken to be the z-axis. Note the *phase* signs indicated on the lobes in the contour diagrams. The linear combinations derived from "sideways" as opposed to "head-on" overlap of p-orbitals change *sign* across the internuclear axis. These are the π-orbitals mentioned on p. 294.

Unfortunately, a complication must now be faced. There is no reason why the $2s$-orbitals which were implicated in formation of one σ_b and σ^* molecular orbital cannot also be involved in linear combinations with the p_z orbitals which form σ_b- and σ^*-orbitals as well. The $2s$- and $2p_z$-derived orbitals are not orthogonal. In fact, an unrestricted molecular-orbital theory would form four linear combinations from the four atomic orbitals entering into σ molecular orbitals. Neglecting any possible combination reflects prejudice as to which combine with which. We should adjust the contributions of each atomic orbital

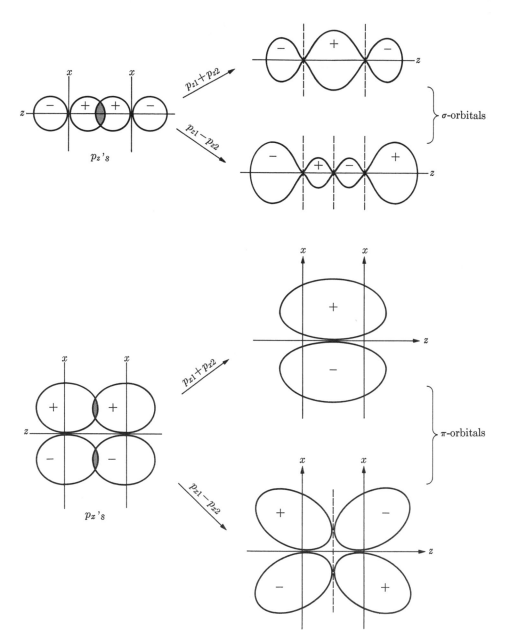

Fig. 9.10. Linear combination of atomic orbitals (LCOA) for p-orbitals. The combinations of p_x-orbitals are identical, except for a rotation through 90°, to those constructed from p_y.

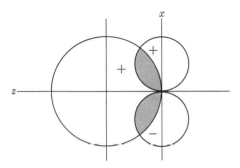

Fig. 9.11. Overlap of an *s*-orbital with p_x-orbital. Note the cancellation of the effects above and below the internuclear axis.

to each molecular orbital in such a way as to reveal the lowest energy (most stable) orbitals. The objective will always be to form those linear combinations that give the *lowest* energy molecular orbitals because we are trying to explain the stability of the molecule. Nature can be expected to "discover" the best arrangement. We must build a theory to find it too. We can, however, get a pretty good picture in *many cases* by using the simple pictures we constructed when we *neglected* the interaction of 2s with $2p_z$-orbitals.

The 2s-orbitals are lower energy orbitals for electrons *near a nucleus*. The linear combinations of 2s-orbitals must then lie at lower energy than the linear combinations of $2p_z$-orbitals. Mixing $2p_z$-orbitals into the molecular orbital will raise the energy for the electron when it is near the nucleus of one atom. If the energy difference between 2s- and 2p-orbitals on an atom is sufficiently large, the lowest molecular orbitals will be purely composed of linear combinations of 2s-orbitals and the higher ones purely of combinations of $2p_z$-orbitals. This can be assumed for pictorial purposes but not quantitative purposes.

There is, fortunately, no σ-π mixing which need be considered. The reason for this is suggested by Fig. 9.11, which shows the overlap between an *s*-orbital on one atom and a p_x-orbital on another. Note that the overlap of *s*- and *p*-orbitals above the internuclear axis where both phases are *positive* is exactly canceled by overlap below the axis where the *s* is *positive* and the *p* is *negative*. Such compensating overlap cannot make any contribution to bonding or antibonding effects.

Now we are prepared to report a typical energy-level diagram for molecular orbitals of diatomic molecules of second row elements. The dashed lines in Fig. 9.12 indicate the mixed origins (*s* and p_z) of the σ levels. Figure 9.12 shows the ordering of molecular orbitals which is most commonly satisfactory for interpretation *in detail* of spectroscopic and magnetic properties of diatomic molecules. Note that the σ bonding level derived "principally" from the $2p_z$-orbitals lies above the π bonding levels. This order will vary with internuclear distance. At short distances π overlap is very favorable, but the positive lobe of the p_z-orbital on atom **1** may extend significantly beyond the nucleus of **2**,

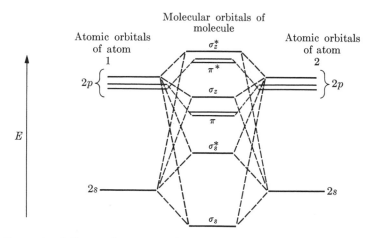

Fig. 9.12. Energy-level diagram for homonuclear diatomic molecules. Each line represents an orbital which may "contain" an electron pair. Recall that the internuclear axis is z.

where the p_z-orbital on that atom is negative. This produces a destructive interference reducing the overall overlap buildup.

If the difference in energy of the 2s- and 2p-orbitals is great, the simplification of ignoring contributions of 2s orbitals to the molecular orbital derived from $2p_z$ and vice versa becomes a good approximation. This, to repeat, is because the electrons moving in the neighborhood of the nucleus *at the lower energy cannot be very much like 2p-electrons*. If they were, they would be at higher energy. An energy-level scheme of this type is shown in Fig. 9.13.

The choice between Figs. 9.12 and 9.13 depends on a knowledge of energies of atomic orbitals, which must be derived from detailed examination of atomic spectra to find the differences of energy levels. The "energy" of each type of atomic orbital is related to how far below ionization that level lies, or the energy required to remove an electron from that orbital. This is closely related to an ionization potential both in concept and in the way it must be derived from atomic spectra. The orbital energies are, as a result, called *valence-state ionization potentials*. A table of these appears as Table 9.2. The values were used in placing atomic orbitals on the previous diagrams.

The energy-level diagrams now constructed are appropriate for the description of diatomic molecules (actual or hypothetical): Li_2, Be_2, B_2, C_2, N_2, O_2, F_2, Ne_2. Some of these exist only under exotic conditions because there are other bonding patterns for the elements, some are very familiar molecules, and some don't exist as stable species. Li_2 with two valence-shell electrons (one from each Li) is the second row analog of H_2 with one bond. Be_2 (Be atom $2s^2$) doesn't exist. The first two electrons occupy the σ_s bonding orbital but the second pair occupies σ_s^*, and the net "bonding" effect is essentially zero. Let's look at nitrogen (N atom $2s^2 2p^3$). The ten electrons will occupy *five* molecular orbitals. These should be σ_s, σ_s^*, π_x, π_y, and σ_z, four bonding and one antibonding, for a *net* bonding effect of *three bonds*. This result is quite

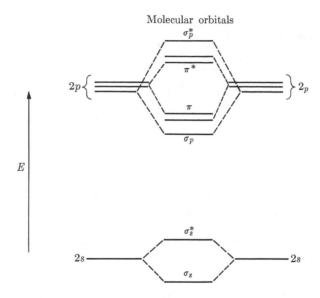

Fig. 9.13. Energy-level diagram for homonuclear diatomic molecules with large 2s and 2p atomic orbital separation. In this case the internuclear distance is assumed large enough for σ_p to lie below π.

reminiscent of the triple bonded Lewis structure

$$: N \equiv N :$$

The O_2 molecule has two more electrons than N_2. The next two levels of Fig. 9.12 are the *two* π^* levels π_x^* and π_y^*. Since electrons *repel* each other, a lower energy arrangement should arise if *one electron is put in each of π_x^* and π_y^**,

Table 9.2

VSIP's (valence-state ionization potentials)
(Orbital energies in various atoms)

Atom	1s	2s	2p
H	110	–	–
He	198	–	–
Li	–	44	–
Be	–	75	–
B	–	113	67
C	–	157	86
N	–	206	106
O	–	261	128
F	–	374	151
Ne	–	391	174

Revealing the traces of their experimental origin in spectroscopy, the energies are given in thousands of wave numbers (cm^{-1}).

rather than both in one. The electrons may (and do) therefore remain unpaired. This has the experimental consequence that the magnetic effects of unpaired electrons remain uncompensated and, in contrast to the majority of simple small molecules, oxygen is *paramagnetic* (attracted into a magnetic field). The prediction of the unusual magnetic properties of oxygen by simple molecular-orbital theory was one of its early great successes. Note that having put two electrons into π^* orbitals should cancel a bond, and O_2 should have *two bonds* net as suggested by:

$$: \overset{\cdot\cdot}{O} = \overset{\cdot\cdot}{O} :$$

Problem 9.21 O_2 has a bond length (O—O distance) of 1.207 Å. In O_2^- the bond length is 1.26 Å, and in O_2^{2-} the bond length is 1.49 Å. Rationalize these changes on adding electrons using the molecular-orbital levels in Fig. 9.12. Describe the change in bond length you would expect on going to O_2^+. N_2 has the very short bond length 1.10 Å; describe the change you would expect on going to N_2^+.

9.11 MOLECULAR ORBITALS FOR HF

Let's turn our attention, now, to a very simple heteronuclear diatomic molecule and examine the new features introduced by having different atoms at the two ends of the molecule. The energy-level diagram may be constructed using the principles derived in the last section. Hydrogen has a valence orbital $1s$ with a valence-state ionization potential, according to Table 9.2, of $-110,000$ cm^{-1}. Fluorine has $2s$- and $2p$-orbitals, both at substantially *lower* energies; for F, $2p = -151,000$ cm^{-1} and $2s = -374,000$ cm^{-1}. Since the $2p$ of F is so much closer in energy to the $1s$ of H, we shall consider only the interaction of these two for construction of molecular orbitals. This *approximation* is based on essentially the same reasoning as that behind Fig. 9.13 where s-p interaction was neglected. The $2s$ orbital of F is such a favorable site for the electron that it cannot be favorable to shift the electron cloud *much* toward H. The resulting situation is shown in Fig. 9.14. Note that p_x and p_y that have π "symmetry" properties cannot interact with $1s$ of H. They are like $2s$ of F *but for a different reason:* molecular orbitals that are unmodified atomic orbitals. Such orbitals are sometimes called *nonbonding* levels.

Now, what do the orbits σ_b and σ^* look like? Since the F atomic $2p_z$-orbital is lower in energy than the H atomic $2s$-orbital, the low-energy (bonding) σ_b might be expected to be more like the $2p$-orbital of F. That is, the mixing should be *uneven*, the sharing favoring the F atom in the bonding level (recall the earlier remarks about sharing between atoms of unequal electronegativity). Uneven mixing is in no way prohibited by the linear combination idea. Superpositions with *any* constant coefficients for the component atomic orbitals are allowed. The wave functions for the bonding molecular orbital would be written:

$$\psi_{\sigma b} = N[1s(\text{H}) + \lambda_1 2p(\text{F})],$$

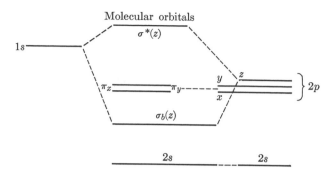

Fig. 9.14. Molecular-orbital energy levels for HF.

where N is the normalizing constant and λ_1 is a constant *greater than* one to weight the mixture in favor of the F-based orbital. The antibonding molecular orbital, would be:

$$\psi_\sigma^* = N[1s(\text{H}) - \lambda_2 2p(\text{F})],$$

where λ_2 is a constant *less than* one to weight the antibonding orbital toward the H-based atomic orbital. That is, most of the $2p$-orbital of F is "used" in the bonding molecular orbital and most of the $1s$-orbital of H is used in the σ^* molecular orbital. Note that a limiting case could be envisioned where the bonding molecular orbital was *all* $2p$ of F. This would represent an ionic state of complete electron transfer. There would then be a coulombic binding force *only*. Thus the ionic model appears as a natural limit of molecular orbital theory.

Problem 9.22   Using the VSIP values of Table 9.2 as a guide, construct an energy-level diagram for LiH. Will the bonding electrons lie "closer" to Li or to H?

9.12 HYBRIDIZATION OF ATOMIC ORBITALS

As we approach polyatomic molecules, the process of forming molecular orbitals becomes more complex. Consider the simple triatomic linear H—Be—H. Sigma bonding here will involve four linear combinations of (calling the two hydrogens H_a and H_b) $2s(\text{Be})$, $2p_z(\text{Be})$, $1s(\text{H}_a)$, and $1s(\text{H}_b)$. The set of molecular orbitals so obtained will be *delocalized* over the whole molecule and will not reveal directly why we have successfully considered such a molecule as held by *two bonds* localized between Be and H_a on the one hand and Be and H_b on the other. Since we saw in Sections 9.2 through 9.5 that directed bonds thought of as shared electron pairs (Lewis and VSEPR theory) form the basis of a very powerful predictive theory, we would like to phrase molecular-orbital arguments so that they reveal these features. Experimentally, BeH_2 is a *linear* H—Be—H structure. It is usefully thought of as constituted from two Be-H bonds at a 180° angle. For our purposes it will be important to see how molecular-orbital theory can encompass and illuminate results of the earlier theory.

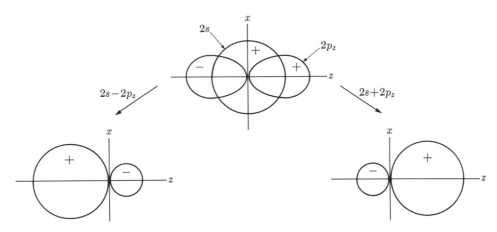

Fig. 9.15. Hybridization to produce two *sp* hybrids. The side built up and the side diminished are reversed as the sign with which $2p_z$ enters the combination is reversed.

At the price of a further approximation, we can obtain a concrete molecular-orbital picture of the familiar bonds. There is no restriction on the order in which we combine orbitals to form molecular orbitals, and if we *combine atomic orbitals on the* Be *first* before mixing in H orbitals, we can get the desired picture. Consider the linear combinations of the 2s- and $2p_z$-orbitals on Be which will yield *two precisely equivalent* combined orbitals called *hybrid* atomic orbitals. They are $(2s + 2p_z)$ and $(2s - 2p_z)$. These combinations are shown pictorially in Fig. 9.15. The resulting *sp*-hybrids are identical to each other, but they point their main lobes in opposite directions so that *one* overlaps *one hydrogen's* orbital strongly and the other hydrogen's orbital very weakly. If we have reconstructed the atomic orbitals this way, it is now very appealing to consider the overlap of *one hybrid* with *one* H orbital and the *other hybrid* with the *other* H orbital so that *bonding* orbitals *localized* in each bond result. When electron pairs are fed into these, they are explanatory of our familiar bonds. How good is this *localized molecular-orbital* approximation based on hybridization? In BeH_2 we construct two bonding orbitals via hybridization. We might well have constructed two bonding orbitals via unrestricted combinations of the four atomic orbitals available for sigma bonding. But, once we have hybridized, we have *forced* s- and p-orbitals to play exactly equivalent roles in the bonding orbitals. This will not always be desirable (e.g., if s and p are widely separated in energy). But it is reasonable if s and p are close in energy, and the energy decreases and increases associated with molecular-orbital formation are large compared to the s-p separations.

In order to develop the desired "wave-mechanical" picture of the chemical "bond," consideration of polyatomic molecules will begin using *hybridized* atomic orbitals on central atoms to form molecular orbitals *localized* in particular bonding regions between particular pairs of atoms. In addition to *sp*-hybrids,

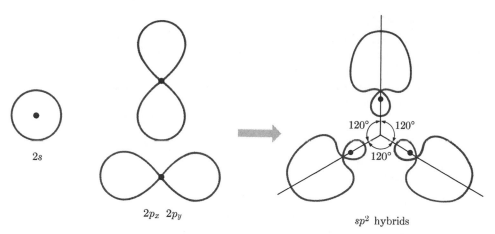

Fig. 9.16. Schematic representation of the formation of sp^2 hybrid orbitals. The orbitals are displaced from their common origin for clarity.

two other hybridization patterns will be important for small molecules formed from light atoms. If we wish to consider a *planar* trigonal molecule like BF_3, we will require orbitals on the central boron directed to the corners of a triangle in one plane. In a single plane, the $2s$ and *two* of the $2p$-orbitals (the two in that plane) can be involved in σ-bonding. (The fourth cannot because it *changes sign* on going through the plane in question and must be π with respect to the plane.) The $2s$ and two $2p$'s may be hybridized to form *three* equivalent sp^2 hybrid atomic orbitals as suggested pictorially in Fig. 9.16. Similarly for tetrahedral molecules, the s and *all three* p orbitals of the central atom may be involved in *sigma* bonding. Hybridization of the central atom (sp^3) can produce *four* equivalent *hybrid* atomic orbitals directed principally to the four corners of a tetrahedron as shown in Fig. 9.17.

Before launching into localized molecular-orbital descriptions we must note one more limitation on this model. In discussion of molecules like N_2 we carried

Fig. 9.17. Schematic representation of the boundary surfaces of the four sp^3 hybrid orbitals.

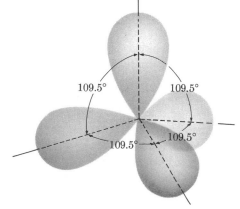

the discussion in a problem (9.21) over to the $N_2{}^+$ ion without changing the molecular orbitals. This can't be done with a *localized* molecular orbital. In BeH_2 an electron cannot be removed from one bond leaving the other intact. The molecule is symmetrical and won't favor one side. To describe $BeH_2{}^+$ with localized molecular orbitals we would have to begin all over again at the beginning of the problem and construct new molecular orbitals that *preserved* the full symmetry of the ion $BeH_2{}^+$ (i.e., gave both H's equivalent roles).

9.13 LOCALIZED MOLECULAR ORBITAL MODELS OF SOME POLYATOMIC MOLECULES

It is now easily seen how to write molecular-orbital descriptions of simple hydride molecules. The two bonds (Be—H) in BeH_2 are each represented by a bond molecular orbital constructed by forming the linear combination of the sp hybrid orbital on Be with the s-orbital on H. This is $[sp(Be) + 1s(H)]$. The CH_4 molecule is tetrahedral:

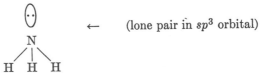

as are the four sp^3 hybrids derived from the $2s$ and three $2p$ orbitals on C.

$\leftarrow$ (simplified sp^3 contour)

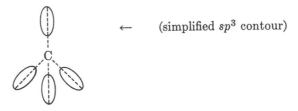

The four bonds in CH_4 are, then, each representable by the bonding molecular orbital localized between one carbon and one hydrogen and written $[sp^3(C) + 1s(H)]$. To the extent that NH_3 is approximately tetrahedral,

$\leftarrow$ (lone pair in sp^3 orbital)

its three N—H bonds can be represented by $[sp^3(N) + 1s(H)]$ localized bonding molecular orbitals. The lone pair is then to be considered as in an sp^3-orbital. H_2O would be described by two bonding molecular orbitals, $[sp^3(O) + 1s(H)]$. The two lone pairs would occupy sp^3 orbitals. The picture is faithful to Lewis theory. An interesting simple extension might be NF_3. Again the N is hybrid-

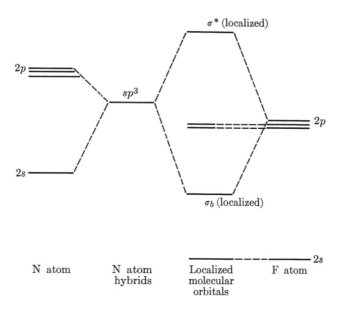

Fig. 9.18. Localized molecular orbitals for an N—F bond in NF_3.

ized sp^3, and a bond is formed with each of the three F atoms by taking the linear combination $[sp^3(N) + p_z(F)]$. The energy-level diagram, Fig. 9.18, shows why neglect of hybridization on the F is acceptable. The $2s$ of F is still well below other orbitals. (Refer once more to the table of valence-state ionization potentials.) The three lone pairs of each F are accommodated in $2s$, $2p_x$, and $2p_y$. The lone pair of N is in an sp^3 hybrid.

Problem 9.23 BF_3 is *planar trigonal* and BF_4 is *tetrahedral*. Describe appropriate hybridization schemes for B in these molecules and identify the orbitals of both B and F involved in forming BF bonds in the simplest model.

A next interesting case introduces the double bond. Consider ethylene, $H_2C{=}CH_2$. Recall (Chapter 3) that van't Hoff argued from simple considerations of a tetrahedral carbon that the two C's and the four H's all lie in a *plane*. This led to the important discovery of geometrical isomers. Can molecular-orbital theory also predict geometrical isomers? We might begin hybridizing each C to give sp^2 atomic orbitals, the p-orbital perpendicular to the plane of the molecule being left out of the hybridization. Sigma bonds may then be formed by combining two sp^2 orbitals of each C with $1s$-orbitals on H and the third sp^2 orbital on one C with the third sp^2 orbital on the other C. The pattern of overlaps is shown in Fig. 9.19(a) by sketching the most important orbital contours. The carbon p-orbitals perpendicular to the molecular plane which were omitted from hybridization remain to be disposed. They are, of course, *precisely correctly oriented* to overlap positive lobes above and negative lobes below the molecular plane to form π_b and π^* molecular orbitals. The overlap

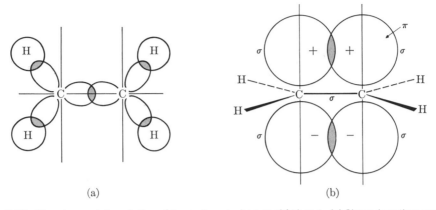

(a)   (b)

Fig. 9.19. The sigma-pi description of the carbon-carbon double bond. (a) Sigma bonding overlap pattern. (b) Pi bonding overlap pattern above and below the molecular plane.

pattern is shown in Fig. 9.19(b). In this description, the carbon-carbon double bond is seen as a sigma bond plus a pi bond. Geometrical isomers are predicted to be separable. Conversion of *cis* to *trans* would require breaking the π-bond.

Now, it is possible to generate an alternative picture beginning with sp^3 hybridization at carbon. This is worth doing to warn us not to become too rigid about the interpretation of simple molecular-orbital pictures. The overlap pattern between the two carbons is shown in Fig. 9.20. This would have two sp^3 hybrids overlapping strongly above the molecular plane and two others below to produce a bent-bond or "banana" orbital description of the carbon-carbon double bond. How can we get the right answer both ways? So long as all the bonding orbitals generated are occupied by electrons, the two are equivalent and equally close to what would be obtained if *no* hybridization had been imposed and molecular orbitals were generated from unrestricted linear combination of all the available C and H atomic orbitals. The differences in the "pictures" are the result of the simplifications in our contour diagrams. Pictures not derived from a complete, unrestricted molecular-orbital theory must be treated with caution.

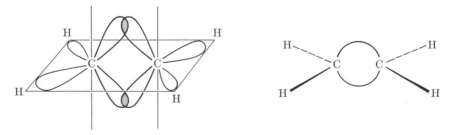

Fig. 9.20. Bent-bond description of the carbon-carbon double bond based on sp^3 hybridization of carbons.

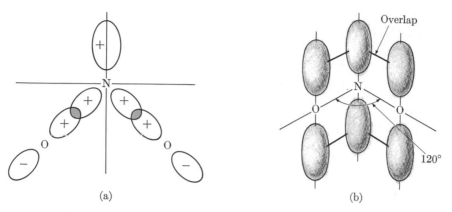

Fig. 9.21. Orbital overlaps for NO_2^- ion. (a) Sigma overlap pattern for NO_2^- ion. Note the orbital available for N lone pair electrons. (b) Pattern of pi overlaps between N and two O's. Note the difficulty of choosing one oxygen over the other, which forces a delocalized treatment.

Problem 9.24 Derive the overlap-pattern diagrams for the sigma-pi and bent-bond descriptions of the planar molecule

$$\begin{array}{c} H \\ \diagdown \\ \diagup \\ H \end{array} C{=}O, \qquad \text{formaldehyde.}$$

Problem 9.25 You might now like to try development of theories of $HC{\equiv}CH$. The molecule is linear. (Recall the use of sp hybridization for the linear BeH_2 molecule.)

As a final example, let's now look at a structure which has required application of resonance theory. Perhaps the simplest is the ion NO_2^- which is described by the Lewis structures (where lines are interpreted as electron pairs):

$$\left[\begin{array}{c} \overset{..}{N} \\ {:}\overset{.}{O} \diagup \diagdown O: \\ \vdots \quad \cdot \overset{..}{} \end{array} \right]^- \quad \leftrightarrow \quad \left[\begin{array}{c} \overset{..}{N} \\ O \diagup \diagdown \overset{..}{O} \\ \ddots \quad \cdot \ddots \end{array} \right]^-$$

Again, the three nuclei lie in one plane, and there is a double bond in each resonance structure. The main concern here will be the effects of different choice of hybridization on the central N and, for simplicity, the oxygens will be left unhybridized. If the central N is hybridized sp^2, sigma bonds to each of the oxygens may be formed by combination of an sp^2 orbital from N with the p orbital from O directed toward the N. The overlap pattern is shown in Fig. 9.21(a). The third sp^2 orbital is available to accommodate the lone pair located on nitrogen. If we insist on a *localized* molecular-orbital picture, the p-orbital on nitrogen directed above and below the plane must be allowed to overlap with the corresponding p-orbital of one of the oxygens to form a π bonding orbital to accommodate two electrons localized between the N and one O nucleus to complete the N-O double bond. The p-orbital of the other oxygen is to accommodate a lone pair as a Lewis structure suggests. But, which oxygen is to play which role? Why single out one for the double bond? Since both are equidistant from the N, their p-orbitals overlap the p-orbital of N equally! We could allow both choices and have *localized* molecular-orbital descriptions of each of our resonance forms. Alternatively, and more economically, we could *drop* the localized molecular-orbital model and form unrestricted linear combinations of the three p-orbitals (one from N, one from each O) which are directed above and below the molecular plane. These three atomic orbitals would form *three* linear combinations (*three* molecular orbitals) of the π type. The electron pair for the second N-O bond of the double bond would go into the lowest energy level. This orbital would *not* be located clearly between either of the N-O pairings. In fact, it has the form $[p(N) + \lambda p(O_1) + \lambda p(O_2)]$, where λ is the *same* constant for both oxygens. This case has a very important point to make. It is exactly in the circumstances where resonance has been involved that a localized molecular-orbital model is cumbersome and the *unrestricted*, "delocalized" model becomes simpler.

9.14 FIRST STEPS TOWARD A THEORY OF METAL LATTICES

One structure type of some importance which has been largely overlooked up to now in this book is the solid metallic lattice. There is some good reason for this. The classical concepts of "bond" and "ion" introduced in Chapters 2 and 5 and explored from an electron viewpoint in this chapter are not too easily applicable to metal lattices. They are, after all, "electron deficient" with respect to the octet rule. The atoms have relatively few s and p valence electrons, and their numbers will be insufficient to complete the "shared octets" of the Lewis theory. Putting this point another way, we may say that the relatively small number of valence electrons could form only a *small* number of *localized* molecular-orbital bonds. The theory of *localized* molecular orbitals will not be appropriate. Experimentally, one of the most interesting properties of metals is electrical conductivity (via electron motion). It should be clear that localized molecular orbitals that tie electron density to particular regions will contribute little toward understanding of electron migration in the electrical conduction process.

One very simple theory of metal structure that meets with some reasonable success imagines each metal atom to be ionized to form a positive ion; the electrons so produced are considered free to move throughout the metal lattice like the molecules of a gas. In such an arrangement, it is clear that average electron-positive metal ion distances will be shorter than the positive ion-positive ion distances in the lattice so that the attractive (negative) contribution to the potential energy will exceed the repulsive (positive) contribution, and the lattice will be stable. This "electron gas" model clearly also allows electron migration under the influence of an electric field.

Problem 9.26 The diagrams show the displacement of ions in a lattice for an ionic crystal and a metal (electron gas model) which would correspond to the fundamental process associated with malleability (ability to be shaped by hammering). Does the model of a metal imply greater malleability? Why?

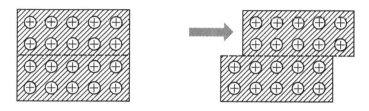

Shift of a metallic crystal lattice along a plane with no resultant strong repulsive forces.

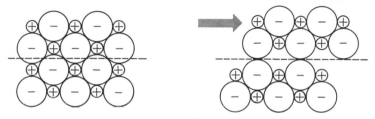

Shift of an ionic crystal lattice along a plane with resultant strong repulsive forces and lattice distortion.

A more sophisticated and probably more satisfactory theory can be constructed using the *delocalized* molecular-orbital approach. Let's consider metallic Li as a very simple example. Each Li atom in the lattice has one 2s *valence orbital*. Since 2s-orbitals are spherically symmetrical (have no directional preferences), each may overlap equally with 2s-orbitals on all of its lattice neighbors. In a Li crystal of the size of a mole (6.9 g), the approximately 10^{23} equivalent 2s atomic orbitals can enter into formation of 10^{23} delocalized molecular orbitals which will be spaced over a relatively narrow range of energies. These can be thought of as forming a "band" of closely spaced levels as indicated in Fig. 9.22. Since each Li atom contributes *one* electron, there will be 10^{23} elec-

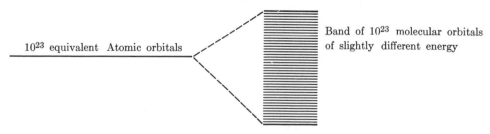

10^{23} equivalent Atomic orbitals

Band of 10^{23} molecular orbitals of slightly different energy

Fig. 9.22. Band of energy levels in a metal.

trons to fill into this band. Pairwise occupancy of the lower levels will require only *half* filling of the band, but there will be quite easily accessible *low* lying empty orbitals. Without pursuing details, it's clear that such a structure is quite "flexible" and can find new arrangements easily when perturbed (either by an electric field or a hammer).

9.15 RESIDUAL INTERMOLECULAR INTERACTIONS

We have now discussed the forces which hold atoms together in a molecule and those that bind ions in a salt lattice, ions in solution, and finally atoms in a metal. Relatively little has been said of the forces which are involved in keeping molecules together in liquids and solids. Clearly such forces must exist, otherwise all substances composed of discrete molecules would be expected to be gaseous. Now molecular substances are more easily vaporized than those of other types but the range of boiling points between, say, He which boils at $4°K$ and H_2O which boils at $373°K$ suggests quite a range in cohesive forces. Why is one liquid so much more easily vaporized than another?

A first important factor may be readily disposed of as an extension (hopefully obvious) of earlier arguments. Consider a liquid or solid composed of dipolar molecules. Electrostatic forces will operate to *orient* the dipoles as shown:

When such orientation occurs, the system is in a low potential-energy state, and energy will be required to achieve the randomized orientation and *large* intermolecular distances characteristic of the gas phase.

Problem 9.27 Consider the following pairs of substances. Could the dipole interaction factor account for the difference in boiling points indicated in each case? (Where dipole moments, D, are not quoted, you should be able to infer relative values on the basis of Lewis structures, electronegativities, and VSEPR theory.)

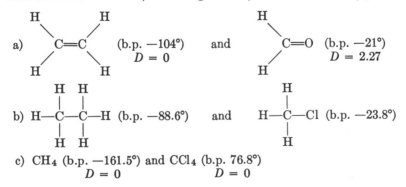

But, we may still wonder about the difference of boiling point of nonpolar molecules, and why closed-shell species like He or Ne condense at any tempera-

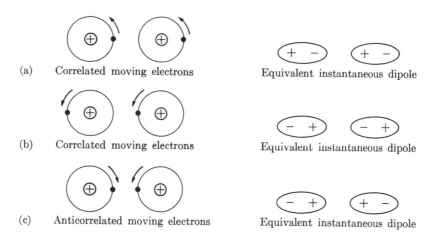

(a)   Correlated moving electrons   Equivalent instantaneous dipole

(b)   Correlated moving electrons   Equivalent instantaneous dipole

(c)   Anticorrelated moving electrons   Equivalent instantaneous dipole

Fig. 9.23. The origin of induced dipole attractions.

ture above absolute zero. Actually, theoreticians have been able to show that there is a residual electrostatic interaction arising from electron movement even in the case of molecules which have no permanent dipole moment. This arises from what may be visualized as "correlation" of electron motion in neighboring free atoms or molecules. Let's allow ourselves the temporary simplification of thinking about electrons as moving on circular paths about atomic nuclei. Figure 9.23(a) and (b) show two electrons on neighboring atoms with correlated motion. Figure 9.23(c) shows the situation where the electrons' motion is "anticorrelated." When the electrons are moving together so that they form instantaneously correctly aligned dipoles, Fig. 9.23(a) and (b), there will be an attraction force and a lowering of potential energy associated with bringing the partners together. When the electrons move in the manner called "anticorrelated" the instantaneous dipoles repel. Of course there will be a tendency for the correlation to develop in the first manner. The forces so arising are called London forces, dispersion forces, or van der Waals forces. They lead to an attraction between any two partners. Of course, the partners will not approach indefinitely close together. As the electron clouds of the partners begin to overlap strongly, compensating repulsions develop in a situation where covalent bonds cannot form.

Dispersion forces, which depend on the correlation of electron motion, will increase with the number of electrons in the molecule and the "flexibility" of the electronic orbitals. This accounts for the frequently encountered increase of boiling point with molecular weight.

Problem 9.28 Predict the order of boiling points for a series of similar nonpolar molecules with central atoms belonging to a single family of the periodic table. Consider, for example, CH_4, SiH_4, GeH_4, SnH_4. Look up the boiling points in a handbook and check your predictions.

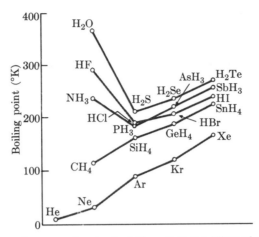

Fig. 9.24. Boiling points of some molecular hydrides. Note the effect of hydrogen bonding on the properties of NH_3, H_2O, and HF.

9.16 HYDROGEN BONDING

Consider the boiling-point trends in the families of simple hydrides shown in Fig. 9.24. It is clear that the first members, HF, H_2O, and NH_3, have anomalously high boiling points but that CH_4 conforms to the trend defined by the heavier compounds. It is not possible to rationalize the remarkably high boiling points of HF, H_2O, and NH_3 by simple recourse to consideration of molecular dipole moment. The situation requires closer examination. These special cases have a common feature. The molecules which are interacting strongly all involve hydrogen bound to a highly electronegative element and all have unshared electron pairs on the electronegative atom. The significance of the first of these facts is that a hydrogen will have only a small share in the electrons in the bonding molecular orbital. From outside, it will look very much like a bare hydrogen nucleus, the fundamental nuclear positive charge unit, the proton. Certainly, a nearly bare proton would be expected to interact with lone pair electrons on a neighboring atom as:

$$\left[\; \overset{\cdot\cdot}{\underset{\cdot\cdot}{:F}} : H : \overset{\cdot\cdot}{\underset{\cdot\cdot}{F}} : \; \right]^{-}$$

Salts of the FHF^- ion have been isolated.

Possibly, this phenomenon, which is called *hydrogen bonding*, is a special case of dipolar interaction. This point has been hotly debated by theoreticians, some making strong contentions for a covalent component in hydrogen bonding. Whatever the detailed nature of the origin of hydrogen bonding, it is certainly a sufficiently important phenomenon to deserve separate attention. Hydrogen bonding is responsible for the special characteristics of water which make it a medium supportive of life. It is also hydrogen bonding which is crucial to determining the three-dimensional structure of proteins and nucleic acids, respectively the building blocks and genetic information carriers of living systems.

READINGS AND EXTENSIONS

The first problem at this point is to list books extending the electronic theory of chemical binding. There have been many in recent years. We shall mention only a few that are available in paperback and that happen to have served us well. This list does miss good books. Again, we refer to Dewar's book listed at the end of Chapter 8; also, to D. K. Sebera, *Electronic Structure and Chemical Bonding*, Blaisdell, Waltham, Mass., 1964, and H. B. Gray, *Electrons and Chemical Bonding*, W. A. Benjamin, New York, 1964. The Gray book is a detailed application of molecular orbital theory to a wide variety of cases.

Second we want to say a few words about where one might go from here. The simple answer is "lots of places." The focus of the last nine chapters has been structural chemistry. Chemical dynamics might now draw attention. For example, there is the book by J. B. Dence, H. B. Gray, and G. S. Hammond, *Chemical Dynamics*, W. A. Benjamin, New York, 1968. This book correlates the behavior of a great many chemical systems in terms of chemical energetics and reaction rates and mechanisms. Another interesting book on rate and mechanism is E. L. King, *How Chemical Reactions Occur*, W. A. Benjamin, New York, 1964. Another line of investigation might be to explore the application of structural principles over the range of elements in the periodic table. See, for example:

R. C. Johnson, *Introductory Descriptive Chemistry*, W. A. Benjamin, New York, 1967
W. L. Jolly, *Chemistry of the Non-Metals*, Prentice-Hall, Englewood Cliffs, N. J., 1966
E. M. Larsen, *Transitional Elements*, W. A. Benjamin, New York, 1965
H. H. Sisler, *Electronic Structure, Properties, and the Periodic Law*, Reinhold, New York, 1963.

Then there is the chemistry of life:

V. H. Cheldelin and R. W. Newburgh, *The Chemistry of Some Life Processes*, Reinhold, New York, 1964
R. J. Light, *A Brief Introduction to Biochemistry*, W. A. Benjamin, New York, 1968.

Seminar: "Electrostatic" or "Ionic" Theory of Complexes

The Seminar of Chapter 5 explored the structure of the "complex" compounds of Co(III) and led to a theory which suggested that they were well represented by an octahedral distribution of six *ligands* bonded to a *central metal*. In the discussion of *Lewis theory* it was suggested that the Co^{3+} ion forms complexes by electron sharing (covalence), using the empty $3d$, $4s$, and $4p$ orbitals of its valence shell, and that the criterion of a *ligand* was the *availability* of an unshared electron pair that could be shared with these empty orbitals.

About 1950, a theory emerged which suggested that electron sharing could be *neglected* as a small effect in holding complexes together. It suggested that the complex was held together by the electrostatic attraction that the positive central metal ion had for the negative *charges* of the ligands or the negative ends of *dipoles* if the ligands were neutral molecules rather than anions. The theory, called *crystal field theory*, was more elaborate than the theory of "ionic" binding in NaCl crystals *only* in the respect that it did not regard the metal ion as a charged *sphere* but looked in detail at the spatial distribution of electron charge in the outermost or valence shells of the metal ion. It considered explicitly the interaction of these outer electron (negative) charge clouds with the ligands.

In your discussion, develop the ionic–crystal-field explanation for the stability of octahedral complexes. Enumerate all of the important interactions, attractive and repulsive. Pay particular attention to the role played by (a) the size of the ligand ion or (b) the dipole moment of the ligand molecule in determining the stability of the complex. How does the theory suggest that stability should vary with (a) metal ion size, (b) number of metal d electrons?

Look at the experimental evidence about the role of ligands. Especially, consider the so-called "spectrochemical series." Try to show that the crystal field theory has serious defects. In a few lines, you might like to offer a suggestion as to why the theory is taken seriously when it has readily identifiable defects. (Note that crystal field theory played an important role in stimulating the tremendous growth of the field of inorganic chemistry in the 1950's and early 1960's. This is not an answer to the last question. It simply adds to the puzzle!)

Useful Reading

F. Basolo and R. C. Johnson, *Coordination Chemistry*, W. A. Benjamin, New York, 1965. Chapter 2.

D. K. Sebera, *Electronic Structures and Chemical Bonding*, Blaisdell, Waltham, Mass., 1965, pp. 207–226.

B. Douglas and D. McDaniel, *Concepts and Models of Inorganic Chemistry*, Blaisdell, Waltham, Mass., 1965, pp. 345–353.

APPENDIX 1

Introduction to the Dynamics of Chemical Reaction

A1.1 INTRODUCTION

The main theme of this book has been the development of ideas of chemical structure, the means of interpreting experimental results in terms of the molecular or ionic structures of reactants and products. This method has resulted in an almost entirely static viewpoint. We have asked about what molecules exist in stable form but not how a reaction proceeds from one stable form to another. In this appendix we shall attempt a preliminary treatment of the "process" of reaction to indicate a little of the flavor of this kind of concern which figures at least as prominently in modern chemical research as the study of structures.

Where should we begin in the analysis? The question is, how do chemical reactions happen? Perhaps it's not unreasonable to begin theoretically with one seemingly obvious postulate.* Herewith we state the first postulate of chemical dynamics:

For two species to react, they must come into collision with each other.

First, this postulate can be applied to gases, where the kinetic theory of gases has things to say about collisions. We shall attempt to calculate the *rate* (or velocity or speed) of a chemical reaction between two partners A and B according to the pattern:

$$A + B \rightarrow products.$$

An example of this might be

$$H_2 + I_2 \rightarrow 2HI.$$

It is by no means *obvious*, it must be warned, that the reaction between H_2 and I_2 to produce HI molecules occurs in a simple collision between an H_2 and I_2

---

* Which recent research has shown to be less true than obvious of a few special reactions— such as transfer of an electron from one atom to another in gases, which appears to occur over very large distances.

molecule. It might actually require several steps, a so-called *complex mechanism* like:

$$I_2 \rightarrow I + I \quad \text{(two I atoms formed)},$$
$$I + H_2 \rightarrow HI + H \quad \text{(one HI plus an H atom formed)},$$
$$H + I \rightarrow HI \quad \text{(the second HI formed)},$$

which is one of a number of sets of steps that would give the overall result of converting one H_2 and one I_2 to two HI. Of course, the desire to find ways of deciding whether a reaction occurs by a complex or simple (one-collision) *mechanism* leads to interest in chemical dynamics and reaction rates.

Problem A1.1 See if you can invent another possible complex mechanism for the H_2-I_2 reaction.

A1.2 GAS MOLECULES COLLISION FREQUENCY

If the problem is to approach a prediction of the number of molecules of A reacting with B each second, the first step is to calculate the number of collisions that should occur between A and B in one second. Consider an A molecule of average velocity $\bar{v}$ moving through a sample of gas containing B molecules as shown in Fig. A1.1(a). For the simplest approximation, imagine that the B molecules are not moving.

In one second, an A molecule moving at a velocity of $\bar{v}$ cm/sec traverses v cm. If the A molecule has a radius r_A and a B molecule has a radius r_B [see Fig. A1.1(b)], whenever the center of the A molecule comes within the distance $r_A + r_B$ of the center of a B molecule, they will collide. The simplest way then to find the number of collisions in a second is to consider the volume swept out in a radius $r_A + r_B = \sigma_{AB}$ about the A molecule as it moves through the gas in one second. This volume swept out will be a *cylinder* of radius σ_{AB} and height $\bar{v}$, the number of centimeters A moves per second. (We shall call σ_{AB} the collision diameter of the A-B pair.) If one can now find the average number of B molecule centers which lie within a volume equal to that swept

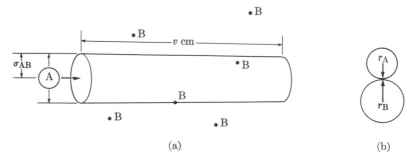

(a)   (b)

Fig. A1.1. (a) The "collision cylinder" swept out by an A molecule moving with velocity *v*. (b) Derivation of radius of collision cylinder.

out by the "collision cylinder" around A, that number will be the number of collisions per second of the A molecule. Of course, the number of B's in any unit of volume can be simply calculated if the *concentration* of B is known. (In the gas phase, recall concentration is proportional to pressure.)

Let's proceed now to calculation of the number of collisions. The volume of the collision cylinder is the base area $\pi\sigma_{AB}^2$ multiplied by the height $\bar{v}$. The number of B molecules in the volume is the volume $\pi\sigma_{AB}^2\bar{v}$ times the number of B molecules per unit volume, or the concentration of B expressed in units of molecules per cubic centimeters, n_B. The number of collisions of *one* A molecule may then be written $\pi\sigma_{AB}^2\bar{v}n_B$. To calculate the total collision number, which we shall call Z_{AB}, we multiply the number of collisions of one A molecule by the number of A molecules per unit volume. This is n_A, so that

$$Z_{AB} = \pi\sigma_{AB}^2\bar{v}n_An_B. \tag{A1.1}$$

Before we can compare the total collision number with any experiment we must decide on a value for $\bar{v}$. Suppose as a first approach we choose the "root mean square" found in the kinetic theory of gases, $\sqrt{\bar{v^2}} = v_{rms} = \sqrt{3RT/Nm}$, where R is the gas constant, T is the absolute temperature, and Nm is Avogadro's number times the mass of a molecule, that is, the molecular weight of A, MW_A. Then we might write

$$Z_{AB} = \pi\sigma_{AB}^2\sqrt{3RT/MW_A}\,n_An_B. \tag{A1.2}$$

Note that we have now reaped the consequences of our initial oversimplification of allowing only A molecules to move and taking B at rest. If we had made the *opposite* choice, we would have the velocity of B in Eq. (A1.2), i.e., $\sqrt{3RT/MW_B}$. If the molecular weights of A and B are different, two different values of Z_{AB} can be calculated using the two equally valid starting approximations! Clearly, what is needed in the derivation is the *relative velocity* of A molecules *with respect to* B. This is a more subtle quantity to evaluate; it turns out to involve the so-called reduced mass, $M_AM_B/M_A + M_B$. It is useful, however, to proceed to *approximate* Z_{AB} using Eq. (A1.2).

Problem A1.2 Argue that it is a *reasonable approximation* to treat B molecules as being, *on the average*, at rest, and choose an average velocity of A molecules to calculate the volume of the collision cylinder.

If [A] is the conventional concentration of A in moles liter^{-1}, then n_A is given by

$$n_A = N[A]/1000,$$

where N is Avogadro's number and the factor of 1000 converts from liters to cubic centimeters. Writing a similar expression for n_B, we may explicitly evaluate Z_{AB}. Taking the O_2 molecule as a typical molecular weight and 300°K (27°C) as the temperature, we get $Z_{AB} = 2.5 \times 10^{32}$ molecules per second if A and B are at concentrations of 1 mole per liter and $\sigma_{AB} = 3.7 \times 16^{-8}$ cm (O_2).

The value of σ_{AB} is derived from the b constant in the van der Waals equation for gases which is presumed to represent the volume occupied by the molecules themselves in the gas.

Problem A1.3 Carry out the indicated calculation of the typical numerical value of Z_{AB} from Eq. (A1.2). How would Z_{AB} change if T were raised from 300° to 310°? If n_A were doubled, if n_B were doubled, if both n_A and n_B were doubled?

If we are interested in chemical reactions, we will observe quantities on the mole scale. The collision number Z_{AB} can be reconverted to the mole level as the number of *moles per liter* of A's colliding with B's per second, converting molecules per cubic centimeter per second to moles per liter per second $Z_{AB} = 4.2 \times 10^9$ moles per liter per second. *If* (and it's a big if) a mole of gas A could be mixed with a mole of gas B very rapidly and then reaction occurred at *every* collision, the reaction would be essentially over in a time of the order of 10^{-9} seconds!

A1.3 FIRST COMPARISONS WITH EXPERIMENT

It is now possible to formulate approaches to experiments on gas reactions to see what the collision theory has accomplished so far. From Eq. (A1.2) we see that Z is proportional to concentrations. In fact, for a *given temperature* we could write

$$Z = \pi\sigma_{AB}^2\sqrt{3RT/MW}\, n_A n_B \tag{A1.2}$$

as

$$Z = k[A][B], \tag{A1.3}$$

where k is a constant (In fact the value of k has already been calculated as 4.2×10^9 for O_2 at 300°K, since it is the value of Z when $[A] = [B] = 1$.) and $[A]$ and $[B]$ are molar concentrations of A and B. Reducing the expression to form A1.3 is very important. It gives the form needed to compare to simple experiments where we measure the changes in reactant concentration as time passes. Note that what this says is that Z *decreases* as the concentrations of A and B *decrease*. Of course, that is just what is happening as the reaction progresses. The reaction $A + B \rightarrow$ products obviously implies changing A and B concentrations. But Z is given for each *particular* value of $[A]$ and $[B]$. If reaction rate is to be related to collision frequency, it must be done through the rate at which a reactant, say A, is disappearing when $[A]$ has a particular value. This is a question of instantaneous rate. It is analogous to asking, if a car can brake to a stop in a period of 10 sec, what is its speed exactly 5 sec after the brakes are applied? The problem is connected with the fundamental idea of differential calculus. The instantaneous rate, the rate at which concentrations are changing at some particular concentration value, is the time derivative of the concentration. We shall *define* the *rate* of a reaction $A + B \rightarrow$ products as

$$-\frac{d[A]}{dt} = -\frac{d[B]}{dt} \equiv \text{reaction rate.}$$

The negative sign is chosen for convenience since reactant concentrations are decreasing. Note that the rate may be defined with respect to any reactant equally well. The only caution here is that if a reaction involves different numbers of molecules (moles) of different reactants, the various ways of defining the rate will differ by simple constants. For example for

$$A + 2B \rightarrow \text{products}, \qquad \text{rate} \equiv -\frac{d[A]}{dt} = -\frac{1}{2}\frac{d[B]}{dt}$$

because two moles of B disappear for each mole of A.

Now if the rate is proportional to the number of collisions, it should be possible to write, for our A-plus-B reaction from Eq. (A1.3),

$$-\frac{d[A]}{dt} \propto Z \propto k[A][B].$$

Or for any concentrations [A] and [B] the rate divided by [A][B] is a *constant* at a given temperature. The experimental constant so obtained is called a *rate constant*.

Problem A1.4   The following data describe changes in concentration of reactants in the reaction $OCl^- + I^- \rightarrow OI^\pm Cl^-$ which is run in 1 M aqueous NaOH solution. Make a graph of $[OCl^-]$ vs. time. Recall that the derivative $d[OCl^-]/dt$ has the geometric interpretation of the *slope* of the line tangent to the curve at the point in question. For times 2, 5, and 8 sec, find $d[OCl^-]/dt$ graphically by drawing tangents and measuring slopes. Note that the three rates are different. Show that $(d[OCl^-]/dt)/[OCl^-][I^-]$ is constant. What are the values of the rate constant for this reaction? What are its units?

$[OCl^-] = [I^-]$, M	0.00200	0.00158	0.00147	0.00135	0.00123
Time (seconds)	0	3	4	5	6

$[OCl^-] = [I^-]$, M	0.00101	0.00073
Time (seconds)	8	14

If reaction occurs at every collision, the rate constant for the A-B reaction in gases at 25°C should be about 4×10^9 M^{-1} sec^{-1}. This is the part of Z independent of concentrations according to our calculation:

$$-\frac{d[A]}{dt} \overset{?}{=} Z = \pi\sigma_{AB}^2 \sqrt{\frac{3RT}{MW}} [A][B] = 4 \times 10^9 [A][B].$$

There are many reactions which have the experimental rate described by

$$-\frac{d[A]}{dt} = k[A][B],$$

but few of them have k as large as 4×10^9 or anything like that value. Some reactions are listed in Table A1.1.

Table A1.1

**Reactions of the type A + B → products which have
rates describable by −d[A]/dt = k[A][B]**

A	B	Products	$k(M^{-1}sec^{-1})$	$T, °K$
HI	HI	$H_2 + I_2$	7.04×10^{-7}	556
H_2	I_2	2HI	4.45×10^{-5}	556
NO_2	NO_2	$2NO + O_2$	0.498	592
H	D_2^*	HD + D	6.9×10^2	400

* D represents the heavy isotope of hydrogen of atomic weight 2 called deuterium.

Let's ask why there should be so many reactions of the A + B type showing the right experimental form but with a *low* rate constant. A simple answer would be that *only a fraction* of collisions lead to reaction. The rate may then be *proportional* to Z but *not equal* to Z. This should be expected. For a stable molecule to react, its structure must be disturbed at an expenditure of energy. One certainly need not expect every molecular collision to sufficiently disturb the internal structure of molecules. Tentatively let's suggest that it is collisions that account for the way rates depend on concentration but that k, the rate constant, depends on other factors, especially whether enough energy is available in a collision.

A1.4 ENERGETIC COLLISIONS IN MOLECULAR BEAMS

Only very recently has technical sophistication of machinery made possible a straightforward exploration of the question of reaction probability as a function of collision energy. Our earlier arguments have relied entirely on taking average molecular velocities, but techniques of forming "molecular beams" of controlled velocities have evolved recently.

Suppose that a gas is contained in a box with a pinhole opening into an evacuated chamber. Gas molecules which come up to the pinhole heading in the right direction pass through, forming a directed molecular beam in the vacuum. The beam is rendered sufficiently accurately rectilinear by additional "collimator" slits. Now if the beam passes through slits in a pair of rotating disks where the slits are offset at an angle from each other, only those molecules traveling at just the right velocity to go from the first disk to the second in the time required for the second slit to rotate into the correct orientation to allow the molecules to pass, will go through. This is illustrated in Fig. A1.2. By variation of the (known) rate at which the velocity-selecting disks rotate, it is possible to *control* the velocity of the molecular beam passing the second disk. If a "target gas" at comparatively low temperature (slow velocity) can be admitted to the chamber and then unreacted and/or reacted molecules coming through it detected, the probability of reaction as a function of the velocity, and hence energy, of the beam may be determined. Figure A1.3 shows

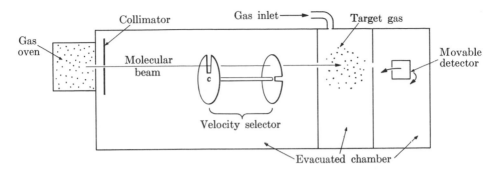

Fig. A1.2. Highly oversimplified schematic diagram of "velocity-selected molecular-beam' apparatus.

such a curve for the *very fast* reaction of potassium atoms with methyl iodide molecules. This is a reaction which has a rate constant k close to Z. Even in this case, beam energies somewhat higher than average energies of molecules are much more reactive. Although the technical difficulties of getting enough molecules in the beam for easy detection have limited the number of studies similar to the one shown in Fig. A1.3, the general form of such curves is now fairly clear. A more typical slower reaction is illustrated by the curve in Fig. A1.4. The reaction probability remains low until the relative kinetic energy of the collision partners is well above the average kinetic energy of molecules at room temperature. Consider the dashed line in Fig. A1.4 as an arbitrary cutoff of energy E_a. If we mix two reactant gases at room temperature, what fraction of molecules will have an energy greater than or equal to E_a?

The velocity selector can be used to determine the fraction of molecules with kinetic energies greater than or equal to E_a. Simply measure the tempera-

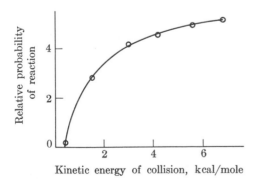

Kinetic energy of collision, kcal/mole

Fig. A1.3. Reaction probability as a function of energy from molecular-beam study of the reaction of K atoms with CH_3I molecules by Airey, Green, Reck, and Ross. [*J. Chem. Phys.* **46** 3295 (1967).] Recall that the average kinetic energy of a molecule at room temperature is about 0.9 kcal/mole.

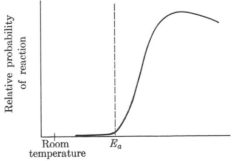

Kinetic energy of collision, kcal/mole

Fig A1.4. Molecular-beam result for a typical slow reaction. It is to be understood that probability is very low until collision energies reach values well above the average kinetic energies of molecules at room temperature.

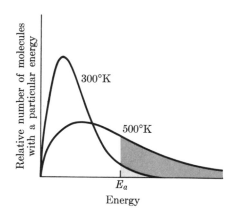

Fig. A1.5. The distribution of molecular kinetic energies at two temperatures. The number of molecules with energy E_a or greater is proportional to the shaded area for each temperature.

ture in the gas "oven" on the left, remove the "target gas," and use the detector to determine the number of molecules passing through the velocity selector for each rate of rotation. This information allows calculation of the fraction of molecules falling in each velocity range. This kind of experimental determination of velocities and molecular kinetic energies has been carried out in recent years. Actually, the experiment simply confirms results that were derived on theoretical grounds by the pioneers in the kinetic theory in the nineteenth century: Maxwell and Boltzmann. (We refer to the mathematically simpler but practically difficult experimental approach to avoid some sophisticated calculus.) Figure A1.5 shows the distribution of molecular kinetic energies in a gas at two temperatures differing by a factor of less than two. A line is drawn at E_a, the "cutoff" energy of Fig. A1.4. All molecules with kinetic energies *larger* than E_a are represented by the size of the shaded area under the curve above E_a. Careful examination of these curves shows that the areas representing the number of molecules of kinetic energy greater than E_a *change very rapidly with temperature!* In fact, the number is an *exponential function* of the temperature. That is, the number of such energetic molecules is *proportional* to e raised to the power E_a/RT, where e is the base of the natural logarithms (2.718 . . .):

$$\text{Fraction of molecules of } E \geq E_a \propto e^{-E_a/RT}. \qquad (A1.4)$$

Recall the results of Problem A1.3. There it emerged that the collision frequency Z changes only slightly as the temperature changes. If it is true, however, that only molecules of kinetic energy $E \geq E_a$ can react, reaction rates should be quite sensitive to temperature.

Problem A1.5 According to Eq. (A1.4), the ratio of energetic molecules at temperature T_2 to the number at temperature T_1, should be given by $e^{-E_a/RT_2}/e^{-E_a/RT_1}$. This may be written $e^{-E_a/R(1/T_2-1/T_1)}$. Taking E_a to be 15 kcal/mole, find the ratio of molecules of kinetic energy 15 kcal/mole or more at 310°K compared to 300°K. [*Hint:* This calulation is most conveniently made using logs. Log to the base e of e^x is x and log to the base e is related to common logs to the base 10 by 2.3 $\log_{10} x = \log_e x$.]

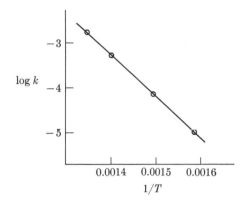

Fig. A1.6. Plot of the log of the rate constant for the reaction $2HI \rightarrow H_2 + I_2$ vs. the reciprocal of absolute temperature. [Data from J. H. Sullivan, *J. Chem. Phys.* **30**, 1292 (1959).]

A1.5 THE ARRHENIUS EQUATION

Experiments of even the most casual sort shows that rates of chemical reactions *are* very sensitive to temperature. The interesting question is how does the *rate constant*, the concentration-independent part of the expression for the rate, depend on temperature. Figure A1.6 shows a typical plot of the variation of a rate constant with temperature. As usual a form of plotting which results in a straight line has been sought. In this case it's the log of the rate constant vs. the reciprocal of the absolute temperature ($\log k$ vs. $1/T$).

This straight line can be translated into an equation:

$$\log_e k = \log_e A - B/T,$$

which may be rewritten

$$k = Ae^{-B/T}, \tag{A1.5}$$

where A and B are two constants to be interpreted (B is obtained from the slope of the line in Fig. A1.6, A is evaluated from the $1/T = 0$ intercept, i.e., $T = \infty$). The preceding section pointed out a very good reason to expect the temperature dependence of Eq. A1.5. If that argument was sound, the interpretation of B is $B = E_a/R$, where E_a is the minimum energy for reaction; E_a is usually called the *activation* energy of the reaction.

Problem A1.6 Using Fig. A1.6, find the activation energy of the HI reaction. Keep in mind that $\log k$ in the figure is given to the base 10 and that Eq. (A1.5) uses base e. (R may be taken to be 2.0 cal mol^{-1} deg^{-1}.) *Answer:* 44 kcal mole^{-1}.

Figure A1.7 collects data on a number of reactions in both fast and slow reactions to illustrate the generality of the exponential increase of reaction rate with increasing temperature. Each reaction has its own characteristic activation energy (slope) but all obey Eq. (A1.5).

One might now suppose that the theory for a simple reaction between two molecules is now complete. A could be identified with the only very slightly

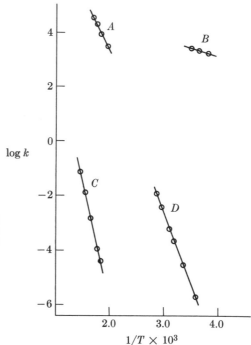

Fig. A1.7. Plot of log of rate constant against $1/T$ for several reactions, fast and slow, to show generality of the fit of the straight line predicted by the Arrhenius equation.

A. $Br + H_2 \quad \rightarrow HBr + H$
B. $Br_2 + 2NO \rightarrow 2NOBr$
C. $H_2 + I_2 \quad \rightarrow 2HI$
D. $2N_2O_5 \quad \rightarrow 2N_2O_4 + O_2$

temperature-dependent collision frequency. That is, we might interpret

$$k = Ae^{-B/T}$$

in terms of collision theory as

$$k = Ze^{-E_a/RT}.$$

If this is correct, the constant A should be of the order of 10^9. For the HI reaction, A is 2.0×10^9, which is to be compared to a rigorously calculated collision frequency, Z, of 2.9×10^9. Unfortunately this very satisfying agreement holds for only some reactions of molecules of only very simple structure. Experimental values of A are compared to theoretical values of Z in Table A1.2.

Table A1.2

Collision frequency, experimental values of A, and correction factors, p

Reaction	A	Z	p
$2HI \rightarrow H_2 + I_2$	2.0×10^9	2.9×10^9	~ 1.0
$2NO_2 \rightarrow 2NO + O_2$	2.6×10^8	4.6×10^9	0.06
$H + D_2 \rightarrow HD + H$	1×10^9	2×10^{10}	0.05
$CH_3 + H_2 \rightarrow CH_4 + H$	1.8×10^7	2.3×10^{10}	0.0008

Formally, the difficulty can be reconciled by introducing a correction factor p. We write

$$A = pZ,$$

and this equation defines p. But what is the significance of p and, especially, why is it always *less than one?* The value $p < 1$ means that not even all collisions with sufficient energy lead to a reaction. Some are still ineffective. Actually, it has to be admitted that this is not a surprise. Consider the two collisions illustrated in Fig. A1.8. The collision (b) could certainly be argued to be more likely to lead to the atom-partner exchange than the "end-on" collision shown in (a). There should be an orientation factor considered in determining whether a collision will be fruitful. The p factor, sometimes called a steric factor, is related to orientations in collision. But it seems unlikely that simple orientation factors can account for p values of one-thousandth or smaller. Yet such small p factors are not uncommon.

Fig. A1.8. Two collision orientations for HI molecules.

To understand very small p, it is necessary to realize that rearrangement of chemical bonds must require that energy associated with collisions must be transferred into appropriate *internal* motions in the molecule, e.g., into extreme vibrations. Recall that internal modes of motion in molecules are subject to quantum restrictions. It is not unreasonable to suppose that the probability that a collision will be fruitful is often quite small. The existence of discrete separate energy states can put severe "matching" restrictions in the transfer of energy.

We are now speculating about an issue which is still a matter of demanding but rewarding research. The fact that the calculated Z does define the experimental upper limit on rates and correctly describes concentration dependence in simple cases, coupled with the success of collision theory in explaining rate variation with temperature, encourages the continuing work.

A1.6 THE STERIC FACTOR

Let's reflect a little more on the origins of a steric factor very much less than one such as the latter examples in Table A1.2. If a species like NO_2 is to break an N—O bond, the NO_2 molecule is required to do more than just accumulate the requisite energy. It must accumulate it in the proper N—O bond in order to make that bond undergo a violent, "pathological" vibration. Not every collision will result in the *transfer* of energy into the *right part* of the molecule, even if *enough* energy is available. Actually, we have seen one aspect of this before. In the plot of reaction probability as a function of energy of colliding molecules (Fig. A1.4), we noted that there was a maximum. It was possible for "too much" energy to be associated with a collision for it to be maximally effective in leading to reaction. The *steric* factor p then includes the correction for the orientation factor in collisions *and* the probability that a given collision leads to delivery of energy efficiently to the right part of a molecule.

The final collision-theory interpretation, then, of the rate constant has three factors: collision frequency, steric factor, and activation energy. We write

$$k = pZe^{-E_a/RT}. \tag{A1.8}$$

Problem A1.7   Consider the reaction between two simple spherical species, an argon ion, Ar^+, and a xenon atom, Xe:

$$Ar^+ + Xe \rightarrow Xe^+ + Ar.$$

Compare this to the reaction

$$\underset{H}{\overset{H}{\diagdown}} C = C \underset{H}{\overset{H}{\diagup}} + HBr \rightarrow H-\underset{\underset{H}{|}}{\overset{\overset{H}{|}}{C}}-\underset{\underset{H}{|}}{\overset{\overset{H}{|}}{C}}-Br.$$

Which do you expect to have the smaller value of p? Why?

A1.7 THE TRANSITION-STATE THEORY

An interesting question we might now raise is whether our collision theory indicates that a reaction occurring in an ordinary gas sample with the normal distribution of energies has one path or many. That is, are there significantly *different* collisions that occur leading to *reaction?* Is the reaction sometimes achieved by a relatively low-energy collision with very favorable "orientations" and sometimes achieved by a higher-energy collision with relatively less favorable "orientation"? The answer to this question comes from comparing the energy-distribution curve (Fig. A1.5) with the curve of the reaction probability as a function of energy (Fig. A1.4). First, it must be realized that very few molecules in the *normal* energy distribution have *enough energy to react at all* in a typical "slow" reaction. We are interested in the "tail" of the curve in Fig. A1.5 and that tail is falling very rapidly. The number of molecules with a little more than the "minimum energy" necessary to react will be considerably smaller than the number with just that minimum energy. The reaction, therefore, must occur mainly by favorably oriented collisions of those molecules with just the necessary minimum energy. Figure A1.4 shows that this tendency to exclude the higher-energy collisions as an important contribution to the reaction is further accentuated by the fact that the reaction-probability curve goes down again at higher energies. Thus, in a normal sample of gas "reacting as best it can" using the available "thermal energy," the reaction can be characterized as going via one best path which involves an energy very near the minimum required and "orientation" factors very near the most favorable. This is suggested by Fig. A1.9, which shows the "superposition" of the energy-distribution curve and the reaction-probability curve to give the number of molecules reacting as a function of energy. This shows that those molecules which actually get to the product state fall in a very *narrow* energy range.

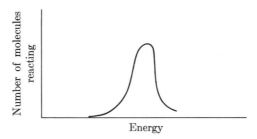

Fig. A1.9. Plot of the number of molecules reacting as a function of energy in a sample of gas at ordinary temperatures. This is the "superposition" of Figs. A1.4 and A1.5.

The result that the reaction may be treated as proceeding via one path which is essentially the one most favorable in energy terms suggests that there might be an alternative approach to developing a theory of reaction rates. We can consider the approach effectively in terms of an exemplary simple reaction, $D + H_2 \rightarrow HD + H$, the reaction of a "heavy hydrogen" or deuterium atom with a normal H_2 molecule to produce an HD molecule and a hydrogen atom.

In this simple reaction, three nuclei are involved. Suppose that a series of molecular-orbital calculations were carried out for various arrangements of these three. We would certainly expect that the most stable configurations of the system would correspond to two nuclei *close* together forming a molecule and the third sufficiently far away that it would not interact significantly, since these are the observed "stable" reactant and product states of the system, an atom and a molecule. As the "atom" is allowed to approach the "molecule," we would expect *less stable* three-atom complexes, some worse than others. If all possible configurations of the three nuclei were examined, it should be possible to pick out a series leading from reactants to products that is the best path in the sense that it goes through the least unstable of the possible unstable states that could lead from reactants to products. We can call this best path a *reaction coordinate*. Figure A1.10 shows a hypothetical plot of the energy of the system at the points along the reaction coordinate. There is a maximum point on this best path where the energy is the highest it is required to get.

Fig. A1.10. Energy at points along the reaction coordinate. The minimum at the left is the stable system $H_2 + D$. The minimum at the right is the stable system HD + H. The maximum in between is the transition state corresponding to an activated complex $D \cdots H \cdots H$.

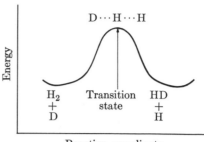

This point is called the transition state, and the configuration of the three nuclei at this point is called the *activated complex*.

If we could calculate the probability that a mixture of D atoms and H_2 molecules in a gas at some temperature could form the activated complex (i.e., reach the transition state) we would be in a good position to calculate the rate of the reaction, since it is reasonable to assume the reaction does go by this best path. The required probability calculation is possible in the framework of a theory called "statistical mechanics." As a result, the approach being described can lead to a complete theory of reaction rate called the *transition-state theory*.

The argument we gave to show that this approach is reasonable should suggest that the energy difference between the "ground" (stable) *reactant state* and the *transition state* is closely related to the activation energy, E_a, of the collision theory. It turns out to be possible to express the results of transition-state theory in the language of thermodynamics, which we introduced in Chapter 6 to describe the relationship between energy quantities and the equilibrium position of a chemical reaction. The activation energy, E_a, is closely related to a quantity called the *enthalpy of activation*, $\Delta H^{\ddagger}$, of transition-state theory. The other factor in determining equilibrium, entropy, which we argued in Chapter 6 is related to the number of configurations available to a system in a certain state and the probability of reaching that state, is clearly related to the "probability of favorable collision" factor pZ of collision theory. Transition-state theory introduces an *entropy of activation*, $\Delta S^{\ddagger}$, which is closely related to the pZ factor of collision theory. Just as the equilibrium constant for reaction is determined by ΔG^0, the standard free energy change, the rate constant is determined by $\Delta G^{\ddagger}$, the *free energy of activation*. $\Delta G^{\ddagger} = \Delta H^{\ddagger} - T \Delta S^{\ddagger}$ in complete analogy with equilibrium thermodynamics.

Problem A1.8 It is sometimes said that transition-state theory fails for reactions of activation energies lower then about 5 kcal/mole occurring at ordinary temperatures. Consider Fig. A1.4 and Fig. A1.5 and try to decide what would be the appearance of Fig. A1.8 for a reaction with a small activation energy. Why does transition-state theory fail in this case?

In most cases, it is a matter of free choice whether one chooses to discuss the rate of a reaction in terms of collision theory or transition-state theory. Why, then, bother with transition-state theory? One important reason is that the main application of reaction-rate study is to the understanding of the mechanisms of reaction, the pathways along which chemical change occurs. We describe reactants and products in terms of molecular or ionic structures, and it is very helpful to discuss the transformation of these structures in terms of a mediating structure, the activated complex which "exists" at the transition state even though the activated complex is not something that lasts long enough to be observed directly in any way.

A1.8 REACTION MECHANISMS

Now, we can turn to the application of the theory of reaction rates. We can ask questions about the nature of atomic and bond motions by which reactant structures are converted into product structures. We can study reaction mechanisms. Perhaps *the* most important principle on which this enterprise depends is the postulate that reaction requires collision. We have established (p. 315) that the *explanation* for the appearance of the product of reactant *concentrations* in the rate equation is the collision probability. If a reaction requires a collision of A and B, its rate will be governed by the equation $k[A][B]$. *Thus we should be able to look at the rate equation for a reaction and decide the collision partners.* Of course, matters are not quite so simple and wouldn't be interesting if they were. The rate equation found *experimentally* for a reaction does involve products of reactant concentrations, but frequently an odd collection of these that bear little relation to the stoichiometric chemical equation for the overall reaction. *Still, our clue is that concentration dependence reveals collision pairing.* How the clue is used is perhaps best shown by example.

Consider the reaction of decomposition of hydrogen peroxide in an aqueous solution containing iodide ion. The reaction is

$$2H_2O_2 \rightarrow 2H_2O + O_2 \text{ (gas)}.$$

Iodide ion is not consumed, and its concentration remains unchanged. The rate of this reaction is conveniently followed by collection of the O_2 gas formed and measurement of its volume. (Clearly, every mole of O_2 collected must correspond to two moles of H_2O_2 decomposed.)

The simplest mechanistic suggestion that might be made is that two molecules of H_2O_2 collide with each other leading to the formation of an activated complex involving four hydrogen atoms and four oxygen atoms that breaks down directly into the products, $2H_2O$ and O_2. If the reaction went along this path, we would predict that its rate would depend on $[H_2O_2][H_2O_2]$ or $[H_2O_2]^2$ because the probability of forming the activated complex would depend on the probability of *two* H_2O_2 molecules colliding. We would predict

$$-\frac{d[H_2O_2]}{dt} = k[H_2O_2]^2. \tag{A1.9}$$

But, this equation *does not summarize the experimental facts*. If the H_2O_2 concentration is doubled, the rate of oxygen evolution is *doubled*, whereas the $[H_2O_2]^2$ factor implies a *fourfold* increase. Also, the rate of oxygen evolution depends on the *concentration of iodide ions* in the solution. (Doubling iodide concentration doubles the rate of oxygen evolution.) The experiments confirm a rate equation:

$$-\frac{d[H_2O_2]}{dt} = k[H_2O_2][I^-]. \tag{A1.10}$$

Now, Eq. (A1.10) implies that iodides are involved in collisions important to the reaction and that only *one* H_2O_2 molecule is critical to the *rate* of the process, which consumes *two*. Consider, now, the mechanism

$$H_2O_2 + I^- \xrightarrow{\text{slow}} HOI + OH^-,$$
$$H^+ + OH^- \xrightarrow{\text{fast}} H_2O, \tag{A1.11}$$
$$H_2O_2 + HOI \xrightarrow{\text{fast}} H_2O + O_2 + H^+ + I^-.$$

If all the steps written are added, they add up to the initial reaction $2H_2O_2 \rightarrow 2H_2O + O_2$. Along the way, however, an iodide is incorporated into HOI and then regenerated as iodide. Looking at the first step of the proposed sequence we find that it requires collision of *one* iodide with *one* H_2O_2 and hence would lead to the predicted rate law:

$$-\frac{d[H_2O_2]}{dt} = k[H_2O_2][I^-], \tag{A1.12}$$

which is, of course, exactly the experimental form. *Moreover*, the suggested mechanism in Eq. (A1.11) includes the proposal that the step which predicts the correct form of the rate law is *slow* compared to the others.

A chemical reaction proceeding in several steps shares characteristics with a stream of water flowing through pipes of different diameter. The narrowest pipe will determine the rate at which water flows. A large pipe following a narrow one will not help increase the rate. Similarly, the overall rate of the chemical reaction (the experimental rate) is determined by the slowest step. Thus, if we write the rate expression for the slowest step correctly, it gives the experimental form. There is only one serious difficulty in reaction-mechanism analysis. The mechanism in Eq. (A1.11) correctly accounts for the experimental rate equation, but there is no assurance that we could not invent others that would do so also if we were sufficiently imaginative. To approach a "proof" that one mechanism is the best account of a reaction requires considering all the available parameters of the reaction (e.g., the activation energy) against experimental or theoretical estimates of the behavior of the proposed intermediate species. This enterprise draws on many areas of chemistry.

Problem A1.9   Write the predicted equation for the rate of disappearance of the reactant for each of the following reactions on the *assumption* that the reaction occurs by a *single-step* mechanism.

a) $CO + N_2O \rightarrow CO_2 + N_2$
b) $2H_2 + O_2 \rightarrow 2H_2O$
c) $2NO_2 \rightarrow 2NO + O_2$
d) C_5H_8 (cyclopentene) $\rightarrow C_5H_6$ (cydopentadiene) $+ H_2$
e) $2N_2O_5 \rightarrow 4NO_2 + O_2$
f) $(CH_3)_3CCl + OH^- \rightarrow (CH_3)_3COH + Cl^-$

Problem A1.10   Reaction (c) of Problem A1.9 goes at low temperature according to the equation

$$-\frac{d[NO_2]}{dt} = k[NO_2]^2.$$

What is the probable mechanism?   At higher temperatures in the presence of added argon gas, the rate law goes over to

$$-\frac{d[NO_2]}{dt} = k[NO_2][Ar].$$

Show that the following mechanism accounts for the high temperature result:

$$NO_2 + Ar \xrightarrow{\text{slow}} NO + O + Ar,$$
$$O + O \xrightarrow{\text{fast}} O_2.$$

From the fact that the second reaction path becomes important only at high temperatures, prove that it has a higher activation energy.

Let us consider one more reaction which introduces another feature of analysis of rate equations.   The reaction is

$$OCl^- + I^- \rightarrow OI^- + Cl^-. \tag{A1.13}$$

It occurs in basic aqueous solutions.   The rate equation is found experimentally to be

$$-\frac{d[OCl^-]}{dt} = \frac{k[OCl^-][I^-]}{[OH^-]}. \tag{A1.14}$$

The puzzle is what the concentration of OH^- has to do with the reaction.   A good guess would be that OCl^- reacts with water to produce small (possibly even undetectable) amounts of neutral $HOCl$ acid molecules.   If $HOCl$ is more reactive than OCl^-, the reaction path might well be expected to involve the acid. Let's consider the initial steps:

$$OCl^- + H_2O \rightleftharpoons HOCl + OH^- \quad \text{(fast, equilibrium)},$$
$$HOCl + I^- \rightarrow HOI + Cl^- \quad \text{(slow)}. \tag{A1.15}$$

The first is a transfer of H^+, which must be reversible if HOCl is present only in traces. Also, it is reasonable on the basis of extensive experience to assume that a transfer of H^+ from combination with one oxygen atom to combination with another occurs *very* rapidly. If the second step is slow, its rate equation will govern the rate at which OCl^- disappears. We can write

$$-\frac{d[OCl^-]}{dt} = k[HOCl][I^-]. \tag{A1.16}$$

Our problem is to calculate the *small* concentration of [HOCl] in terms of the *easily observed* concentrations $[OCl^-]$ and $[OH^-]$. We can write the *equilibrium-constant* expression for the first step of Eq. (A1.15). If the reactions involved in the H^+ transfer are fast, the reaction should stay *very close* to its equilibrium position as the slow step changes the overall concentrations. As HOCl is consumed in its reaction with I^-, the fast reaction of OCl^- with H_2O can reestablish equilibrium. The equilibrium constant for the H^+ transfer is

$$K = \frac{[HOCl][OH^-]}{[OCl^-]}. \tag{A1.16}$$

Rearrangement of this equation gives the concentration of [HOCl] that we need to find the rate:

$$[HOCl] = K[OCl^-]/[OH^-]. \tag{A1.17}$$

Substituting the right-hand side of (A1.17) into (A1.16) gives

$$-\frac{d[OCl^-]}{dt} = kK[OCl^-][I^-]/[OH^-]. \tag{A1.18}$$

This is the experimental rate equation. We see that the *experimental rate constant* must be *interpreted* as the product of the rate constant for the slow step k times the equilibrium constant K for the fast "preequilibrium" step.

Problem A1.11   Consider the reaction

$$H_2O_2 + 2H^+ + 2I^- = I_2 + 2H_2O.$$

The rate law consists of *two terms* representing two parallel pathways of reaction:

$$\frac{d[I_2]}{dt} = k_1[H_2O_2][I^-] + k_2[H_2O_2][H^+][I^-].$$

Show that the k_1 path can be interpreted by the mechanism:

$$H_2O_2 + I^- \rightarrow OH^- + HOI \quad \text{(slow)},$$
$$H^+ + OH^- \rightarrow H_2O \quad \text{(rapid)},$$
$$HOI + H^+ + I^- \rightarrow I_2 + H_2O \quad \text{(rapid)}.$$

Show that the k_2 path can be interpreted by assuming that the *slow* step is the displacement of H_2O from $H_3O_2^+$:

$$H_3O_2^+ + I^- \rightarrow H_2O + HOI \quad \text{(slow)}$$

by consideration of the transfer of H^+ to H_2O_2 as a fast equilibrium step.

A1.9 A CASE STUDY: SOLVOLYSIS OF $(CH_3)_3CX$

In Chapter 2, we considered three functional groups which can be seen to be closely related when reaction mechanisms are considered. These are alkyl halides, RX, alcohols, ROH, and ethers, ROR. In talking about reaction mechanisms, it turns out to make *more* difference whether we are talking about reactions of CH_3OH and $(CH_3)_3COH$ than about reactions of CH_3OH and CH_3CI. In the new similarities and the new differences revealed, reaction-mechansim study lends an altogether new perspective to theoretical organic chemistry. In this section one case of correlation of rate data with other chemical information to solve a mechanistic problem is presented. The reactions chosen are those of $(CH_3)_3CX$ (X = CI, Br, I), which are quite different in mechanism from those of CH_3X or CH_3CH_2X. The reactions are substitutions at a carbon center.

In aqueous ethyl alcohol solutions (mixtures of H_2O and C_2H_5OH), the tertiary butyl halides, $(CH_3)_3CX$, react to produce *tert* butyl alcohol, $(CH_3)_3COH$ and ethyl *tert*-butyl ether $(CH_3)_3COC_2H_5$ plus H^+ and X^-. The rate of disappearance of $(CH_3)_3CX$ follows the simple rate equation

$$- \frac{d[(CH_3)_3CX]}{dt} = k[(CH_3)_3CX]. \tag{A1.19}$$

The simplest interpretation of this rate law suggests that only $(CH_3)_3CX$ is involved in the slow step. An ionization mechanism can be proposed:

$$(CH_3)_3CX \rightarrow (CH_3)_3C^+ + X^- \quad \text{(slow)},$$

$$(CH_3)_3C^+ + H_2O \rightarrow (CH_3)_3COH + H^+ \quad \text{(fast)}, \tag{A1.20}$$

or

$$(CH_3)_3C^+ + C_2H_5OH \rightarrow (CH_3)_3COC_2H_5 + H^+ \quad \text{(fast)}.$$

This is an interesting notion. It suggests that carbon compounds, almost straitlaced in their insistent molecularity, undergo transposition through *ionic* states. The species $(CH_3)_3C^+$ has been called a "carbonium ion." Unfortunately, path (A1.20) cannot be surely inferred from the *experimental* rate equation (A1.19). It is entirely possible that the rate equation for the formation of, for example, the alcohol $(CH_3)_3COH$ should be written

$$-\frac{d[(CH_3)_3CX]}{dt} = k[(CH_3)_3CX][H_2O]. \qquad (A1.21)$$

We cannot tell experimentally whether or not there is a term in the rate equation for the concentration of water because the solvent is present in such *large* excess that its concentration *does not change appreciably in the course of the reaction.* Of course, we could try changing the concentration of water in the ethyl alcohol-water solvent mixture. But this would not really help. Changing the solvent mixture changes not only the concentration of water, but also properties of the solvent that can affect the value of the *rate constant k*. For example, the pathway of Eq. (A1.20) requires *ionization* of $(CH_3)_3CX$ in the slow step. The incipient ions in the transition state must be solvated. As the alcohol-water mixture is varied, the dielectric constant of the solvent mixture changes and as a result the energy of ions must change. Thus it would be expected that *at least* the activation energy for the ionization reaction would change as the dielectric constant changes. The point of this argument is that it is not possible to distinguish clearly between factors that affect k, the rate constant, and variation of the H_2O concentration. How, then, is the scheme (A1.20) to be defended?

The most persuasive argument for (A1.20) is external to the question of rate itself. Suppose that the carbonium ion $(CH_3)_3C^+$ is formed in the slow step. Then, what happened afterward (to produce the product) would not depend on the *nature* of the leaving group. X^- could be either Cl^-, Br^-, or I^- and this would not affect the question of the product being either $(CH_3)_3COH$

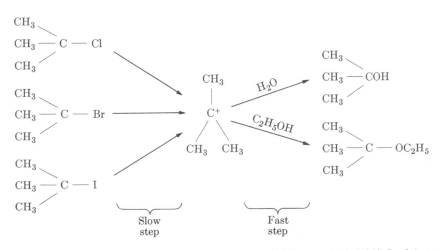

Fig. A1.11. The mechanism of solvolysis of tertiary butyl halides in C_2H_5OH-H_2O mixtures.

or $(CH_3)_3COC_2H_5$. The question of product is settled by the fast reaction of the carbonium ion, $(CH_3)_3C^+$. At that stage, the X^- ion is already gone. It is observed that the rate of disappearance of $(CH_3)_3CX$ *decreases* along the series $(CH_3)_3CI$, $(CH_3)_3CBr$, and $(CH_3)_3CCl$. But despite the change of rate, the *ratio* of alcohol produced to ether produced remains *constant* for a given solvent mixture. Figure A1.11 reiterates the interpretation of these facts. The combination of rate and product considerations gives strong support to the concept of *ionic* intermediates in reactions of these *molecular* substances.

Problem A1.12 It is found in basic media that the rate equation for the reaction $CH_3I + OH^- \rightarrow CH_3OH + I^-$ is

$$-\frac{d[CH_3I]}{dt} = k[CH_3I][OH^-].$$

Argue that this implies an important difference of mechanism for the $CH_3I + OH^-$ reaction as compared to the $(CH_3)_3CI + H_2O$ reaction. Explain the significance of the observation that reactions of $(CH_3)_3CI$ are independent of OH^- concentration.

APPENDIX 2

Brief Review of Some Basic Concepts of Mechanics and Electricity

A2.1 SPEED

Our everyday operations in an automotive culture involve us with the concept of speed. We speak of speed limits such as 40 mph. Basically, 40 mph can be thought of as a distance traveled (40 miles) divided by a time for traveling that distance (one hour). We might think of speed as distance/time, or algebraically, speed $= d/t$. But a little thought indicates that this is too simple. We realize that the speed of the car may *change* a lot in an hour. If we speak of a car traveling 40 mph *now*, it is the *rare* case that the car is exactly 40 miles down the road one hour later. Our remark about a car going 40 mph is a remark about *instantaneous* speed. Clearly the speed at *one moment* cannot be defined via division by an *elapsed* time. To see how this subtlety is dealt with it is important to go back to the basic kind of measurements which can be made about moving bodies. Speed is not a basic parameter.

Consider a ball rolling along a line as in Fig. A2.1. We define a coordinate system by some device such as laying a meter stick next to the line. With this, a number can be assigned to the *position* of the ball. Adding a clock of some sort to our equipment, we can now also read the clock "simultaneously" with observation of positions. We can define a concept of average speed by observing that the ball is at position x_1 at time t_1 and at x_2 at t_2. Then the average speed is

$$\text{average speed} = \frac{|x_2 - x_1|}{t_2 - t_1}. \tag{A2.1}$$

That is, the average speed is the change in position, $|x_2 - x_1|$ (the bars imply absolute value; neglect sign), divided by the time interval $t_2 - t_1$. This is what we called before "dividing the distance traveled by the elapsed time." It should be clear that we can get closer to a measure of "instantaneous" speed if we choose the *small* distance $|x_2 - x_1|$ traveled during a *short* time interval $t_2 - t_1$, but an "instantaneous" speed is the speed for a *zero* time interval $(t_2 - t_1 = 0)$, and this is *undefined*. The mathematical problem here is the

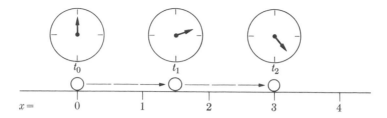

Fig. A2.1. A ball rolling along a line with a position coordinate marked off and a clock.

fundamental one of the calculus. Can we find average speeds for shorter and shorter intervals closer and closer to the time t, and *find* that these *average speeds approach a definite value more and more* closely as the time interval is made smaller? If such a *limit* for indefinitely small intervals can be shown to exist, we call that the "instantaneous" speed at time t, or the speed. In calculus notation, writing Δx for $|x_2 - x_1|$, and Δt for $t_2 - t_1$, we have

$$\text{speed} \overset{\text{def}}{=} \text{limit (as } \Delta t \text{ approaches 0) } \frac{\Delta x}{\Delta t} \overset{\text{def}}{=} \frac{dx}{dt}. \qquad (A2.2)$$

The best verbal form to give the definition of speed is that it represents the *rate of change of position.* Unfortunately, the verbal formulation adds little to the above mathematics because the mathematics showed how to *define* "rate."

A2.2 VELOCITY AND ACCELERATION

In the preceding section we defined speed; but speed, it will turn out, is not the fundamentally important quantity when we try to *explain* motion. Speed has nothing to say about *direction.* Direction of motion is very significant. The quantity which includes direction is *velocity.* If we are considering motion along a straight line as in Fig. A2.1, all that is needed to go from speed to velocity is to include the *algebraic sign* of $x_2 - x_1$. If the ball is rolling so that it goes toward higher position numbers x as time passes, then $x_2 - x_1$ is *positive.* Conversely, if the ball is rolling the other way, $x_2 - x_1$ is *negative.* When $x_2 - x_1$ is *positive,* the *velocity,* v, is *positive.* When $x_2 - x_1$ is negative (ball rolling back toward $x = 0$), the *velocity,* v, is *negative.* When the motion is not in a single straight-line direction, velocity must be defined by a *vector.* But an important *experimental* result is worth keeping in mind. Any motion in three-dimensional space may be *treated* as *three independent* motions in three orthogonal directions as, for example, x, y, and z of the Cartesian coordinate system shown in the sketch (Fig. A2.2). This result is illustrated by the famous story of the monkey who falls from a tree just as a gun aimed straight (and level) at him is fired. Since the bullet and the monkey are both falling with the same

Figure A2.2

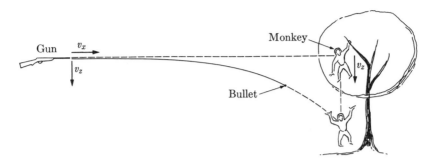

Fig. A2.3. "Shooting the monkey." The bullet hits the monkey because the downward component of its velocity, v_z, is the same as v_z for the monkey and *independent* of the forward velocity component v_x.

velocity, and that velocity component is independent of the forward velocity of the bullet, the bullet hits the monkey. This is shown in Fig. A2.3.

Since velocities can be resolved into one-directional (one-dimensional) components, we may describe velocities in terms of three components v_x, v_y, and v_z, which are simply numbers with *algebraic signs* ($+$ or $-$) to indicate directions.

Now, it is clear that our problem with the defining of instantaneous speed and velocity is due to the fact that *velocities may change* in the course of motion. We need a quantity to describe the change of velocity just as velocity is a description of rate of change of position. This quantity is called *acceleration, a*, and is defined by a process very similar to the definition of velocity. Acceleration is the *rate of change of velocity*. In calculus notation

$$a \stackrel{\text{def}}{=} \frac{dv}{dt}. \tag{A2.3}$$

Positive acceleration indicates velocity *becoming* more positive or less negative. Negative acceleration indicates velocity becoming less positive *or* more negative.

A2.3 LAWS OF MOTION

In ancient Greece, the model for motion on which physics was based was essentially a cart pulled by a horse. When the horse quit pulling, the cart stopped. Thus, it was thought that *rest* was the "natural" "state of motion" for any object and that any moving object must be "caused" to move. The key to development of modern physics (since Galileo and Newton) has been the recognition that it is the interaction of the cart with the road, called *friction*, that stops the cart. The crucial case now is considered to be the case of a can like that shown in Fig. A2.4, which has a small hole in the bottom and a stopper in the top. With dry ice (solid CO_2) in the can, a gas stream comes out of the hole in the bottom so that the can slides on a cushion of gas. If this can is given a small "push," it slides freely, slowing down *very little* as it goes. It is what is called "nearly frictionless." The experiment suggests that the ideal

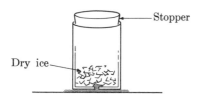

Fig. A2.4. A "nearly frictionless" dry ice puck. The dry ice vaporizes out the hole, so that the puck slides on a gas cushion.

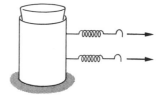

Fig. A2.5. Puck pulled by two like springs stretched to the same extent.

case should be considered to be an object in motion at *constant velocity not necessarily rest*. This is Newton's first law:

A body in motion tends to remain in motion at constant velocity* unless acted upon by an unbalanced force.

There is a term in this law that must be further defined: *force*.

Consider a "dry ice puck" being pulled by a thread attached to the puck through a spring. The idea, force, which we wish to define precisely is closely related to the notion of stretching springs. Suppose that we assume that pulling so that our chosen spring is "stretched" by 0.2 cm produces a certain *force* on the puck. If this constant force is applied, it is *observed* that the puck undergoes a *constant acceleration*. Now suppose a second *equivalent* spring is added so that the puck is pulled as shown in Fig. A2.5. In this case, the *acceleration* is just *twice* the acceleration produced by the *one* spring. In general, if we use a given spring stretched to a specified degree as a force unit, we find that the *acceleration* of a specified object is proportional to the force applied.

Now we must consider different objects. Using the same springs (same forces) on a larger puck, we find that although the accelerations are *all* less they are proportionately so. It appears that there is a property of bodies, roughly related to size, that determines how readily they'll accelerate. This is the property called *mass*. If we combine the observations about springs with the observations about size, we can formulate a second law (Newton's second law) which quantifies the idea of force. The law is:

Acceleration is proportional to force and inversely proportional to mass.

Alternatively, we write

$$F = ma. \tag{A2.4}$$

Equation (A2.4) is Newton's second law, which says that accelerations are produced by unbalanced forces and (in a sense) resisted by masses. Of course, the two terms *force* and *mass* are precisely defined only in the framework of Newton's first and second laws.

---

* This includes the case of "rest," where the velocity is zero.

There is one more important law. Anyone who has tried to "push" something while standing on a platform with little friction has experienced it. If you stand in a boat and try to "push" a dock away from you, you'll find you move and the dock doesn't. This exemplifies Newton's third law:

Every action has an equal and opposite reaction.

A2.4 TYPES OF FORCES

Before developing a force catalog, we should mention that we don't always measure a force by observing an acceleration. More often we *prevent* an acceleration by balancing the *unknown* force by a *known* force. The idea is that forces are additive (vectorially) and that a force directed one way will exactly prevent acceleration by a force directed the opposite way. We should revise Eq. (A2.4) to

$$F_{\text{net}} = ma, \qquad (A2.5)$$

where F_{net} designates the *net* force, the sum of all forces acting on the body.

Let's now consider some types of forces. We have already seen forces arising from compression and distension: springs. Also we have had to recognize that objects moving along a surface experience friction. Perhaps the next most familiar ones are gravitational forces, which reflect the interactions among "all masses in the universe." The law is

$$F_g = G\frac{m_1 m_2}{r^2}. \qquad (A2.6)$$

Here, F_g is the gravitational force, G is a units conversion constant, m_1 and m_2 are the masses of the interacting bodies (e.g., you and the earth), and r^2 is the square of the distance between the centers of gravity of the two masses. We usually measure gravitational forces by a weighing procedure in which the gravitational force on one object is balanced against the gravitational force on another.

Another sort of force arises when a rod of hard rubber (or, historically, amber) is rubbed with fur. Such a rod will *pick up* small objects like bits of paper against the action of the force of gravity. The force counteracting gravity is an *electrical* force. The rod is said to be *charged*. A very similar phenomenon can be produced by rubbing a glass rod with silk. This rod, too, will pick up small objects against the force of gravity. If each of these rods is touched to a small object like a *pith ball* hung from a silk thread, it transfers electrical charge to the pith ball (see Fig. A2.6). The funny thing is that once a rod has *touched* the pith ball and transferred charge to it, the rod repels the ball. Interestingly, the pith balls are now charged in such a way that they *attract each other or the rod that touched the other ball*. Clearly, there are *two kinds* of electrical forces, attractive, like gravity, *and also repulsive*. There must be two kinds of charge! Repulsions arise when *two* objects have been charged the same way. Thus we formulate the rule that *like charges repel*. Conversely, *unlike*

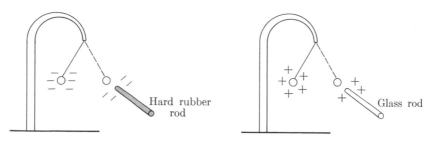

Fig. A2.6. A pith ball touched by a charged rod is subsequently repelled by that rod.

charges attract. Interestingly enough, these forces can be shown (for example, by balancing them against gravity or springs) to obey a mathematical law very similar to the gravitational one:

$$F = K \frac{q_1 q_2}{r^2}. \tag{A2.7}$$

In this equation, K is a unit conversion constant, q_1 and q_2 are the magnitudes of the charges, and r^2 is the square of the distance as in the gravitational law. The difference between the gravitational and "electrostatic" force laws is that gravity is always an *attractive* force between *masses*, whereas electrostatic forces may be both attractive (opposite charges) and repulsive (same charges). Both positive and negative signs must be assigned to the forces. The assignment of a positive algebraic sign to one type of charge (that produced by rubbing glass with silk) and a negative algebraic sign to the other type of charge (that produced by rubbing fur on hard rubber) results in the force having a different sign (direction) when the charges are unlike from the one it has when they are like. Equation (A2.7) is called *Coulomb's law.*

Another word or two about electricity is worthwhile. There is a touch of circularity in the above argument. Equation (A2.7) is involved in the measurement of the magnitude of charges because the magnitude of a charge q is defined in terms of the force it exerts. This "semicircularity" is an almost inevitable component of physical theory.*

There is another thing to be said about electrical effects. If objects having like charges are connected by a silk thread or plastic fiber (a nonconductor), there is no change. But, if two objects of unlike charges are connected by a copper wire, the charges "discharge"; electrical effects disappear. The only way to preserve electrical effects when there are conductors connecting the charge centers is to have a continuous source of charge like a battery (see Chapter 6) or generators. Generators involve the first cousins of electrical forces, magnetic forces. We shall note about these only that they arise *whenever there are moving electric charges.*

---

* On this point see D. Hawkins, *The Language of Nature,* W. H. Freeman, San Francisco, 1964, Chapter 4 on measurement.

A2.5 ENERGY, WORK, AND VOLTAGE

Let's turn now to the notion of energy. The word enjoys considerable use in ordinary language to denote *activity*. But, it is a word borrowed by ordinary speech from physics, where it has a most precise meaning.

As long ago as 1668, Huygens, a Dutch physicist, summarized the laws governing collisions of hard (elastic) objects like billiard balls. He included the rule that the sum of the quantities mv^2 for the balls was the *same* before and after collision. That is, he noted that there was a quantity *conserved* in ideally elastic collisions, the quantity mv^2. Values of the quantity for individual balls may change but the collision doesn't affect the *sum* for the group.

Huygens had recognized the tendency for certain systems to *conserve* what we now call *kinetic energy*, KE. In modern usage we denote the kinetic energy of a body as $\frac{1}{2}mv^2$. But the elastic collision in which KE is neither lost nor gained by the system of objects is a very special case. Let's examine some instances where KE-changes do occur. Clearly, they will require velocity *changes* and Newton's first law guarantees that outside forces are acting.

Consider a "superball," the recently developed toy ball that bounces back from the floor to almost the same height from which it was dropped. As it *falls*, the force of gravity (constant near the earth's surface) accelerates the ball, and its downward velocity increases. It *gains kinetic energy*. Now it *bounces* and starts up. Its velocity is reversed in direction but it still has about the same value of $\frac{1}{2}mv^2$, or KE. As it rises on the bounce, however, it slows down and it is *losing* KE as it rises *against* the force of gravity. In the ideal case, the kinetic energy gained in the fall with gravity will have been *lost* in the rise *against* gravity when the ball returns to the height from which it started. This suggests the following language. "The force acting along the line of fall over a certain distance produces a given KE. The KE is lost again when the body moves through the same distance against the force." The product of "the constant force acting along a line" and "the distance the body moves under its action" is called *work*. The final "KE" acquired by a body is equal to the *work* done on it by a *force*. This idea has its negative counterpart; when the body moves against the force it must *do work* and it loses KE. Equation (A2.8) summarizes the situation:

$$F \cdot \Delta x = -W = \Delta(\tfrac{1}{2}mv^2). \qquad (A2.8)$$

The product of a force along the direction of motion and the distance moved, Δx, equals work, W, and equals the *change* in the KE of the body, $\Delta(\frac{1}{2}mv^2)$.

Now, let's examine the consequences of the loss of KE as the ball rises doing work against the force of gravity. Do we really want to say that the *energy* is "gone" when the ball has reached the top of its bounce? In an important sense we do not. The ball loses its KE in the process of just getting to a height at which it can *recover it again in a fall*. That is, as the ball does work against the force of gravity and loses KE, it reaches positions where the force of gravity can do work on it to restore its KE. Under the action of gravity there are positions where there is a potentiality to gain energy. We might speak of the values of the potentiality here as *potential energies*, PE. Since the bouncing ball losing KE as it rises will rise just enough to recover that KE on falling, we can easily see that the energy associated with its final position, its PE, is just equal to the amount of work done against the force:

$$-\Delta(\tfrac{1}{2}mv^2) = W = \text{PE}. \tag{A2.9}$$

Of course we could eliminate work from our analysis above and simply note that the ball losing KE is gaining PE and, conversely, the ball gaining KE is losing PE. Algebraically,

$$\text{KE} + \text{PE} = \text{constant}. \tag{A2.10}$$

This equation is one form of the *law of conservation of mechanical energy*. Note explicitly that the force of gravity near the earth's surface may be written $F = ma = mg$ since the acceleration due to gravity is easily shown to be constant. We can then analyze the energy situation of the falling superball as shown in Fig. A2.7.

Now we know, in fact, that no "real" ball obeys Eq. (A2.10). Each bounce is a little lower than the one before. Mechanical energy is being lost *especially at the collision of the ball with the floor*. But are the effects of mechanical energy entirely disappearing? Could we find some other effect which is proportional to the energy loss? In fact, if a very sensitive thermometer were placed in the floor at the point of impact of the ball, it would show that a *temperature* rise *proportional* to the amount of mechanical energy lost in the bounce. This leads

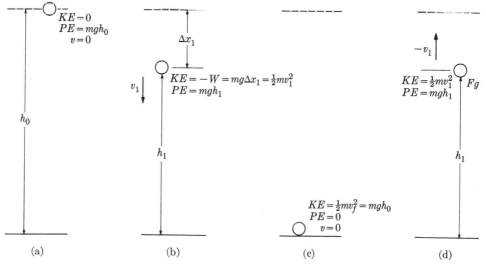

Fig. A2.7. The bouncing superball, an energy analysis.

us to a further generalization of the notion of energy. When mechanical energy is lost, heating effects proportional to the loss arise. These correspond, in fact, to work against *frictional* forces. We write an even more elaborate law of conservation of energy:

$$KE + PE + q = \text{constant.} \tag{A2.11}$$

That is, as mechanical energy is lost, heat appears proportionately. Of course, it is a matter of *experiment* to determine the proportionality constant that relates ordinary units of mechanical energy ($g \times (cm/sec)^2 = $ ergs or kg $\times$ $(m/sec)^2 = $ joules) to usual units of heat (calories). This factor (first determined by James Joule) is expressed by the statement that 4.18 joules of mechanical energy produce 1.00 calories of heat.

To complete our review of energy notions as they are needed in elementary chemistry, we need comment on only one other point. In the above, only PE of position with respect to gravitational forces was considered. Clearly, since Coulomb's electrostatic force law is very like the gravitational force law, a charged particle in the field of other charged particles can have potential energy associated with its positions. The potential energy will depend upon the body's position with respect to other charges *and its own charge*. The effect of all the other charges can be expressed for an arbitrarily charged object by discussing the *work per unit* charge, W/q, involved in moving a charged object from one place to another. The quantity W/q would be the change of PE for an object carrying a unit charge (one coulomb), and a knowledge of W/q allows us to calculate PE-changes for any charged object simply by multiplication of W/q by the q on the object. This quantity, W/q, the work per unit charge for moving a charged object subjected to electrical forces, is given the name electrical potential or *voltage*.

A particularly important case of potential energy analysis with respect to electrical forces arises when we consider the energy of one charged sphere moving under the influence of another which is fixed (see the H atom problem). Suppose that the moving and fixed charged spheres have opposite charge. When the two spheres are effectively "infinitely" separated, the attractive force between them is essentially zero. It is quite reasonable and convenient to choose this arrangement as the *zero* of a scale of electrical potential energy. Now consider what happens as the moving charged object approaches the fixed one. The attractive force does *work* on the moving charged sphere, and it acquires *kinetic energy* in precise analogy to a falling body. Thus its *potential energy* is decreasing. If the zero of the potential energy scale is set at infinite separation of the two objects, all closer distances correspond to *negative* potential energies. (Convince yourself that the potential energies would be positive if the two spheres had like charges.) An analysis of the work done as the moving sphere approaches the fixed one, in which Coulomb's law is used for the force acting, gives the following expression for the potential energy of the movable sphere

$$PE = -K\frac{q_1 q_2}{r}$$

where q_1 and q_2 are the *magnitudes* of the charges and r is the distance between them. Again, PE is negative if the charges are opposite and positive if they are like.

PROBLEMS

A2.1 What force is required to impart a 5 m/sec^2 acceleration to a 5-kg body in the force units of *newtons* ($= \text{kg} \cdot \text{m/sec}^2$); in the force units of *dynes* ($= \text{g} \cdot \text{cm/sec}^2$)?

A2.2 An automobile weighing 2000 kg slows down from 60 km/hr to 30 km/hr in 5 sec. Assuming that the braking force is constant, calculate it in newtons.

A2.3 If you were set down on a frozen lake so smooth that it offered *no* frictional resistance at all on the surface, how could you get off?

A2.4 Show from

$$F_g = G\frac{m_1 m_2}{r^2}$$

that any object of any mass m_2 will experience the same constant acceleration g when it falls near the surface of the earth.

A2.5 Argue in your own words that there are *two* kinds of electricity (charge) but no more.

A2.6 A charge of $+3 \times 10^{-9}$ C (C = coulombs) is located 0.5 cm from a charge of -5×10^{-9} C. If the constant K of Eq. (A2.7) is 9×10^9 N m^2/C, what is the force between these charges? What is its direction?

A2.7 Two small spheres are given identical positive charges. When they are 1 cm apart the force between them is 0.001 N. What would be the force if

a) the distance was increased to 2 cm?
b) one charge is doubled?
c) both charges are doubled?
d) one charge is doubled and the distance increased to 2 cm?

A2.8 An object of a mass of 1 kg has a PE above the ground of 1 J. How high is it? What will be its velocity when it hits the ground if it is released from this height (the acceleration due to gravity, g, may be taken as 9.8 m/sec^2)?

A2.9 The earth revolves about the sun in an elliptical orbit. At what part of the orbit is its PE greatest? At what part of the orbit is its KE greatest? Explain your answers. (Incidentally, what is the direction of the force that holds the earth in orbit?)

A2.10 Between the plates of a parallel-plate capacitor (a device with two charged plates separated by a nonconducting medium) the electrical force on a charge remains constant. Suppose it is 0.2 N on an object bearing a charge of -5×10^{-9} C. How much work will be required to move the charge from the positive plate to the negative plate if they are 0.2 m apart? What is the electrical potential difference, or *voltage*, between the two plates of the capacitor?

Waves*

A. B. Arons

A3.1 INTRODUCTION

When we drop a stone into a quiet pond or sharply displace the end of a rope or long helical spring (Fig. A3.1), we produce motions that are transmitted from one point to another in the water or on the rope, eventually affecting regions very remote from the location of the initial disturbance.

In such instances, water is not displaced bodily to outer reaches of the pond from the point at which the stone fell; coils of the helical spring in Fig. A3.1 are not displaced from one end to the other with the disturbance that moves along the spring. What we see are motions that are communicated from one layer of the medium to the next, and our eye observes the propagation of a "pulse"—a shape or a disturbance. This shape moves along the medium with a finite velocity, and, in the illustrations cited above, it is apparent that two very different velocities can be distinguished.

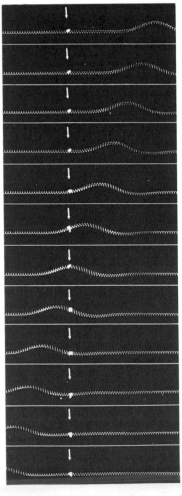

Fig. A3.1. Successive photographs showing propagation of pulse on a rope or helical spring. Initial disturbance is produced by imparting a sharp transverse (back and forth or up and down) deflection of the end of the spring. Note that motion of particle on spring is *up and down* while pulse shape moves from left to right. (From *PSSC Physics*, D. C. Heath, Boston, 1960.)

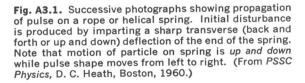

* Reprinted with the kind permission of Professor Arons from A. B. Arons, *Development of Concepts of Physics*, Addison-Wesley, Reading, Mass., 1965.

One velocity is that of the pulse as a whole; we call this *propagation velocity*. The other velocity is that of the particles of the medium; this is referred to as *particle velocity*. The particles of water and of rope move up and down or back and forth in a direction *perpendicular* to that in which the pulse moves, and within the region of the moving pulse, some layers of particles are moving up while others are moving down.

The propagation of a pulse or disturbance in a medium represents a new kind of motion, distinctly different from the bodily displacement of particles or rigid bodies. We call this phenomenon wave motion or wave propagation.

Our concern in this chapter will be to build on everyday experience and simple laboratory observations in an effort to develop some feeling for the nature of wave motion, and for the way in which waves behave when they arrive at boundaries or interact with each other. Our approach will be essentially kinematic and phenomenological: we shall describe the essential properties and characteristics of wave behavior without carrying out a mathematical analysis of dynamical effects and relationships. It is perfectly possible to apply Newtonian mechanics to wave motion and to predict the effects of forces and accelerations, but an analysis at this level will be left for more advanced courses and texts.

It is characteristic of the pattern of scientific thought that insights and conceptions which are developed or clarified at some simple, easily approachable level of perception frequently serve as models or analogies for explanation of phenomena at more complex, more remote levels. In the physical situation we have been describing, both particle and wave motion are directly perceptible to the senses of sight and touch. Although we can neither see nor feel mechanical effects in the air when we experience phenomena of sound, we soon become aware that sounds always originate with some kind of motion. Sounds are heard at a distance from the source when a blow is struck, when an explosion occurs, when a string or other object vibrates sufficiently rapidly. We quickly seize upon the plausible analogy, comparing the invisible spreading of an acoustic disturbance in the air with the spread of other waves we know from visual experience. At one point in the First Day of the *Two New Sciences* [of Galileo], Salviati illustrates this mode of thought and connects an action producing sound with that simultaneously producing ripples on a water surface:

That undulations of the medium are widely dispersed about a sounding body is evinced by the fact that a glass of water may be made to emit a tone merely by the friction of the fingertip upon the rim of the glass; [and in the] water is produced a series of regular waves. The same phenomenon is observed to better advantage by fixing the base of the goblet upon the bottom of a rather large vessel filled nearly to the edge of the goblet; for if, as before, we sound the glass by friction of the finger, we shall see ripples spreading with utmost regularity to large distances. . . . I have often remarked, in thus sounding a rather large glass nearly full of water, that at first the waves are spaced with uniformity, and when, as sometimes happens, the tone of the glass jumps an octave higher, I have noted that at this moment each of the aforesaid waves divides into two; a phenomenon which shows clearly that the ratio involved in the octave is two.

In the historical development of wave concepts, the qualitative insights came long before the dynamical, mathematical theory of acoustic waves provided quantitative theoretical validation. After the middle of the seventeenth century, scientific investigation moved rapidly to new ranges of physical phenomena: inquiry was directed toward the nature of light and heat, the nature of electricity and magnetism, the structure of matter. To help visualize, explain, and explore these subtle, nonmechanical levels of experience, scientists made use of models and analogies based on familiar mechanical phenomena of particle behavior on the one hand, and of wave motion on the other. Such models have had a profound influence on the evolution of modern physical concepts.

A3.2 TYPES OF WAVES

Pulses on a string and ripples on a water surface exemplify a class of waves described as *transverse*, in the sense that the direction of particle motion is perpendicular, or transverse, to the direction of propagation of the wave form. Transverse waves, when excited within the body of a solid medium such as the earth or a steel beam, are called *shear waves*.

In contrast, consider the wave illustrated in Fig. A3.2, in which pulses of compression or rarefaction are propagated among the loops of a helical spring. The particle velocity of the individual loops is parallel to the direction of propagation, and this type of wave is characterized as longitudinal. Sound waves in air or water are longitudinal waves of compression and rarefaction, completely analogous to the waves illustrated in Fig. A3.2. Only longitudinal waves can

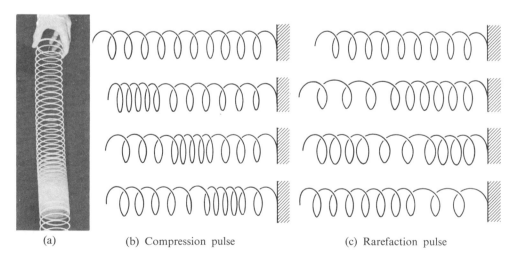

(a)   (b) Compression pulse   (c) Rarefaction pulse

Fig. A3.2. Propagation of compression and rarefaction pulses among the loops of a helical spring. A region of increased or decreased "density" of coils is produced by a longitudinal deflection parallel to the direction of propagation of the disturbance. A wave of this type is called *longitudinal*. Sound waves in air or water are also longitudinal waves of compression and rarefaction. (Photo (a) Courtesy Educational Services, Inc.)

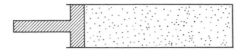

Figure A.3.3

propagate within the body of a fluid medium; shear waves cannot propagate because a fluid cannot transmit appreciable transverse stress from one layer to the next.

Problem A3.1 Visualize air contained in a tube and represent by means of dots the particles of air distributed uniformly throughout the tube (Fig. A3.3). Suppose that the piston is moved abruptly toward the right and then toward the left. Sketch, in the manner of Fig. A3.2(b) and (c), the propagation down the tube of pulses of compression and rarefaction.

Figure A.3.4

Problem A3.2 A long solid rod is lying at rest on a table (Fig. A3.4). Suppose you apply a push **P** at end A hard enough to displace the rod. Does end B move to the right at the instant you touch A? Describe what happens in this case. In what ways is it similar to the situation in Figs. A3.2 and A3.3? In what ways different?

Consider the suspension illustrated in Fig. A3.5. Weighted rods are attached to a wire or flexible metal tape and suspended as shown. Suppose that the bottom dumbbell is quickly rotated through an angular displacement in the horizontal plane. The twist is communicated to the next rod and the next, and a disturbance, called a *torsional wave*, is propagated up the suspension. In the system of Fig. A3.5 the disturbance travels relatively slowly and can be followed by eye. If we twist the end of an iron bar or the crankshaft of an engine, the resulting torsion wave travels very rapidly and involves very small angular displacements; we cannot see it visually, but we could detect its effects with sufficiently sensitive, rapidly responding electrical instruments.

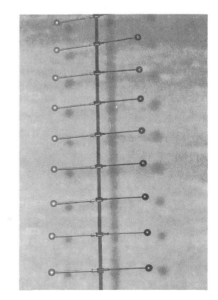

Fig. A3.5. Torsional wave is propagated up the suspension if bottom dumbbell is rotated through an angular displacement in horizontal plane.

In some cases a given disturbance may entail several different types of waves simultaneously. For example, earthquakes involve a complicated superposition of both longitudinal and shear waves, together with other types that are confined to surfaces of discontinuity such as the surface of the earth itself or to layers within the earth. Each type of wave travels at its own characteristic velocity, with the longitudinal waves (designated as "P" waves by seismologists) arriving first at a point of observation and the shear ("S" waves) arriving afterward.

The velocity of waves of various kinds depends, in general, on properties of the medium and on the size or amplitude of the disturbance in the wave itself. Very large deflections on a string or very large pressure pulses in the air (explosion waves or sonic booms from aircraft) propagate more rapidly than very small disturbances. Small disturbances propagate with a velocity that depends only on properties of the medium. The velocity of such a wave on a stretched string is given by $V = \sqrt{T/\mu}$, where T is the tension in the string and μ is the mass per unit length. In the language of physics, *sound* is defined as referring to small-amplitude waves, and the velocity of sound in a fluid is given by $V = \sqrt{(dp/d\rho)_{\text{adiabatic}}}$, where the derivative describes the rate of change of pressure with density under conditions such that there is no time for appreciable heat flow from compressed to rarefied regions in the wave.

It is frequently convenient to represent a wave form by means of a picture or graph. We can readily imagine taking a photograph of a wave on a string or on a water surface and plotting a scaled version of such a photograph on a set of coordinates in which the ordinate represents the transverse displacements and the abscissa represents corresponding horizontal positions. This would be an instantaneous picture of the wave in a quite literal sense, as in Fig. A3.1. For the torsional wave of Fig. A3.5 we might plot *angular* displacement as a function of vertical position. This would no longer be a photographic likeness of the wave form but would convey the basic idea on a two-dimensional graph.

For the wave on the helical spring (Fig. A3.2) we might plot instantaneous longitudinal displacements from equilibrium position of points on the spring and obtain a diagram such as Fig. A3.6, where the crest represents a region of

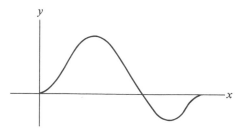

Fig. A3.6. Hypothetical graph that might represent longitudinal displacements versus normal horizontal position, *x*, of coils of a helical spring, or of pressure versus horizontal position in a sound wave.

compression or bunching of the coils, and the trough a region of rarefaction. To delineate a sound wave we might plot instantaneous pressures as a function of horizontal position and obtain a diagram very similar to Fig. A3.6, with essentially similar implications of compression and rarefaction. These last representations are certainly not photographic likenesses, but they nevertheless convey simple and readily interpretable information about the "shapes" of the wave forms.

Another graphical representation of a wave disturbance can be given by plotting a graph of displacements or pressure variations, etc., at a *fixed* point in the medium as a function of time. In this text, we shall make little use of this representation, but both forms are widely used in the literature and are, of course, closely related mathematically. Each has its own advantages in the appropriate context.

A3.3 WAVELENGTH AND FREQUENCY OF PERIODIC WAVE TRAINS

Wave disturbances can assume an infinite variety of forms. Several principal types, together with the vocabulary used to describe them, are illustrated in Fig. A3.7. Regular periodic wave trains, such as those in Fig. A3.7(c) and (d), are excited by sources having a regular, oscillatory motion of some definite frequency—a vibrating string or tuning fork, a vibrating rod producing ripples at a water surface. If the source makes ν (Greek letter "nu") complete oscillations in one second, it must emit ν complete waves. If we take up a point of observation in the medium, we observe ν crests or ν complete waves pass by

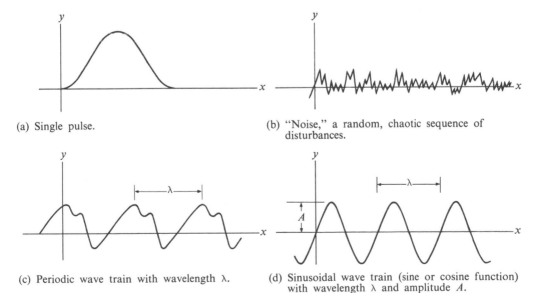

(a) Single pulse.

(b) "Noise," a random, chaotic sequence of disturbances.

(c) Periodic wave train with wavelength λ.

(d) Sinusoidal wave train (sine or cosine function) with wavelength λ and amplitude A.

Fig. A3.7. Illustrations of several different types of wave

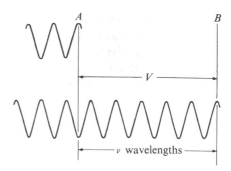

Fig. A3.8. Wave crest, initially at point A, advances distance V in one second. Length $AB = V$ contains ν wavelengths.

in each second. We therefore associate a definite frequency ν with any periodic wave train, and we recognize this frequency to be identical with that of the source. The period T (the duration of the complete cycle of the wave) is given by $T = 1/\nu$.

We have an intuitive sense that for any particular kind of wave, a low frequency should be associated with a long wavelength and a high frequency with a short wavelength. Let us explore the arithmetical relationship. A periodic wave train of frequency ν (Fig. A3.8) is moving at propagation velocity V. If we watch a crest that passes point of observation A at $t = 0$, one second later the crest will have advanced to position B a distance V away from A. The space AB will contain ν wavelengths. Therefore the individual wavelength λ must be given by

$$\lambda = \frac{V}{\nu} = VT. \tag{A3.1}$$

Problem A3.3 Interpret Eq. (A3.1): A vibrating source in a ripple tank oscillates at 3.6 cycles/sec, and the ripples in the given depth of water propagate at 4.8 cm/sec. Calculate the wavelength of the ripples. The musical note middle C has a frequency of about 260 cycles/sec, and the velocity of sound in air is about 1100 ft/sec. Calculate the wavelength of this sound wave. Suppose a 260-cycle/sec sound wave is excited in water, will the wavelength be larger, smaller, or equal to the wavelength in air? On what do you base your answer?

A3.4 RELATIVE PHASES OF SINUSOIDAL WAVE TRAINS

The wave forms produced by a tuning fork vibrating in air or by a bar vibrating in a ripple tank are very nearly sinusoidal; i.e., we can to a very good approximation represent the experimentally observed wave forms by a sine or cosine function.

Figure A3.9 illustrates how this curve fitting is achieved. We are given a sinusoidal wave form of amplitude A and wavelength $\lambda = 8$ ft. We can choose the origin of coordinates anywhere we please, and we have elected to place it at point 0. This immediately suggests that the curve is to be represented by a sine function since $\sin \theta = 0$ when $\theta = 0$ and increases as θ increases. The

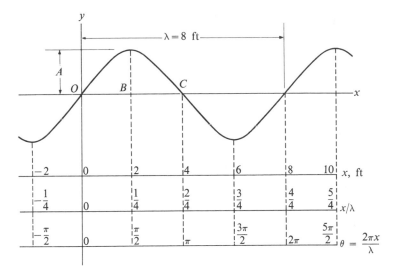

Fig. A3.9. Sinusoidal wave form; origin of coordinates taken at point of zero deflection, with deflections increasing toward the right. Equation of curve: $y = A \sin (2\pi x/\lambda)$.

maximum value of the sine function is unity, and the maximum deflection in the wave form is A; our first impulse might then be to represent the given curve by the equation $y = A \sin x$. A few simple checks show that this equation has absurd implications. At $x = 4$ ft the deflection in Fig. A3.9 is zero, but the equation would read $y = A \sin 4$, which is not zero. If we measured the horizontal distance in meters instead of feet, $x = 1.22$ m, and the deflection at the same position would be given by $y = A \sin 1.22$, which is not equal to $A \sin 4$. Depending on the units in which we elect to measure horizontal distance x, we can get any number we please.

The inconsistency, of course, lies in the fact that sine and cosine functions were defined with pure numbers as their domain; geometrically, these pure numbers can be interpreted as angles. Horizontal distance x, however, is a dimensional quantity expressed in arbitrary units. If we attempt to relate $\sin x$ to the given wave form, we shall simply get a meaningless jumble. In order to obtain $y = 0$ at $x = 4$, it is necessary to associate a pure number, or angle, θ with $x = 4$ such that the sine of this value of θ is zero. Similarly we must associate another value of θ with $x = 6$ such that the sine of this value is -1, etc.

Figure A3.9 shows how this association can be achieved. First we describe horizontal positions in terms of numbers of wavelengths x/λ, as shown in the second row of numbers along the abscissa. Then we note that a whole wavelength ($x/\lambda = 1$) corresponds to $\theta = (2\pi) \cdot 1$, and that the angle corresponding to one quarter wavelength must be

$$\theta = 2\pi \cdot \tfrac{1}{4} = \pi/2,$$

and that, in general, for x/λ wavelengths $\theta = 2\pi(x/\lambda)$. Therefore the equation describing the sinusoid having the origin located as shown in Fig. A3.9 is

$$y = A \sin \frac{2\pi x}{\lambda}. \qquad (A3.2)$$

Problem A3.4 Suppose that the origin in Fig. A3.9 were located at point B; show that the curve would then be represented by $y = A \cos 2\pi x/\lambda$. If the origin were at point C, show that the equation would be $y = -A \sin 2\pi x/\lambda$.

Problem A3.5 A wave on a string is known to be represented by the equation $y = 0.25 \sin (2.36x)$, where deflection and horizontal distances are measured in feet. What are the amplitude and wavelength of this wave? [*Answer:* $A = 0.25$ ft, $\lambda = 2.66$ ft.]

Figure A3.10 illustrates several sinusoidal wave trains, of identical amplitude and wavelength, shifted relative to each other along the x-axis. We speak of these waves as "differing in phase" or as being "shifted in phase" relative to each other, and the phase difference is measured quantitatively by the *angle* of shift or by the number of wavelengths. There are, of course, many different phase shifts that would reproduce the patterns illustrated in Fig. A3.10. For example, the out-of-phase pattern of (c) relative to (a) can be re-established by any angular shift, either right or left, of an odd multiple of π or an odd number of half-wavelengths. Similarly, the in-phase pattern of (e) and (a) can be re-established by any angular shift of 2π or by an integral number of whole wave-

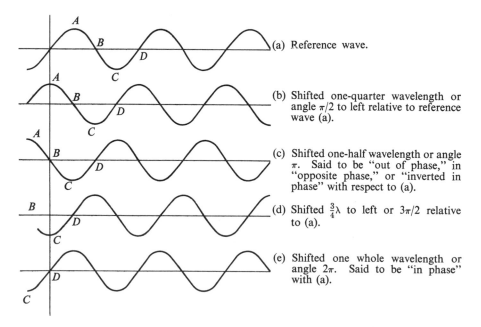

(a) Reference wave.

(b) Shifted one-quarter wavelength or angle $\pi/2$ to left relative to reference wave (a).

(c) Shifted one-half wavelength or angle π. Said to be "out of phase," in "opposite phase," or "inverted in phase" with respect to (a).

(d) Shifted $\frac{3}{4}\lambda$ to left or $3\pi/2$ relative to (a).

(e) Shifted one whole wavelength or angle 2π. Said to be "in phase" with (a).

Fig. A3.10. Sinusoidal waves of a given wavelength in various different phases relative to each other.

lengths. We shall be particularly concerned with these two special cases when we deal with superposition and interference of sinusoidal wave trains in Sections A3.11–A3.13.

A3.5 WAVES IN TWO DIMENSIONS

One of the most distinctive aspects of a wave disturbance is the manner in which it spreads out in either two or three dimensions from a source. If we touch a quiet water surface with a pencil point or our fingertip, we see a circular wave pulse spread out from the origin of the disturbance. By vibrating the pencil point up and down we can send out a continuous train of waves, successive crests and troughs forming concentric circles around the source. In an exactly analogous manner, a vibrating tuning fork sends out a continuous train of sound waves. At distances larger relative to the size of the fork itself, the fork appears to be a point source, with successive regions of compression and rarefaction forming concentric spherical shells around it.

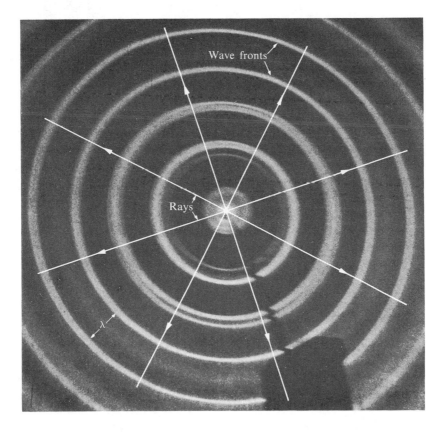

Fig. A3.11. Wave fronts and rays of waves spreading out from a point source. (From *PSSC Physics*, D. C. Heath, Boston, 1960.)

A specific vocabulary is developed for the description of waves propagating in two or three dimensions. A region of constant phase (a crest, a trough, a locus of zero deflection, or a locus of any arbitrarily chosen deflection from $-A$ to $+A$) is called a *wave front*. Lines drawn in such a way as to be everywhere *normal* to the wave fronts they cross are called *rays*. For ripples from a point source on a water surface, wave fronts form concentric circles, as illustrated in Fig. A3.11, and rays are radial lines originating at the point source. Figure A3.11 would serve equally well as an illustration of wave fronts and rays lying in a plane cross section of a spherical sound wave emitted by a vibrating tuning fork.

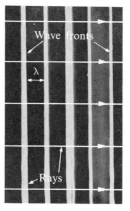

Fig. A3.12. Straight waves emitted by bar vibrating in ripple tank. Both the wave fronts and the rays are straight lines. (From *PSSC Physics,* D. C. Heath, Boston, 1960.)

Waves may also be generated by the motion of an extended source. If we vibrate a long bar up and down at a water surface, it will generate a wave having the same phase along its entire length. At distances relatively close to the bar, the successive wave fronts have the form of parallel straight lines. Such waves are called "plane" or "straight" waves; their rays are also parallel to each other and do not diverge as do the rays of circular waves. Figure A3.12 illustrates this case, and could equally well apply to a plane cross section taken through a sound wave emitted by a large diaphragm vibrating, as a whole, back and forth along the x-axis. At distances very large relative to the size of the bar or diaphragm, the wave fronts would appear to be circles of very large radius centered at the source. On the other hand, small sections of circular wave fronts of very large radius, far from the source, appear to be very nearly plane and are frequently treated to a very good approximation as plane waves with essentially parallel rays. The basic simplicity of the plane and circular waves resides, in part, in the fact that the rays are stright lines in both cases.

A3.6 DIFFRACTION

Figure A3.13 illustrates three characteristic patterns formed when a plane wave encounters an obstacle pierced by a single opening. Waves continue to propagate through the opening, but the wave pattern on the other side is determined

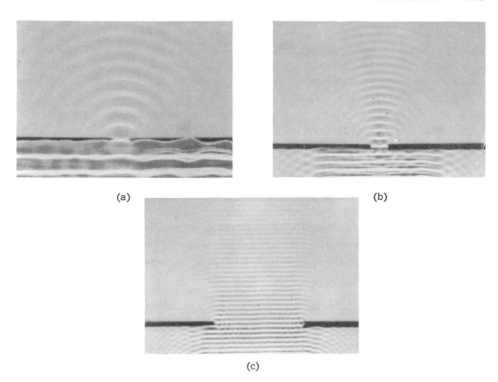

(a)   (b)

(c)

Fig. A3.13. Plane waves incident on slits of various widths, D. (From *PSSC Physics*, D. C. Heath, Boston, 1960.) (a) $\lambda \sim D$. Slit approximates point source of circular waves. (b) $\lambda \sim 0.7D$. Emerging waves form single-slit interference pattern. (c) $\lambda \ll D$. Waves exhibit some diffraction at edges, but remain straight in central portion.

by the relation between the wavelength λ of the wave and the width D of the opening. Such bending and modification of a wave front on passing through an opening is called *diffraction*. Diffraction effects are small when λ is much smaller than D (Fig. A3.13c) and become increasingly pronounced as λ becomes of the order of magnitude of D or larger. The pattern illustrated in Fig. A3.13(b) is called a single-slit diffraction pattern. When λ is larger than the opening (Fig. A3.13a) the slit becomes, to a good approximation, a point source of circular waves.

If a wide "beam" of particles, all traveling parallel to each other, were to arrive at a barrier, as in Fig. A3.13, a narrower parallel beam, of width D, would continue on the other side, with perhaps a few particles deflected (or scattered) out of their original path by the edges of the slit. A wave with $\lambda/D \ll 1$ behaves in a somewhat analogous fashion, but the pronounced diffraction effects of Fig. A3.13(a) and (b) are uniquely characteristic of wave behavior and markedly different from the behavior of a beam of particles. We shall subsequently see this distinction used repeatedly to characterize wave disturbances versus particle motion under circumstances in which it is otherwise not at all obvious what sort of phenomenon one is dealing with.

A3.7 REFRACTION

In many circumstances the velocity of wave propagation may vary from point to point in a medium. The velocity of ripples decreases with decreasing depth of water; if a ripple tank has a sloping bottom, the propagation velocity becomes a function of position in the tank. Sound velocity in air or water increases with increasing temperature; if the temperature varies in the horizontal and vertical directions, as it does in both the atmosphere and the ocean, sound velocity becomes a function of position in the medium.

When a quantity varies in space, it is said to have a "space gradient." The gradient in a particular direction x is measured by the derivative or space rate of change of the quantity in that direction. For example, a temperature gradient in the x-direction at any particular point is measured by the derivative dT/dx. Similarly, a gradient in propagation velocity is measured by dV/dx. Figure A3.14 illustrates the behavior of ripples propagating over a sloping bottom. The water becomes shallower in the positive x-direction, and therefore the gradient dV/dx in propagation velocity is negative.

Any local portion of a wave front propagates forward at the local propagation velocity. In Fig. A3.14(a) the ripples move into water of decreasing depth, with wave fronts aligned parallel to loci of constant propagation velocity. Moving into regions of lower velocity, the fronts remain parallel to each other but the wave length ($\lambda = V/\nu$) decreases, since the frequency is fixed, having been established by the driving source.

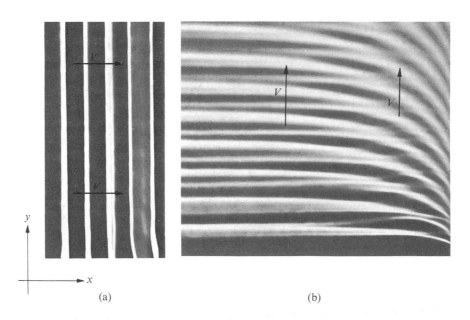

Fig. A3.14. Waves propagating over sloping bottom. Water becomes shallower from left to right: dV/dx negative. Waves propagate in directions shown by arrows. (Photograph from *PSSC Physics*, D. C. Heath, Boston, 1960.)

In Fig. A3.14(b) the wave fronts are not parallel to regions of constant propagation velocity. As a result the left-hand portions of the ripples, in deeper water, move faster than the right-hand portions. The wave fronts tend to bend around as shown in the figure; the rays are no longer straight lines emanating from the source, but curve toward the region of lower propagation velocity.

The phenomenon of distortion of wave fronts and bending of rays under the influence of gradients in the propagation velocity is called refraction. Refraction occurs not only under the conditions illustrated in Fig. A3.14(b) but also when a wave propagates through an interface between two uniform regions having sharply different propagation velocities: when a sound wave in air encounters a water surface or when ripples in a water layer of uniform depth encounters a "shelf" at which the depth changes to another uniform value. In these instances the propagation velocity gradient is zero in each region, but refraction occurs sharply at the boundary, where there is an abrupt change in the propagation velocity.

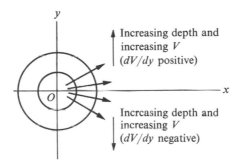

Figure A3.15

Problem A3.6 Consider a ripple-tank situation with velocity gradients as indicated in Fig. A3.15. The bottom slopes away very gradually on either side of the x-axis, so that the axis is a line of minimum (not zero) depth and of minimum propagation velocity. The velocity V increases with distance from the x-axis in both the $+y$ and $-y$ directions. A point source of ripples is located at O. Make a rough qualitative sketch of what will happen to the several rays indicated in the figure as the waves propagate away from O. What happens to a ray if it crosses the x-axis?

A3.8 MOVING SOURCE

Figure A3.16 illustrates the changes of wavelength that arise when a point source of ripples moves over a water surface with a velocity v_s lower than the propagation velocity V in the medium.

Let us calculate wavelengths along the axis of motion of the source. Taking v_s positive when it is in the same direction as V and negative when its direction is opposite to V, making use of the derivation suggested in Fig. A3.8, we find that the wavelength of ripples in the medium is given by

$$\lambda = \frac{V - v_s}{v_s}.\tag{A3.3}$$

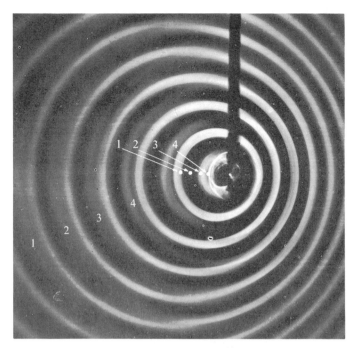

Fig. A3.16. Point source of fixed frequency ν_s moves toward the right at uniform velocity v_s, emitting crest 1 from position 1, crest 2 from position 2, etc.; v_s is smaller than propagation velocity V. Wavelength ahead of moving source is shorter, and wavelength behind is greater, than undisturbed wavelength from stationary source. (Courtesy Educational Services, Inc.)

From a frame of reference stationary relative to the water surface, we observe ripples of wavelength λ to have an apparent frequency ν_{app} given by

$$\nu_{app} = \frac{V}{\lambda}. \tag{A3.4}$$

Eliminating λ from (A3.3) and (A3.4), we can write

$$\nu_{app} = \frac{V}{V - v_s}\nu_s = \frac{1}{1 - (v_s/V)}\nu_s. \tag{A3.5}$$

Note that Eqs. (A3.3) and (A3.5) apply only along the line of motion of the source and do *not* tell us about values of λ and ν_{app} at positions away from the line of motion. A complete analysis of Fig. A3.16 is perfectly possible but will not be undertaken here.

Problem A3.7 How might a figure such as A3.16 be constructed with compass and ruler; i.e., what must the relation be between the unperturbed wavelength and the spacing of positions 1, 2, etc? Construct such a figure for yourself. Justify Eq. (A3.3) by using an argument such as that presented in Fig. A3.8 (or by using any other explanation you feel is clear and logical). Verify that Eq. (A3.3) makes sense physically by testing its agreement with the facts exhibited in Fig. A3.16. What happens as $v_s \to V$?

Problem A3.8 We cannot see the wavelengths of sound waves, but our ears discriminate their apparent *frequency* through our sense of pitch: a note or tone of higher frequency has a higher pitch than one of lower frequency. Interpret Eq. (A3.5) by deducing its predictions concerning the relative pitch we would expect to hear if we listened to a stationary tuning fork and then caused the fork to move toward us or recede from us at velocity v_s. Do the predictions make physical sense? What happens as $v_s \to V$?

If we ourselves, as observers, adopt a frame of reference that moves at velocity v_0 relative to the medium in which the waves propagate, the waves appear to go more slowly if we move in the same direction as the waves and pass us more rapidly if we move in the opposite direction. This also affects the apparent frequency ν_{app} that we would detect, even when the source itself is stationary relative to the medium; i.e., when $v_s = 0$.

The phenomenon in which wavelengths and apparent frequencies are altered by motion of source and observer relative to the medium is called the *Doppler effect*. (Doppler was an Austrian scientist who, in 1842, worked out the theory for sound waves, and then called attention to the relevance of this phenomenon to investigation of the nature and behavior of light. A test of the theory was carried out in Holland in 1845, with trumpeters riding a railroad flat car and musically trained observers estimating, by ear, the apparent change in pitch. The two groups then exchanged positions.)

A3.9 BOW WAVES

Our analysis in Section A3.8 was carefully restricted to the case $v_s < V$. It is perfectly possible, however, for the source velocity to exceed the propagation velocity. This happens when a speedboat runs through the water, when we move a pencil point rapidly along the surface in a ripple tank, when a bullet or airplane exceeds the speed of sound. Under such circumstances, the wave cannot get out ahead of the source; the source is continually at the very leading edge of the disturbance and generates a characteristic pattern called a *bow wave* (Fig. A3.17).

Bow waves formed by the boat and bullet are actually quite complicated phenomena. The disturbance in each case is intense and the amplitude quite high. In the case of the bullet or supersonic* plane, the wave produced by supersonic motion is called a *shock wave*. The velocity of propagation of high-amplitude waves depends not only on the medium but also on the instantaneous amplitude of the wave itself; shock wave velocity, for example, decreases with decreasing amplitude, approaching the local velocity of sound as the amplitude becomes very small. Some of the complex features and curvature apparent in Figs. A3.17(a) and (b) are associated with the effects of large wave amplitude.

---

* The term "supersonic" refers to velocities exceeding that of sound.

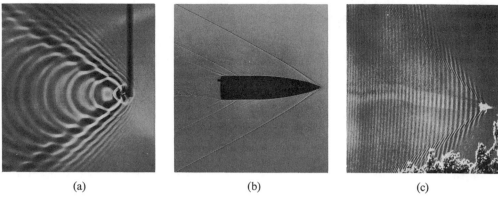

(a)   (b)   (c)

Fig. A3.17. Bow waves generated in various circumstances. (a) Vibrating source moving through ripple tank with velocity of source v_s greater than ripple propagation velocity V. (Courtesy Educational Services, Inc.) (b) Projectile moving through air. (Courtesy U. S. Naval Ordnance Laboratory, White Oak, Md.) (c) Speedboat.

If we idealize the problem, however, and confine ourselves to waves of small amplitude, the bow wave takes the form of two plane wave fronts. We can construct the basic pattern by drawing circles representing wave crests emitted at regular intervals by a vibrating source, as shown in Fig. A3.18. If the source does not vibrate, but emits a single disturbance because of its continuous motion, we see just the bow wave with relatively little disturbance in the wake.

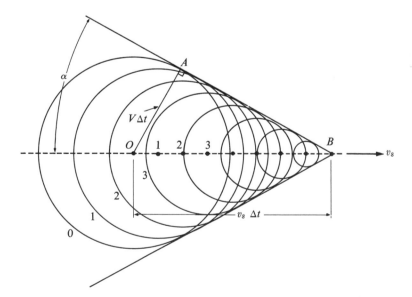

Fig. A3.18. Successive crests emitted by vibrating source moving at velocity v_s, greater than wave propagation velocity V. Crest 0 was emitted when source was at position 0, crest 1 from position 1, etc. "Envelope," or line tangent to the various crests, forms the bow wave.

From Fig. A3.18 it is easy to deduce the dependence of the bow wave angle α on the velocities v_s and V. During the time interval Δt, the source moves from 0 to B, a distance $v_s \Delta t$. During the same interval, the first wave crest, emitted at 0, moves a distance $OA = V \Delta t$. Then

$$\sin \alpha = \frac{V \Delta t}{v_s \Delta t} = \frac{V}{v_s}. \tag{A3.6}$$

We shall find this result to be very useful in analyzing the reflection and refraction of wave trains at sharp interfaces.

A3.10 SUPERPOSITION

A dramatic aspect of wave behavior is exhibited in the manner in which wave pulses, traveling in different directions, can literally pass through each other without any appreciable effect on their form and propagation. The deflections at any given location add algebraically from instant to instant, and the pulses

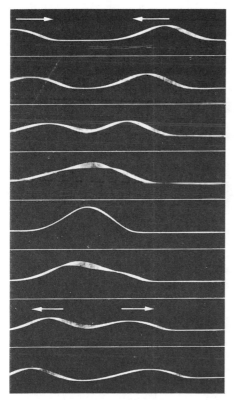

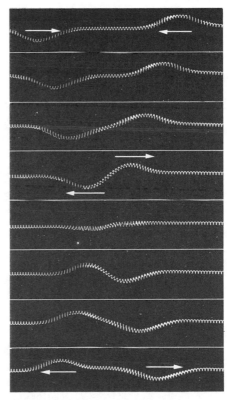

Fig. A3.19. Sequence showing two pulses, both of upward deflection, traveling in opposite directions on a string. Pulses superpose, pass through each other, and continue propagating. (From *PSSC Physics,* D. C. Heath, Boston, 1960.)

Fig. A3.20. Sequence showing superposition of two pulses of opposite deflection traveling in opposite directions. (From *PSSC Physics,* D. C. Heath, Boston, 1960.)

(a) Rack consists of metal bars that are free to slide up and down. Their tops are cut to form a sinusoidal wave form. Wooden template, cut to the same wave form, can be "superposed," lifting the bars from below. In this picture wave trains are in phase with each other.

(b) Constructive interference. Wave trains are superposed crest to crest and trough to trough. Resultant wave form has twice the amplitude of the individual wave trains.

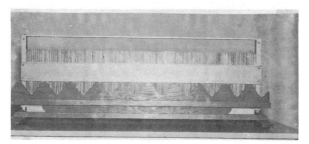

(c) Destructive interference. Wave trains are superposed crest to trough, in opposite phase. Resultant wave (tops of bars) has zero amplitude.

(d) Wave forms with same amplitude but slightly different wavelength. If they are out of phase at one point, they will be in phase at another.

(e) Superposition of wave forms in (d) leads to succession of regions of destructive and constructive interference. This phenomenon is called "beating" or "beats."

Figure A3.21

continue propagating as though each were present alone. This phenomenon is illustrated for pulses on a spring in Figs. A3.19 and A3.20, and constitutes another example of the broad concept of superposition referred to in a variety of different contexts in earlier chapters. It is apparent that the waves illustrated obey a simple superposition principle.

Since we are able to discern different sounds simultaneously, we infer that sound waves also superpose, traveling through each other without alteration.

Our descriptions of the formation of bow waves have already utilized the superposition concepts implicitly. In the following sections we shall pursue additional consequences of this aspect of wave behavior.

A3.11 INTERFERENCE

If we contrive to make two periodic trains of the same wave form and wavelength overlap while traveling in very nearly the same direction, we can anticipate two extremes that might occur in the superposition (Fig. A3.21). This special case of superposition is referred to as *interference*, and the examples illustrated in Fig. A3.21(b) and (c) are called constructive and destructive, respectively.

Figure A3.22 shows patterns observed when circular ripples originate from two point sources, oscillating in phase with each other, and placed fairly close together in the ripple tank. The geometry of the pattern changes as either the wavelength or the distance between the sources is changed. On examining the photographs, we can readily identify lines (called nodal* lines) along which the

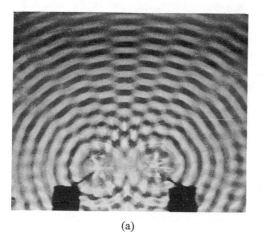   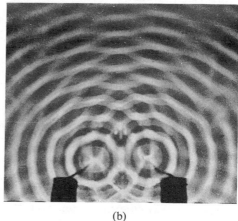

(a)   (b)

Fig. A3.22. Ripple-tank photographs showing interference pattern formed by superposition of circular waves from two point sources. Spacing between sources is identical in (a) and (b), but wavelength is larger in (b). (From *PSSC Physics*, D. C. Heath, Boston, 1960.)

---

* The term "node" refers to a point or region of permanently zero deflection or disturbance.

resultant wave amplitude appears to be zero. Here the waves from the two sources overlap systematically crest to trough and interfere destructively, as illustrated in Fig. A3.21(c). In the sectors between the nodal lines, we can identify constructive interference as the waves superpose crest to crest and trough to trough.

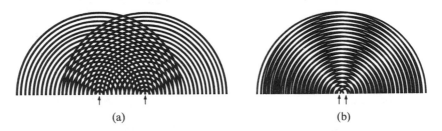

(a)   (b)

Fig. A3.23. Lines represent crests; spaces between them represent troughs. When source spacing d is much larger than the wavelength λ, waves crisscross each other as in (a) without forming a pronounced pattern. When d is only slightly larger than λ, the waves intersect at a more oblique angle, and a pattern is formed as in (b). The dark loci are regions where crests cross troughs and simulate nodal lines of destructive interference. Light regions are loci where crests cross crests and troughs cross troughs, simulating constructive interference.

The reader will find it very illuminating to produce diagrams such as those in Fig. A3.23 by drawing circles with a compass carrying a very blunt pencil, making a wide line. Taking a fixed wavelength as shown in the figure and changing the spacing between the sources, he can develop a feeling for how the pattern of dark loci, simulating destructive interference, fans *out* as the spacing d between the sources is made smaller and more nearly equal to λ, and how it contracts into the crisscross pattern of Fig. A3.23(a) as d is made much larger than λ. Figure A3.24 shows a greatly expanded drawing of the region near the two sources in Fig. A3.23(b).

Examining Fig. A3.24, we note that the principal axis is itself a locus of constructive interference. Everywhere along it the waves from the two sources arrive in phase, having traveled equal distances $r_1 = r_2$. The first locus of destructive interference to the right of the axis is one along which the waves are everywhere exactly out of phase; the wave from S_1 has traveled a distance r_1 which exceeds r_2 by exactly one-half wavelength; i.e., the locus is defined by the relation $r_1 - r_2 = \lambda/2$. The other loci are defined in a similar way; for example, point R is in a region of constructive interference because the seventh crest from S_1 superposes on the fifth crest from S_2, the phase lag being exactly two wavelengths; that is, $r_1 - r_2 = 2\lambda$.

It has been implied in the preceding discussion that we are concerned with interference patterns formed by waves from two separate vibrating sources. We could, of course, produce an exactly equivalent situation by allowing a plane wave train to impinge on a barrier with two small openings or slits a distance d apart. If the openings themselves are narrower than the wavelength, they act

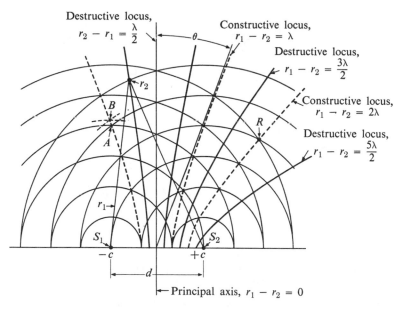

Fig. A3.24. Greatly expanded drawing of region near sources in Fig. A3.23(b). Loci of constructive and destructive interference are hyperbolas with foci at S_1 and S_2. At relatively large distances from the origin the hyperbolas are indistinguishably close to their asymptotes, and appear to be straight lines radiating from the origin at angles θ to the principal axis. The loci are fixed in space even though the waves are continually moving. As the crests which intersect at point A both move outward to the dotted positions, their intersection moves to B along the fixed locus of constructive interference.

as point sources (Fig. A3.13a), diffracting the incident wave and emitting circular wavelets exactly in phase with each other. The resulting interference pattern is precisely like that of two separate vibrating sources.

Problem A3.9   Using a compass, construct figures similar to Figs. A3.23 and A3.24, and locate the loci of constructive and destructive interference. Explain in your own words the definitions of several of these loci by accounting for the phase difference between the two waves in each case.

A3.12 ELEMENTARY ANALYSIS OF THE TWO-SOURCE INTERFERENCE PATTERN

Examination of Figs. A3.22, A3.23, and A3.24 reveals that the loci of constructive and destructive interference are curved lines but that the curvature is pronounced only in the immediate neighborhood of the sources. At distances of the order of two or three times the source spacing d, the loci are indistinguishable from straight lines radiating from an origin located at the midpoint between the sources. This property suggests a very simple way of obtaining an algebraic description of the angles θ (Fig. A3.24) between various particular loci and the principal axis.

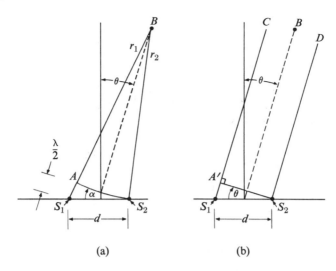

(a)   (b)

Fig. A3.25. (a) Point B lies relatively far out on the first nodal line, the locus $r_1 - r_2 = \lambda/2$. With B as a center, we draw arc S_2A. The length S_1A is then equal to $\lambda/2$ and, since S_2A is very nearly a straight line, we can think of α as an angle whose sine is $\lambda/2d$. (b) Point B lies on first nodal line, but instead of radii r_1 and r_2, we draw lines S_1C and S_2D parallel to the nodal line. Line S_2A' is then drawn perpendicular to S_1C. Angle $A'S_2S_1$ is then equal to θ. When point B is relatively far from the origin, triangle AS_2S_1 in (a) is very nearly identical with triangle $A'S_2S_1$ in (b). S_1A' is then very nearly equal to $\lambda/2$ and θ is very nearly equal to α. Thus $\sin \theta = \lambda/2d$, to an extremely close approximation.

In Fig. A3.25 and its accompanying caption, it is shown that for points relatively far out along the first nodal line the inclinations of radii r_1 and r_2 toward each other can be neglected; the radii can be treated as though they were essentially parallel to each other. On this basis a simple relation among θ, λ, and d can be inferred from the small right triangle $A'S_2S_1$ in Fig. A3.25(b). It is immediately apparent that

$$\sin \theta_1 = \frac{\lambda}{2d}, \tag{A3.7}$$

where θ_1 denotes the angular position of the first nodal line on either side of the principal axis. Other nodal lines will occur at larger angles, for which the distance S_1A' in Fig. A3.25(b) is equal to three half-wavelengths, five half-wavelengths, etc. Expressing this result more compactly, we note that nodal lines occur at those angles θ_n for which:

$$\sin \theta_n = (2n - 1) \frac{\lambda}{2d}, \qquad n = 1, 2, 3, 4, \ldots, \tag{A3.8}$$

$(2n - 1)$ being an expression that gives us the successive odd numbers when we substitute integer values for n.

In an exactly similar fashion, we note that constructive interference will occur along lines oriented in such a way that the path difference S_1A' in

Fig. A3.25(b) will be equal to one, two, three, or any other whole number of wavelengths. Thus we conclude that loci of constructive interference will lie at angles θ_n for which

$$\sin \theta_n = n\frac{\lambda}{d}, \qquad n = 1, 2, 3, 4, \ldots . \qquad (A3.9)$$

Thus, if we know λ and d, we can immediately predict the angular positions of the various constructive and destructive loci. If we do not know λ but determine the source spacing and measure, say θ_2, the angular position of the second nodal line, we can compute the wavelength of the wave train.

Problem A3.10 Interpret Eqs. (A3.8) and (A3.9), and connect them explicitly with various loci in Figs. A3.22, A3.23, and A3.24. Note that for any particular source spacing d and wavelength λ, the number n ceases to have physical significance after exceeding a certain value. Why? For any given pattern, how do you determine the highest meaningful value of n? What is this value in Fig. A3.24? What happens when λ becomes larger than d? Larger than $2d$? (Note that if $\lambda \gg 2d$, the two sources cannot be distinguished from a single source emitting circular waves into the medium.)

Problem A3.11 In a ripple-tank experiment with a spacing of 3.0 cm between the sources, a point on the first nodal line is found to be located as shown in Fig. A3.26. Compute the wavelength of the ripples. [*Answer:* $\lambda = 1.2$ cm.]

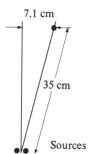

7.1 cm

35 cm

Figure A3.26

Sources

Problem A3.12 Suppose that Fig. A3.24 represents a ripple-tank experiment in which $d = 6.0$ cm and $\lambda = 2.0$ cm. Suppose further that you cannot see the pattern directly for lack of projection of a shadow pattern, but that you are equipped with a small probe that is sensitive to wave amplitude. You start at a point on the y-axis at a large radius R from the source and move the probe around the circular arc of radius R. Predict what the probe will indicate, giving numerical values of angles at which you expect a maximum or a minimum response. [*Answer:* minima at $\theta = 9.6°$, 30°, 56°; maxima at $\theta = 19°$, 43°.]

Problem A3.13 Suppose that you were to perform an experiment exactly like that in Problem A3.12, except that the sources are two tuning forks having a frequency of 440 cycles/sec, placed 7.0 ft apart and vibrating exactly in phase with each other. This time the sensitive "probe" is your own ear as you walk around the arc of a circle at radius R from the source. Describe what you would hear, giving numerical values of appropriate angles. (Take the velocity of sound to be 1100 ft/sec.)

Problem A3.14 Suppose that the two point sources vibrate unreliably, randomly shifting their phase relative to each other. What would happen; would you expect to observe an interference pattern? Suppose that in Fig. A3.24 the vibration is steady but the phase relation between the sources is changed, so that S_2 emits a trough at precisely the moment S_1 emits a crest (i.e., the phase lag would be equivalent to $\lambda/2$). What would happen to the interference pattern?

A3.13 PLANE WAVE TRANSMITTED THROUGH A GRATING

An important modification of the interference pattern is produced when, instead of just two sources, we have a large array of sources, uniformly spaced along a straight line, and all emitting circular waves in the same phase. This condition can be achieved in the ripple tank by allowing a plane wave train to impinge on a grating—a barrier pierced by narrow slits uniformly spaced a distance d apart. The slit opening D is smaller than λ. The slits diffract the incident wave and act as coherent point sources. A sketch of what is observed in such a situation is shown in Fig. A3.27. Emerging from the grating are several *plane* wave trains having different directions of propagation relative to the principal axis!

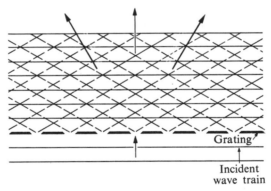

Fig. A3.27. Plane wave train in ripple tank incident on a plane grating. Arrows show directions of propagation of three emerging wave trains.

The ripple-tank photographs in Fig. A3.28 show how this pattern is built up, as more and more openings are added to the array.

We can help ourselves understand Figs. A3.27 and A3.28 by once more constructing wavelets with a compass. Such a construction is illustrated in Fig. A3.29. (The reader will find it easier to follow the discussion if he constructs a diagram such as A3.29 for himself.) In this construction we find that wavelets overlap so as to reconstitute plane wave fronts (as in the illustration of the bow wave in Fig. A3.18), the fronts being pointed up more clearly if we draw tangents to them, as we have at several locations in Fig. A3.29. One reconstituted wave front (Fig. A3.29a) travels in the original direction along the principal axis; the inclination of its ray relative to the axis is zero. Additional plane wave fronts

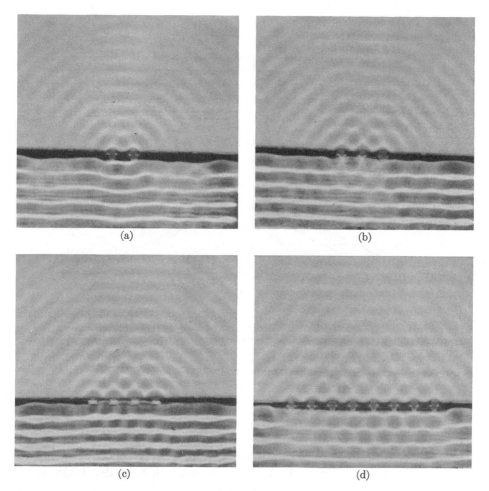

(a)   (b)

(c)   (d)

Fig. A3.28. Ripple-tank photographs showing alteration of transmitted wave pattern as more openings are added to an array so as to form a "grating." (From the film, "Interference and Diffraction," produced by Educational Services, Inc.)

are formed propagating in the particular direction θ_1 (Fig. A3.29b) indicated by the respective rays, but there are *no* organized wave fronts traveling in any direction between 0 and θ_1.

It is evident from Fig. A3.29(b) that

$$\sin \theta_n = \frac{\lambda}{d}$$

and that, in general, plane wave fronts would be formed moving in directions defined by

$$\sin \theta_n = \frac{n\lambda}{d}, \qquad \text{where } n = 0, 1, 2, 3, \ldots, \qquad (A3.10)$$

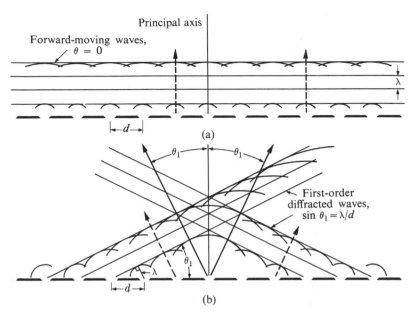

Fig. A3.29. Plane waves diffracted through openings in a uniformly spaced grating. In the forward direction ($\theta = 0$) wavelets superpose after equal distances of travel to form forward-moving plane waves as in (a). In the direction θ_1, wavelets with path differences of one whole wavelength between successive openings superpose to form a bow wave traveling at an angle θ_1 relative to the principal axis. Another wave is to be expected at a larger angle θ_2, such that path differences between waves from successive openings are two whole wavelengths, etc.

the pattern being symmetrical on either side of the principal axis. The case $n = 0$ represents the wave traveling parallel to the principal axis, while the two waves on either side of the axis with $n = 1$ and angle θ_1 are called the first-order waves; the ones for $n = 2$, the second-order waves, etc. Second-order waves will not exist if $(2\lambda/d) > 1$. As d is made larger or λ smaller, θ_1 decreases, and more orders of waves appear in the pattern. A grating used in this fashion is called a *diffraction grating;* the wavelets diffracted through the small openings are responsible for the resulting interference pattern.

Contrast the effect of the grating with the two-source pattern exhibited in the figures of Section A3.11. If you walk around the periphery of a two-source pattern at a large distance from the source, you would observe fairly sharp regions of destructive interference at certain particular angles. Between these angles you would observe broad regions in which the waves do not cancel each other but vary in amplitude from zero at one nodal line, through a maximum of constructive interference, to zero at the next nodal line. In these regions the wave fronts are curved, and rays exist for all angles between those of the two nodal lines that bound the region. If you walk around a circle at a large distance R from the grating, however, you would observe a plane wave arriving along the principal axis and additional plane waves at the sharply defined directions θ_1 and θ_2, but you would observe no waves having rays at intermediate angles. The effect of the grating as opposed to the double source is to produce a tremendous sharpening of the directions of constructive interference and to extinguish the waves in the intermediate directions.

Problem A3.15   Interpret Eq. (A3.10). Describe how you might utilize a diffraction-grating experiment to determine the wavelength of an unknown wave. Answer the questions raised in Problem A3.12, supposing that you use a grating with a slit spacing of 6.0 cm instead of a double source.

A3.14 SINGLE-SLIT DIFFRACTION PATTERN

We return now to the sequence of ripple experiments illustrated in Fig. A3.13. Experience with the double-source and diffraction-grating patterns has taught us that interference effects can be expected to be most pronounced when the wavelength λ is somewhat less than that of a source spacing d. In Fig. A3.13(b) we have a definite interference pattern when the slit opening D is itself some-

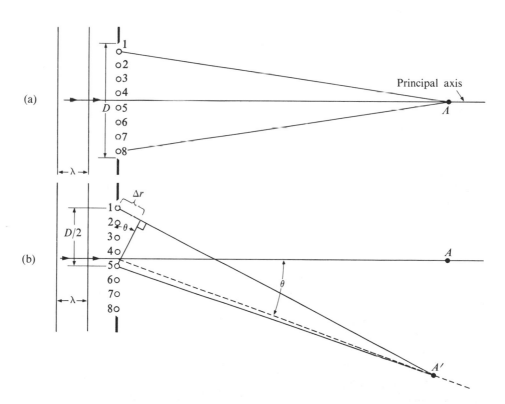

Fig. A3.30. (a) Along the principal axis at points such as A: For every fictitious source point above the axis, there is a symmetrically placed source point at the slit below the axis (1 and 8, 2 and 7, etc.). Waves arriving at A from these pairs of points travel the same distance and interfere constructively. Therefore the net effect is to produce constructive interference everywhere along the principal axis. (b) As the point of observation is moved away from the axis along the arc from A to A', path lengths from the fictitious source points change, and wavelets arrive at A' more and more out of phase with each other. At some angle θ_1, Δr will become equal to $\lambda/2$, and waves from sources 1 and 5 will cancel each other. The other pairs (2 and 6, 3 and 7, etc.) will cancel at the same time. There will be a nodal line at angular position θ_1, given by $\sin \theta_1 = (\lambda/2)/(D/2) = \lambda/D$.

what larger than λ. This pattern is apparently completely fanned out in (a), where $\lambda \sim D$, and the opening behaves more like a point source. The pattern is apparently compressed into a forward-propagating beam of waves when $\lambda \ll D$, as in Fig. A3.13(c).

These observations motivate an attempt to interpret the pattern in Fig. A3.13(b) by visualizing an array of point sources lined up along the slit opening D, and emitting waves in phase with each other, as sketched in Fig. A3.30.

Figure A3.30(a) indicates that we would expect constructive interference all along the principal axis. Figure A3.30(b) indicates that we would expect the wave amplitude to decrease as we move to angular positions θ away from the principal axis, and that we would expect nodal lines on either side of the axis at angular positions given by

$$\sin \theta_n = (2n - 1)\frac{\lambda}{D}, \qquad \text{where } n = 1, 2, 3, \ldots. \qquad (A3.11)$$

Similarly, we would expect lines of greatest amplitude at angles given by

$$\sin \theta_n = 2n\frac{\lambda}{D}, \qquad \text{where } n = 0, 1, 2, \ldots. \qquad (A3.12)$$

Problem A3.16 Extend the explanation given in the caption of Fig. A3.30(b) so as to derive and justify Eqs. (A3.11) and (A3.12). Interpret these equations. At what value of n for a given λ/D do they cease to have physical meaning? What approximations are implicit in the treatment? (Compare with the discussion associated with Fig. A3.25.) Make several sketches which are similar to A3.30(b), but which show how the path lengths differ and sources pair up along the next outlying regions of constructive and destructive interference.

A3.15 ENERGY AND MOMENTUM IN WAVE PROPAGATION

The laws of conservation of energy and momentum are found to apply to wave propagation fully as rigorously as they apply to the simpler particle motions and displacements discussed in earlier chapters. Detailed mathematical calculations substantiating this assertion can be carried out in straightforward fashion, but are beyond the scope of this text. We can, however, readily draw some useful qualitative insights from our direct experience with wave phenomena.

To keep generating a bow wave from a moving object or to keep generating a continuous periodic wave train from a vibrating source, it is obvious that we must maintain a continuous supply of energy in the form of work done on the source. If we fail to do so, the source "runs down": the moving boat coasts to a stop; the wave rippler stops moving up and down; the tuning fork stops vibrating. This running down and coming to a stop, however, is very different from frictional sliding of a block on a table. In the latter case the kinetic energy of the block is totally converted into heat, raising the temperature of the block and its surroundings. In the former case, the running down is accompanied by extremely little temperature rise; instead, a disturbance propagates to regions remote from the source and is capable of accelerating objects, and causing displacements against resisting forces, at these remote points.

In other words, both momentum and energy are lost by the source, but the energy is not immediately dissipated as heat. Nor are the momentum and energy imparted to the medium in such a way that it moves as a unit, like a rigid body. Momentum and energy are *propagated through* the medium from layer to layer and region to region by the wave disturbance at its own finite velocity V.

Any small region of a wave disturbance possesses momentum given by the product of the mass of the region m and the local transverse or longitudinal particle velocity v; similarly, the region possesses kinetic energy $\frac{1}{2}mv^2$. The system as a whole also possesses potential energy by virtue of the position of the region in question. In a water wave, potential energy is associated with the upward or downward displacement of particles of water against the restoring gravitational force. In a transversely vibrating string, potential energy is associated in an exactly similar way with displacements against the restoring effect of the tension in the string. In longitudinal waves, as in a helical spring or sound in air, potential energy is associated with the compressions and rarefactions of the medium.

Thus the work that we do to keep the source emitting waves is propagated as both kinetic and potential energy to distant points, and, after a time lag determined by the wave propagation velocity, is capable of exerting forces and doing work on other objects at these remote positions. But the force exerted on the medium by the vibrating source is not immediately transmitted to the distant object whose motion is being excited; Newton's third law does *not* apply to a simple pair of forces between the source and the distant object. The two bodies do not act on each other directly at a distance. Rather, we can only think of adjacent particles of the medium exerting equal and opposite forces on each other; for the system as a whole we must think of the relation between initial and final conditions: we must think of energy and momentum as being conserved in the propagation of a disturbance through the medium (or "field") between the source and the receiver of the effects.

A moment's thought will now convince us that transfer of momentum and energy by wave propagation is one of the most common and all-pervasive physical effects that we encounter: If we push a solid object by applying a force to one side, it does not accelerate instantaneously as a whole; the force we exert causes a compression wave that propagates at the velocity of sound through the solid, transmitting the disturbance to the other side of the object in an extremely short, but nevertheless finite, time interval. (See Problem A3.2.) In an elastic collision of hard bodies, there is no such thing as perfect rigidity; the interval of interaction between the bodies is finite and is accompanied by deformation and by the complicated bouncing back and forth of compressions and rarefactions within the colliding bodies.

Problem A3.17   Think of some other everyday experiences in which forces act to accelerate systems of various kinds, and analyze the situation qualitatively by describing the wave propagations that must occur.

The emission of waves, energy, and momentum by a source is frequently referred to as *radiation*. A radiating source runs down unless energy is supplied to it. The radiated energy is transmitted to distant points. Energy propagated through any region is related to the local wave amplitude, but the relation is not immediately obvious; calculation shows that, in general, *the rate at which energy flows past a point of observation in the medium* is proportional to the *square of the amplitude at that point*. As a wave recedes from its source, the amplitude decreases, principally because the energy in a given layer is spread out over an ever-increasing surface. A very small portion of the decrease, however, results from frictional dissipation within the medium as the wave moves along. Gradually, the radiated energy is converted into heat, warming up the medium through which the disturbance passes. Sometimes, as with water waves, the dissipative process may be enhanced at a boundary where waves splash and tumble at a beach or wall instead of being completely reflected.

The concept of radiation will be found to play a very significant role in our discussions of the nature of light and of the structure of matter.

Useful Constants and
Unit Conversion Factors

Avogadro number	N_0	6.02252×10^{-23} mole^{-1}
Gas constant	R	0.08207 l atm mole^{-1}deg^{-1}
		8.314 J mole^{-1}deg^{-1}
		1.98 cal mole^{-1}deg^{-1}
Electron charge	q_e	1.60210×10^{-19} C
		4.80×10^{-10} esu
Electron mass	m_e	9.11×10^{-31} kg
Planck constant	h	6.6256×10^{-27} erg sec
Bohr radius	a_0	0.5292 Å
Faraday constant	$\mathfrak{F}$	96,487 C
Speed of light in vacuum	c	2.998×10^8 m sec^{-1}
Coulomb law constant	K	8.9876×10^9 n m^2 C^{-2}

The following energy conversion factors are given to precision appropriate for ordinary calculations to three significant figures:

$$4.19 \text{ J} = 1.00 \text{ cal}$$
$$1,000 \text{ cm}^{-1} \text{ molecule}^{-1} = 2.85 \text{ kcal mole}^{-1}$$
$$= 11,900 \text{ J mole}^{-1}$$
$$= 0.1239 \text{ eV molecule}^{-1}$$

Index

A CATALOG OF SELECTED
DOVER BOOKS
IN SCIENCE AND MATHEMATICS

A CATALOG OF SELECTED

DOVER BOOKS

IN SCIENCE AND MATHEMATICS

QUALITATIVE THEORY OF DIFFERENTIAL EQUATIONS, V.V. Nemytskii and V.V. Stepanov. Classic graduate-level text by two prominent Soviet mathematicians covers classical differential equations as well as topological dynamics and ergodic theory. Bibliographies. 523pp. 5⅜ × 8½. 65954-2 Pa. $10.95

MATRICES AND LINEAR ALGEBRA, Hans Schneider and George Phillip Barker. Basic textbook covers theory of matrices and its applications to systems of linear equations and related topics such as determinants, eigenvalues and differential equations. Numerous exercises. 432pp. 5⅜ × 8½. 66014-1 Pa. $10.95

QUANTUM THEORY, David Bohm. This advanced undergraduate-level text presents the quantum theory in terms of qualitative and imaginative concepts, followed by specific applications worked out in mathematical detail. Preface. Index. 655pp. 5⅜ × 8½. 65969-0 Pa. $13.95

ATOMIC PHYSICS (8th edition), Max Born. Nobel laureate's lucid treatment of kinetic theory of gases, elementary particles, nuclear atom, wave-corpuscles, atomic structure and spectral lines, much more. Over 40 appendices, bibliography. 495pp. 5⅜ × 8½. 65984-4 Pa. $12.95

ELECTRONIC STRUCTURE AND THE PROPERTIES OF SOLIDS: The Physics of the Chemical Bond, Walter A. Harrison. Innovative text offers basic understanding of the electronic structure of covalent and ionic solids, simple metals, transition metals and their compounds. Problems. 1980 edition. 582pp. 6⅛ × 9¼. 66021-4 Pa. $15.95

BOUNDARY VALUE PROBLEMS OF HEAT CONDUCTION, M. Necati Özisik. Systematic, comprehensive treatment of modern mathematical methods of solving problems in heat conduction and diffusion. Numerous examples and problems. Selected references. Appendices. 505pp. 5⅜ × 8½. 65990-9 Pa. $12.95

A SHORT HISTORY OF CHEMISTRY (3rd edition), J.R. Partington. Classic exposition explores origins of chemistry, alchemy, early medical chemistry, nature of atmosphere, theory of valency, laws and structure of atomic theory, much more. 428pp. 5⅜ × 8½. (Available in U.S. only) 65977-1 Pa. $10.95

A HISTORY OF ASTRONOMY, A. Pannekoek. Well-balanced, carefully reasoned study covers such topics as Ptolemaic theory, work of Copernicus, Kepler, Newton, Eddington's work on stars, much more. Illustrated. References. 521pp. 5⅜ × 8½. 65994-1 Pa. $12.95

PRINCIPLES OF METEOROLOGICAL ANALYSIS, Walter J. Saucier. Highly respected, abundantly illustrated classic reviews atmospheric variables, hydrostatics, static stability, various analyses (scalar, cross-section, isobaric, isentropic, more). For intermediate meteorology students. 454pp. 6⅛ × 9¼. 65979-8 Pa. $14.95

RELATIVITY, THERMODYNAMICS AND COSMOLOGY, Richard C. Tolman. Landmark study extends thermodynamics to special, general relativity; also applications of relativistic mechanics, thermodynamics to cosmological models. 501pp. 5⅜ × 8½. 65383-8 Pa. $12.95

APPLIED ANALYSIS, Cornelius Lanczos. Classic work on analysis and design of finite processes for approximating solution of analytical problems. Algebraic equations, matrices, harmonic analysis, quadrature methods, much more. 559pp. 5⅜ × 8½. 65656-X Pa. $13.95

SPECIAL RELATIVITY FOR PHYSICISTS, G. Stephenson and C.W. Kilmister. Concise elegant account for nonspecialists. Lorentz transformation, optical and dynamical applications, more. Bibliography. 108pp. 5⅜ × 8½. 65519-9 Pa. $4.95

INTRODUCTION TO ANALYSIS, Maxwell Rosenlicht. Unusually clear, accessible coverage of set theory, real number system, metric spaces, continuous functions, Riemann integration, multiple integrals, more. Wide range of problems. Undergraduate level. Bibliography. 254pp. 5⅜ × 8½. 65038-3 Pa. $7.95

INTRODUCTION TO QUANTUM MECHANICS With Applications to Chemistry, Linus Pauling & E. Bright Wilson, Jr. Classic undergraduate text by Nobel Prize winner applies quantum mechanics to chemical and physical problems. Numerous tables and figures enhance the text. Chapter bibliographies. Appendices. Index. 468pp. 5⅜ × 8½. 64871-0 Pa. $11.95

ASYMPTOTIC EXPANSIONS OF INTEGRALS, Norman Bleistein & Richard A. Handelsman. Best introduction to important field with applications in a variety of scientific disciplines. New preface. Problems. Diagrams. Tables. Bibliography. Index. 448pp. 5⅜ × 8½. 65082-0 Pa. $12.95

MATHEMATICS APPLIED TO CONTINUUM MECHANICS, Lee A. Segel. Analyzes models of fluid flow and solid deformation. For upper-level math, science and engineering students. 608pp. 5⅜ × 8½. 65369-2 Pa. $13.95

ELEMENTS OF REAL ANALYSIS, David A. Sprecher. Classic text covers fundamental concepts, real number system, point sets, functions of a real variable, Fourier series, much more. Over 500 exercises. 352pp. 5⅜ × 8½. 65385-4 Pa. $10.95

PHYSICAL PRINCIPLES OF THE QUANTUM THEORY, Werner Heisenberg. Nobel Laureate discusses quantum theory, uncertainty, wave mechanics, work of Dirac, Schroedinger, Compton, Wilson, Einstein, etc. 184pp. 5⅜ × 8½. 60113-7 Pa. $5.95

INTRODUCTORY REAL ANALYSIS, A.N. Kolmogorov, S.V. Fomin. Translated by Richard A. Silverman. Self-contained, evenly paced introduction to real and functional analysis. Some 350 problems. 403pp. 5⅜ × 8½. 61226-0 Pa. $9.95

PROBLEMS AND SOLUTIONS IN QUANTUM CHEMISTRY AND PHYSICS, Charles S. Johnson, Jr. and Lee G. Pedersen. Unusually varied problems, detailed solutions in coverage of quantum mechanics, wave mechanics, angular momentum, molecular spectroscopy, scattering theory, more. 280 problems plus 139 supplementary exercises. 430pp. 6½ × 9¼. 65236-X Pa. $12.95

THE ELECTROMAGNETIC FIELD, Albert Shadowitz. Comprehensive undergraduate text covers basics of electric and magnetic fields, builds up to electromagnetic theory. Also related topics, including relativity. Over 900 problems. 768pp. 5⅜ × 8¼. 65660-8 Pa. $18.95

FOURIER SERIES, Georgi P. Tolstov. Translated by Richard A. Silverman. A valuable addition to the literature on the subject, moving clearly from subject to subject and theorem to theorem. 107 problems, answers. 336pp. 5⅜ × 8½. 63317-9 Pa. $8.95

THEORY OF ELECTROMAGNETIC WAVE PROPAGATION, Charles Herach Papas. Graduate-level study discusses the Maxwell field equations, radiation from wire antennas, the Doppler effect and more. xiii + 244pp. 5⅜ × 8½. 65678-0 Pa. $6.95

DISTRIBUTION THEORY AND TRANSFORM ANALYSIS: An Introduction to Generalized Functions, with Applications, A.H. Zemanian. Provides basics of distribution theory, describes generalized Fourier and Laplace transformations. Numerous problems. 384pp. 5⅜ × 8½. 65479-6 Pa. $9.95

THE PHYSICS OF WAVES, William C. Elmore and Mark A. Heald. Unique overview of classical wave theory. Acoustics, optics, electromagnetic radiation, more. Ideal as classroom text or for self-study. Problems. 477pp. 5⅜ × 8½. 64926-1 Pa. $12.95

CALCULUS OF VARIATIONS WITH APPLICATIONS, George M. Ewing. Applications-oriented introduction to variational theory develops insight and promotes understanding of specialized books, research papers. Suitable for advanced undergraduate/graduate students as primary, supplementary text. 352pp. 5⅜ × 8½. 64856-7 Pa. $8.95

A TREATISE ON ELECTRICITY AND MAGNETISM, James Clerk Maxwell. Important foundation work of modern physics. Brings to final form Maxwell's theory of electromagnetism and rigorously derives his general equations of field theory. 1,084pp. 5⅜ × 8½. 60636-8, 60637-6 Pa., Two-vol. set $21.90

AN INTRODUCTION TO THE CALCULUS OF VARIATIONS, Charles Fox. Graduate-level text covers variations of an integral, isoperimetrical problems, least action, special relativity, approximations, more. References. 279pp. 5⅜ × 8½. 65499-0 Pa. $7.95

HYDRODYNAMIC AND HYDROMAGNETIC STABILITY, S. Chandrasekhar. Lucid examination of the Rayleigh-Benard problem; clear coverage of the theory of instabilities causing convection. 704pp. 5⅜ × 8¼. 64071-X Pa. $14.95

CALCULUS OF VARIATIONS, Robert Weinstock. Basic introduction covering isoperimetric problems, theory of elasticity, quantum mechanics, electrostatics, etc. Exercises throughout. 326pp. 5⅜ × 8½. 63069-2 Pa. $8.95

DYNAMICS OF FLUIDS IN POROUS MEDIA, Jacob Bear. For advanced students of ground water hydrology, soil mechanics and physics, drainage and irrigation engineering and more. 335 illustrations. Exercises, with answers. 784pp. 6⅛ × 9¼. 65675-6 Pa. $19.95

HANDBOOK OF MATHEMATICAL FUNCTIONS WITH FORMULAS, GRAPHS, AND MATHEMATICAL TABLES, edited by Milton Abramowitz and Irene A. Stegun. Vast compendium: 29 sets of tables, some to as high as 20 places. 1,046pp. 8 × 10½. 61272-4 Pa. $24.95

MATHEMATICAL METHODS IN PHYSICS AND ENGINEERING, John W. Dettman. Algebraically based approach to vectors, mapping, diffraction, other topics in applied math. Also generalized functions, analytic function theory, more. Exercises. 448pp. 5⅜ × 8¼. 65649-7 Pa. $9.95

A SURVEY OF NUMERICAL MATHEMATICS, David M. Young and Robert Todd Gregory. Broad self-contained coverage of computer-oriented numerical algorithms for solving various types of mathematical problems in linear algebra, ordinary and partial, differential equations, much more. Exercises. Total of 1,248pp. 5⅜ × 8½. Two volumes. Vol. I 65691-8 Pa. $14.95
Vol. II 65692-6 Pa. $14.95

TENSOR ANALYSIS FOR PHYSICISTS, J.A. Schouten. Concise exposition of the mathematical basis of tensor analysis, integrated with well-chosen physical examples of the theory. Exercises. Index. Bibliography. 289pp. 5⅜ × 8½. 65582-2 Pa. $8.95

INTRODUCTION TO NUMERICAL ANALYSIS (2nd Edition), F.B. Hildebrand. Classic, fundamental treatment covers computation, approximation, interpolation, numerical differentiation and integration, other topics. 150 new problems. 669pp. 5⅜ × 8½. 65363-3 Pa. $15.95

INVESTIGATIONS ON THE THEORY OF THE BROWNIAN MOVEMENT, Albert Einstein. Five papers (1905–8) investigating dynamics of Brownian motion and evolving elementary theory. Notes by R. Fürth. 122pp. 5⅜ × 8½. 60304-0 Pa. $4.95

CATASTROPHE THEORY FOR SCIENTISTS AND ENGINEERS, Robert Gilmore. Advanced-level treatment describes mathematics of theory grounded in the work of Poincaré, R. Thom, other mathematicians. Also important applications to problems in mathematics, physics, chemistry and engineering. 1981 edition. References. 28 tables. 397 black-and-white illustrations. xvii + 666pp. 6⅛ × 9¼. 67539-4 Pa. $16.95

AN INTRODUCTION TO STATISTICAL THERMODYNAMICS, Terrell L. Hill. Excellent basic text offers wide-ranging coverage of quantum statistical mechanics, systems of interacting molecules, quantum statistics, more. 523pp. 5⅜ × 8½. 65242-4 Pa. $12.95

ELEMENTARY DIFFERENTIAL EQUATIONS, William Ted Martin and Eric Reissner. Exceptionally clear, comprehensive introduction at undergraduate level. Nature and origin of differential equations, differential equations of first, second and higher orders. Picard's Theorem, much more. Problems with solutions. 331pp. 5⅜ × 8½. 65024-3 Pa. $8.95

STATISTICAL PHYSICS, Gregory H. Wannier. Classic text combines thermodynamics, statistical mechanics and kinetic theory in one unified presentation of thermal physics. Problems with solutions. Bibliography. 532pp. 5⅜ × 8½. 65401-X Pa. $12.95

ORDINARY DIFFERENTIAL EQUATIONS, Morris Tenenbaum and Harry Pollard. Exhaustive survey of ordinary differential equations for undergraduates in mathematics, engineering, science. Thorough analysis of theorems. Diagrams. Bibliography. Index. 818pp. 5⅜ × 8½. 64940-7 Pa. $16.95

STATISTICAL MECHANICS: Principles and Applications, Terrell L. Hill. Standard text covers fundamentals of statistical mechanics, applications to fluctuation theory, imperfect gases, distribution functions, more. 448pp. 5⅜ × 8½. 65390-0 Pa. $11.95

ORDINARY DIFFERENTIAL EQUATIONS AND STABILITY THEORY: An Introduction, David A. Sánchez. Brief, modern treatment. Linear equation, stability theory for autonomous and nonautonomous systems, etc. 164pp. 5⅜ × 8¼. 63828-6 Pa. $5.95

THIRTY YEARS THAT SHOOK PHYSICS: The Story of Quantum Theory, George Gamow. Lucid, accessible introduction to influential theory of energy and matter. Careful explanations of Dirac's anti-particles, Bohr's model of the atom, much more. 12 plates. Numerous drawings. 240pp. 5⅜ × 8½. 24895-X Pa. $6.95

THEORY OF MATRICES, Sam Perlis. Outstanding text covering rank, non-singularity and inverses in connection with the development of canonical matrices under the relation of equivalence, and without the intervention of determinants. Includes exercises. 237pp. 5⅜ × 8½. 66810-X Pa. $7.95

GREAT EXPERIMENTS IN PHYSICS: Firsthand Accounts from Galileo to Einstein, edited by Morris H. Shamos. 25 crucial discoveries: Newton's laws of motion, Chadwick's study of the neutron, Hertz on electromagnetic waves, more. Original accounts clearly annotated. 370pp. 5⅜ × 8½. 25346-5 Pa. $10.95

INTRODUCTION TO PARTIAL DIFFERENTIAL EQUATIONS WITH APPLICATIONS, E.C. Zachmanoglou and Dale W. Thoe. Essentials of partial differential equations applied to common problems in engineering and the physical sciences. Problems and answers. 416pp. 5⅜ × 8½. 65251-3 Pa. $10.95

BURNHAM'S CELESTIAL HANDBOOK, Robert Burnham, Jr. Thorough guide to the stars beyond our solar system. Exhaustive treatment. Alphabetical by constellation: Andromeda to Cetus in Vol. 1; Chamaeleon to Orion in Vol. 2; and Pavo to Vulpecula in Vol. 3. Hundreds of illustrations. Index in Vol. 3. 2,000pp. 6⅛ × 9¼. 23567-X, 23568-8, 23673-0 Pa., Three-vol. set $41.85

CHEMICAL MAGIC, Leonard A. Ford. Second Edition, Revised by E. Winston Grundmeier. Over 100 unusual stunts demonstrating cold fire, dust explosions, much more. Text explains scientific principles and stresses safety precautions. 128pp. 5⅜ × 8½. 67628-5 Pa. $5.95

AMATEUR ASTRONOMER'S HANDBOOK, J.B. Sidgwick. Timeless, comprehensive coverage of telescopes, mirrors, lenses, mountings, telescope drives, micrometers, spectroscopes, more. 189 illustrations. 576pp. 5⅜ × 8¼. (Available in U.S. only) 24034-7 Pa. $9.95

ROTARY-WING AERODYNAMICS, W.Z. Stepniewski. Clear, concise text covers aerodynamic phenomena of the rotor and offers guidelines for helicopter performance evaluation. Originally prepared for NASA. 537 figures. 640pp. 6⅛ × 9¼.
64647-5 Pa. $15.95

DIFFERENTIAL GEOMETRY, Heinrich W. Guggenheimer. Local differential geometry as an application of advanced calculus and linear algebra. Curvature, transformation groups, surfaces, more. Exercises. 62 figures. 378pp. 5⅜ × 8½.
63433-7 Pa. $8.95

INTRODUCTION TO SPACE DYNAMICS, William Tyrrell Thomson. Comprehensive, classic introduction to space-flight engineering for advanced undergraduate and graduate students. Includes vector algebra, kinematics, transformation of coordinates. Bibliography. Index. 352pp. 5⅜ × 8½.
65113-4 Pa. $8.95

A SURVEY OF MINIMAL SURFACES, Robert Osserman. Up-to-date, in-depth discussion of the field for advanced students. Corrected and enlarged edition covers new developments. Includes numerous problems. 192pp. 5⅜ × 8½.
64998-9 Pa. $8.95

ANALYTICAL MECHANICS OF GEARS, Earle Buckingham. Indispensable reference for modern gear manufacture covers conjugate gear-tooth action, gear-tooth profiles of various gears, many other topics. 263 figures. 102 tables. 546pp. 5⅜ × 8½.
65712-4 Pa. $14.95

SET THEORY AND LOGIC, Robert R. Stoll. Lucid introduction to unified theory of mathematical concepts. Set theory and logic seen as tools for conceptual understanding of real number system. 496pp. 5⅜ × 8¼.
63829-4 Pa. $12.95

A HISTORY OF MECHANICS, René Dugas. Monumental study of mechanical principles from antiquity to quantum mechanics. Contributions of ancient Greeks, Galileo, Leonardo, Kepler, Lagrange, many others. 671pp. 5⅜ × 8½.
65632-2 Pa. $14.95

FAMOUS PROBLEMS OF GEOMETRY AND HOW TO SOLVE THEM, Benjamin Bold. Squaring the circle, trisecting the angle, duplicating the cube: learn their history, why they are impossible to solve, then solve them yourself. 128pp. 5⅜ × 8½.
24297-8 Pa. $4.95

MECHANICAL VIBRATIONS, J.P. Den Hartog. Classic textbook offers lucid explanations and illustrative models, applying theories of vibrations to a variety of practical industrial engineering problems. Numerous figures. 233 problems, solutions. Appendix. Index. Preface. 436pp. 5⅜ × 8½.
64785-4 Pa. $10.95

CURVATURE AND HOMOLOGY, Samuel I. Goldberg. Thorough treatment of specialized branch of differential geometry. Covers Riemannian manifolds, topology of differentiable manifolds, compact Lie groups, other topics. Exercises. 315pp. 5⅜ × 8½.
64314-X Pa. $9.95

HISTORY OF STRENGTH OF MATERIALS, Stephen P. Timoshenko. Excellent historical survey of the strength of materials with many references to the theories of elasticity and structure. 245 figures. 452pp. 5⅜ × 8½. 61187-6 Pa. $11.95

GEOMETRY OF COMPLEX NUMBERS, Hans Schwerdtfeger. Illuminating, widely praised book on analytic geometry of circles, the Moebius transformation, and two-dimensional non-Euclidean geometries. 200pp. 5⅜ × 8¼.
63830-8 Pa. $8.95

MECHANICS, J.P. Den Hartog. A classic introductory text or refresher. Hundreds of applications and design problems illuminate fundamentals of trusses, loaded beams and cables, etc. 334 answered problems. 462pp. 5⅜ × 8½. 60754-2 Pa. $9.95

TOPOLOGY, John G. Hocking and Gail S. Young. Superb one-year course in classical topology. Topological spaces and functions, point-set topology, much more. Examples and problems. Bibliography. Index. 384pp. 5⅜ × 8¼.
65676-4 Pa. $9.95

STRENGTH OF MATERIALS, J.P. Den Hartog. Full, clear treatment of basic material (tension, torsion, bending, etc.) plus advanced material on engineering methods, applications. 350 answered problems. 323pp. 5⅜ × 8½. 60755-0 Pa. $8.95

ELEMENTARY CONCEPTS OF TOPOLOGY, Paul Alexandroff. Elegant, intuitive approach to topology from set-theoretic topology to Betti groups; how concepts of topology are useful in math and physics. 25 figures. 57pp. 5⅜ × 8½.
60747-X Pa. $3.50

ADVANCED STRENGTH OF MATERIALS, J.P. Den Hartog. Superbly written advanced text covers torsion, rotating disks, membrane stresses in shells, much more. Many problems and answers. 388pp. 5⅜ × 8½. 65407-9 Pa. $9.95

COMPUTABILITY AND UNSOLVABILITY, Martin Davis. Classic graduate-level introduction to theory of computability, usually referred to as theory of recurrent functions. New preface and appendix. 288pp. 5⅜ × 8½. 61471-9 Pa. $7.95

GENERAL CHEMISTRY, Linus Pauling. Revised 3rd edition of classic first-year text by Nobel laureate. Atomic and molecular structure, quantum mechanics, statistical mechanics, thermodynamics correlated with descriptive chemistry. Problems. 992pp. 5⅜ × 8½. 65622-5 Pa. $19.95

AN INTRODUCTION TO MATRICES, SETS AND GROUPS FOR SCIENCE STUDENTS, G. Stephenson. Concise, readable text introduces sets, groups, and most importantly, matrices to undergraduate students of physics, chemistry, and engineering. Problems. 164pp. 5⅜ × 8½. 65077-4 Pa. $6.95

THE HISTORICAL BACKGROUND OF CHEMISTRY, Henry M. Leicester. Evolution of ideas, not individual biography. Concentrates on formulation of a coherent set of chemical laws. 260pp. 5⅜ × 8½. 61053-5 Pa. $6.95

THE PHILOSOPHY OF MATHEMATICS: An Introductory Essay, Stephan Körner. Surveys the views of Plato, Aristotle, Leibniz & Kant concerning propositions and theories of applied and pure mathematics. Introduction. Two appendices. Index. 198pp. 5⅜ × 8½. 25048-2 Pa. $7.95

THE DEVELOPMENT OF MODERN CHEMISTRY, Aaron J. Ihde. Authoritative history of chemistry from ancient Greek theory to 20th-century innovation. Covers major chemists and their discoveries. 209 illustrations. 14 tables. Bibliographies. Indices. Appendices. 851pp. 5⅜ × 8½. 64235-6 Pa. $18.95

DE RE METALLICA, Georgius Agricola. The famous Hoover translation of greatest treatise on technological chemistry, engineering, geology, mining of early modern times (1556). All 289 original woodcuts. 638pp. 6¾ × 11.
60006-8 Pa. $18.95

SOME THEORY OF SAMPLING, William Edwards Deming. Analysis of the problems, theory and design of sampling techniques for social scientists, industrial managers and others who find statistics increasingly important in their work. 61 tables. 90 figures. xvii + 602pp. 5⅜ × 8½. 64684-X Pa. $15.95

THE VARIOUS AND INGENIOUS MACHINES OF AGOSTINO RAMELLI: A Classic Sixteenth-Century Illustrated Treatise on Technology, Agostino Ramelli. One of the most widely known and copied works on machinery in the 16th century. 194 detailed plates of water pumps, grain mills, cranes, more. 608pp. 9 × 12.
28180-9 Pa. $24.95

LINEAR PROGRAMMING AND ECONOMIC ANALYSIS, Robert Dorfman, Paul A. Samuelson and Robert M. Solow. First comprehensive treatment of linear programming in standard economic analysis. Game theory, modern welfare economics, Leontief input-output, more. 525pp. 5⅜ × 8½. 65491-5 Pa. $14.95

ELEMENTARY DECISION THEORY, Herman Chernoff and Lincoln E. Moses. Clear introduction to statistics and statistical theory covers data processing, probability and random variables, testing hypotheses, much more. Exercises. 364pp. 5⅜ × 8½. 65218-1 Pa. $9.95

THE COMPLEAT STRATEGYST: Being a Primer on the Theory of Games of Strategy, J.D. Williams. Highly entertaining classic describes, with many illustrated examples, how to select best strategies in conflict situations. Prefaces. Appendices. 268pp. 5⅜ × 8½. 25101-2 Pa. $7.95

MATHEMATICAL METHODS OF OPERATIONS RESEARCH, Thomas L. Saaty. Classic graduate-level text covers historical background, classical methods of forming models, optimization, game theory, probability, queueing theory, much more. Exercises. Bibliography. 448pp. 5⅜ × 8¼. 65703-5 Pa. $12.95

CONSTRUCTIONS AND COMBINATORIAL PROBLEMS IN DESIGN OF EXPERIMENTS, Damaraju Raghavarao. In-depth reference work examines orthogonal Latin squares, incomplete block designs, tactical configuration, partial geometry, much more. Abundant explanations, examples. 416pp. 5⅜ × 8¼.
65685-3 Pa. $10.95

THE ABSOLUTE DIFFERENTIAL CALCULUS (CALCULUS OF TENSORS), Tullio Levi-Civita. Great 20th-century mathematician's classic work on material necessary for mathematical grasp of theory of relativity. 452pp. 5⅜ × 8½.
63401-9 Pa. $9.95

VECTOR AND TENSOR ANALYSIS WITH APPLICATIONS, A.I. Borisenko and I.E. Tarapov. Concise introduction. Worked-out problems, solutions, exercises. 257pp. 5⅜ × 8¼. 63833-2 Pa. $7.95

THE FOUR-COLOR PROBLEM: Assaults and Conquest, Thomas L. Saaty and Paul G. Kainen. Engrossing, comprehensive account of the century-old combinatorial topological problem, its history and solution. Bibliographies. Index. 110 figures. 228pp. 5⅜ × 8½.   65092-8 Pa. $6.95

CATALYSIS IN CHEMISTRY AND ENZYMOLOGY, William P. Jencks. Exceptionally clear coverage of mechanisms for catalysis, forces in aqueous solution, carbonyl- and acyl-group reactions, practical kinetics, more. 864pp. 5⅜ × 8½.   65460-5 Pa. $19.95

PROBABILITY: An Introduction, Samuel Goldberg. Excellent basic text covers set theory, probability theory for finite sample spaces, binomial theorem, much more. 360 problems. Bibliographies. 322pp. 5⅜ × 8½.   65252-1 Pa. $8.95

LIGHTNING, Martin A. Uman. Revised, updated edition of classic work on the physics of lightning. Phenomena, terminology, measurement, photography, spectroscopy, thunder, more. Reviews recent research. Bibliography. Indices. 320pp. 5⅜ × 8¼.   64575-4 Pa. $8.95

PROBABILITY THEORY: A Concise Course, Y.A. Rozanov. Highly readable, self-contained introduction covers combination of events, dependent events, Bernoulli trials, etc. Translation by Richard Silverman. 148pp. 5⅜ × 8¼.
                                                                63544-9 Pa. $5.95

AN INTRODUCTION TO HAMILTONIAN OPTICS, H. A. Buchdahl. Detailed account of the Hamiltonian treatment of aberration theory in geometrical optics. Many classes of optical systems defined in terms of the symmetries they possess. Problems with detailed solutions. 1970 edition. xv + 360pp. 5⅜ × 8½.
                                                                67597-1 Pa. $10.95

STATISTICS MANUAL, Edwin L. Crow, et al. Comprehensive, practical collection of classical and modern methods prepared by U.S. Naval Ordnance Test Station. Stress on use. Basics of statistics assumed. 288pp. 5⅜ × 8½.
                                                                60599-X Pa. $6.95

DICTIONARY/OUTLINE OF BASIC STATISTICS, John E. Freund and Frank J. Williams. A clear concise dictionary of over 1,000 statistical terms and an outline of statistical formulas covering probability, nonparametric tests, much more. 208pp. 5⅜ × 8½.   66796-0 Pa. $6.95

STATISTICAL METHOD FROM THE VIEWPOINT OF QUALITY CONTROL, Walter A. Shewhart. Important text explains regulation of variables, uses of statistical control to achieve quality control in industry, agriculture, other areas. 192pp. 5⅜ × 8½.   65232-7 Pa. $7.95

THE INTERPRETATION OF GEOLOGICAL PHASE DIAGRAMS, Ernest G. Ehlers. Clear, concise text emphasizes diagrams of systems under fluid or containing pressure; also coverage of complex binary systems, hydrothermal melting, more. 288pp. 6½ × 9¼.   65389-7 Pa. $10.95

STATISTICAL ADJUSTMENT OF DATA, W. Edwards Deming. Introduction to basic concepts of statistics, curve fitting, least squares solution, conditions without parameter, conditions containing parameters. 26 exercises worked out. 271pp. 5⅜ × 8½.   64685-8 Pa. $8.95

TENSOR CALCULUS, J.L. Synge and A. Schild. Widely used introductory text covers spaces and tensors, basic operations in Riemannian space, non-Riemannian spaces, etc. 324pp. 5⅜ × 8¼. 63612-7 Pa. $8.95

A CONCISE HISTORY OF MATHEMATICS, Dirk J. Struik. The best brief history of mathematics. Stresses origins and covers every major figure from ancient Near East to 19th century. 41 illustrations. 195pp. 5⅜ × 8½. 60255-9 Pa. $7.95

A SHORT ACCOUNT OF THE HISTORY OF MATHEMATICS, W.W. Rouse Ball. One of clearest, most authoritative surveys from the Egyptians and Phoenicians through 19th-century figures such as Grassman, Galois, Riemann. Fourth edition. 522pp. 5⅜ × 8½. 20630-0 Pa. $10.95

HISTORY OF MATHEMATICS, David E. Smith. Nontechnical survey from ancient Greece and Orient to late 19th century; evolution of arithmetic, geometry, trigonometry, calculating devices, algebra, the calculus. 362 illustrations. 1,355pp. 5⅜ × 8½. 20429-4, 20430-8 Pa., Two-vol. set $23.90

THE GEOMETRY OF RENÉ DESCARTES, René Descartes. The great work founded analytical geometry. Original French text, Descartes' own diagrams, together with definitive Smith-Latham translation. 244pp. 5⅜ × 8½. 60068-8 Pa. $7.95

THE ORIGINS OF THE INFINITESIMAL CALCULUS, Margaret E. Baron. Only fully detailed and documented account of crucial discipline: origins; development by Galileo, Kepler, Cavalieri; contributions of Newton, Leibniz, more. 304pp. 5⅜ × 8½. (Available in U.S. and Canada only) 65371-4 Pa. $9.95

THE HISTORY OF THE CALCULUS AND ITS CONCEPTUAL DEVELOPMENT, Carl B. Boyer. Origins in antiquity, medieval contributions, work of Newton, Leibniz, rigorous formulation. Treatment is verbal. 346pp. 5⅜ × 8½. 60509-4 Pa. $8.95

THE THIRTEEN BOOKS OF EUCLID'S ELEMENTS, translated with introduction and commentary by Sir Thomas L. Heath. Definitive edition. Textual and linguistic notes, mathematical analysis. 2,500 years of critical commentary. Not abridged. 1,414pp. 5⅜ × 8½. 60088-2, 60089-0, 60090-4 Pa., Three-vol. set $29.85

GAMES AND DECISIONS: Introduction and Critical Survey, R. Duncan Luce and Howard Raiffa. Superb nontechnical introduction to game theory, primarily applied to social sciences. Utility theory, zero-sum games, n-person games, decision-making, much more. Bibliography. 509pp. 5⅜ × 8½. 65943-7 Pa. $12.95

THE HISTORICAL ROOTS OF ELEMENTARY MATHEMATICS, Lucas N.H. Bunt, Phillip S. Jones, and Jack D. Bedient. Fundamental underpinnings of modern arithmetic, algebra, geometry and number systems derived from ancient civilizations. 320pp. 5⅜ × 8½. 25563-8 Pa. $8.95

CALCULUS REFRESHER FOR TECHNICAL PEOPLE, A. Albert Klaf. Covers important aspects of integral and differential calculus via 756 questions. 566 problems, most answered. 431pp. 5⅜ × 8½. 20370-0 Pa. $8.95

CHALLENGING MATHEMATICAL PROBLEMS WITH ELEMENTARY SOLUTIONS, A.M. Yaglom and I.M. Yaglom. Over 170 challenging problems on probability theory, combinatorial analysis, points and lines, topology, convex polygons, many other topics. Solutions. Total of 445pp. 5⅜ × 8½. Two-vol. set.

Vol. I 65536-9 Pa. $7.95
Vol. II 65537-7 Pa. $6.95

FIFTY CHALLENGING PROBLEMS IN PROBABILITY WITH SOLUTIONS, Frederick Mosteller. Remarkable puzzlers, graded in difficulty, illustrate elementary and advanced aspects of probability. Detailed solutions. 88pp. 5⅜ × 8½.
65355-2 Pa. $4.95

EXPERIMENTS IN TOPOLOGY, Stephen Barr. Classic, lively explanation of one of the byways of mathematics. Klein bottles, Moebius strips, projective planes, map coloring, problem of the Koenigsberg bridges, much more, described with clarity and wit. 43 figures. 210pp. 5⅜ × 8½. 25933-1 Pa. $5.95

RELATIVITY IN ILLUSTRATIONS, Jacob T. Schwartz. Clear nontechnical treatment makes relativity more accessible than ever before. Over 60 drawings illustrate concepts more clearly than text alone. Only high school geometry needed. Bibliography. 128pp. 6⅛ × 9¼. 25965-X Pa. $6.95

AN INTRODUCTION TO ORDINARY DIFFERENTIAL EQUATIONS, Earl A. Coddington. A thorough and systematic first course in elementary differential equations for undergraduates in mathematics and science, with many exercises and problems (with answers). Index. 304pp. 5⅜ × 8½. 65942-9 Pa. $8.95

FOURIER SERIES AND ORTHOGONAL FUNCTIONS, Harry F. Davis. An incisive text combining theory and practical example to introduce Fourier series, orthogonal functions and applications of the Fourier method to boundary-value problems. 570 exercises. Answers and notes. 416pp. 5⅜ × 8½. 65973-9 Pa. $9.95

THE THEORY OF BRANCHING PROCESSES, Theodore E. Harris. First systematic, comprehensive treatment of branching (i.e. multiplicative) processes and their applications. Galton-Watson model, Markov branching processes, electron-photon cascade, many other topics. Rigorous proofs. Bibliography. 240pp. 5⅜ × 8½. 65952-6 Pa. $6.95

AN INTRODUCTION TO ALGEBRAIC STRUCTURES, Joseph Landin. Superb self-contained text covers "abstract algebra": sets and numbers, theory of groups, theory of rings, much more. Numerous well-chosen examples, exercises. 247pp. 5⅜ × 8½. 65940-2 Pa. $7.95

Prices subject to change without notice.
Available at your book dealer or write for free Mathematics and Science Catalog to Dept. GI, Dover Publications, Inc., 31 East 2nd St., Mineola, N.Y. 11501. Dover publishes more than 175 books each year on science, elementary and advanced mathematics, biology, music, art, literature, history, social sciences and other areas.